AA004094

GEOProcessing 2016

The Eighth International Conference on Advanced Geographic Information Systems, Applications, and Services

April 24 - 28, 2016

Venice, Italy

GEOProcessing 2016 Editors

Claus-Peter Rückemann, Leibniz Universität Hannover / Westfälische Wilhelms-Universität Münster / North-German Supercomputing Alliance (HLRN), Germany
Yerach Doytsher, Technion - Israel Institute of Technology, Haifa, Israel

Printed from e-media with permission by:

Curran Associates, Inc.
57 Morehouse Lane
Red Hook, NY 12571

Some format issues inherent in the e-media version may also appear in this print version.

Printed by Curran Associates, Inc. (2016)

International Academy, Research, and Industry Association (IARIA)
412 Derby Way
Wilmington, DE 19810

Phone: (408) 893-6407
Fax: (408) 527-6351

petre@iaria.org

Additional copies of this publication are available from:

Curran Associates, Inc.
57 Morehouse Lane
Red Hook, NY 12571 USA
Phone: 845-758-0400
Fax: 845-758-2633
Email: curran@proceedings.com
Web: www.proceedings.com

TABLE OF CONTENTS

GEOPROCESSING 1: GEO-MODELING I

Evaluation of CASE Tools with UML Profile Support for Geographical Database Design 1
T. Ferreira, J. Lisboa-Filho, S. Stempliuc

Impact of DEM Processing on the Geotechnical Instability Analysis of Waste Heaps in Wallonia 7
O. Close, N. Stephenne, C. Frippiat

Uncovering User Profiles in Location-Based Social Networks 14
S. Mohamed, A. Abdelmoty

Modelling Urban Expansion: A Multiple Urban-Densities Approach 22
A. Mustafa, I. Saadi, M. Cools, J. Teller

GEOPROCESSING 2: GEO-MODELING II

Spatial Regression in Health: Modelling Spatial Neighbourhood of High Risk Population 26
S. Bertazzon

On Improving Data Quality and Topology in Vector Spatial Data 32
N. Solomakhina, T. Hubauer, S. Becher

Identification of Areas with Potential for Flooding in South America 38
S. Rosim, J. Oliveira, A. Jardim, L. Namikawa, C. Almeida

A Common Land-Use Change Model for Both the Walloon and Flanders Regions in Belgium 43
B. Beaumont, N. Stephenne, E. Hallot, L. Poelmans, O. Close

GEOPROCESSING 3: GEO-SPATIAL DOMAIN APPLICATIONS

Brokered Approach to Federating Data using Semantic Web Techniques 46
J. Fa, G. West, D. McMeekin, S. Moncrieff

Underground Monitoring Systems using 3D GIS for Public Safety 56
K. Kim, D. Park, J. Lee, I. Lee

Image Based-Localization on Mobile Devices Using Geometric Features of Buildings 58
H. Ramanana, A. Rahajaniaina, J. Jessel

GEOPROCESSING 4: SPECIFIC GEO-DATA PROCESSING

A Method for Calculating Shape Similarity among Trajectory of Moving Object Based on Statistical Correlation of Angular Deflection Vectors 63
A. Melo, G. Scheibel, F. Baldo, F. Braz

Advanced Association Processing and Computation Facilities for Geoscientific and Archaeological Knowledge Resources Components 69
C. Ruckemann

Differential Morphological Profile for Threat Detection on Pipeline Right-of-Way: Heavy Equipment Detection 76
K. Stankov, B. Tolton

GEOPROCESSING 5: MANAGING GEO-SPATIAL DATA

SOLAP_Frame: A Framework for SOLAP using Heterogeneous Data Sources 78
T. Silva, D. Leite, C. Baptista

Geographic Metadata Searching with Semantic and Spatial Filtering Methods 85
T. Reed, E. Gulland, G. West, D. McMeekin, S. Moncrieff

A Linear Approach for Spatial Data Integration 93
A. Noskov, Y. Doytsher

GEOPROCESSING 6: GIS

BIM/GIS-based Data Integration Framework for Facility Management 100
T. Kang, S. Park, C. Hong

Context-aware Indoor-Outdoor Detection for Seamless Smartphone Positioning 106
N. Kroll, M. Jager, S. Suß

A Raster SOLAP for the Visualization of Crime Data Fields 109
J. Kasprzyk, J. Donnay

GEOPROCESSING 7: GEO-SPATIAL WEB SERVICES

Comparative Evaluation of Alternative Addressing Schemes 118
K. Clemens

Geospatial Content Services in the Digital Government: CASE: National Data Exchange Layer in the Finnish National Architecture for Digital Services 121
L. Lehto, P. Latvala, T. Sarjakoski, J. Reini

GEOPROCESSING 8: EARTH GEO-OBSERVATION

Mission Exploitation Platform PROBA-V 126
J. Dries, E. Goor, D. Daems

On Feasibility to Detect Volcanoes Hidden under Ice of Antarctica via their "Gravitational Signal" 128
J. Klokocnik, A. Bezdek, J. Kostelecky

We Need to Rethink How We Describe and Organize Spatial Information: Instrumenting and Observing the Community of Users to Improve Data Description and Discovery 131
B. Adams, M. Gahegan

GEOPROCESSING 9: GEO-SPATIAL FUNDAMENTALS

Improving Spatial Data Supply Chains: Learnings from the Manufacturing Industry 137
L. Arnold

Spatial Data Supply Chain Provenance Modelling for Next Generation Spatial Infrastructures Using Semantic Web Technologies 146
M. Sadiq, D. McMeekin, L. Arnold

Geographical General Regression Neural Network (GGRNN) Tool For Geographically Weighted Regression Analysis 154
M. Irfan, A. Koj, H. Thomas, M. Sedighi

Author Index

GEOProcessing 2016

Forward

The Eighth International Conference on Advanced Geographic Information Systems, Applications, and Services (GEOProcessing 2016), was held between April 24 and April 28, 2016 in Venice, Italy, continued a series of international events addressing fundamentals of advances in geographic information systems and the new applications related to them using Web Services. Such systems can be used for assessment, modeling and prognosis of emergencies. As an example, they can be used as assessment of accidents from chemical pollution by considering hazardous chemical zones dimensions represented on a computer map of the region's territory.

Geographical sensors and satellites provide a huge volume of spatial data which is available on the Web. Making use of Web Services, the users are able for provisioning and using these services instead of only for document searching. These services are published in a directory and may be automatically discovered in a given context by software agents. Accessing large digital geographical libraries with geo-spatial information raises some challenges with respect to data semantics, interfaces, data accuracy and updates, distributed processing, as well as with discovery, indexing and integration of geographical information systems; this raise the issue of distributed catalogs forming a federation of spatial databases. Some spatial data infrastructures use service-oriented architecture for accessing these large databases via Web Services.

The event was very competitive in its selection process and very well perceived by the international scientific and industrial communities. As such, it has attracted excellent contributions and active participation from all over the world. We were very pleased to receive a large amount of top quality contributions.

The conference had the following tracks:

- Geo-modeling
- Geo-spatial domain applications
- Specific geo-data processing
- Managing geo-spatial data
- GIS
- Geo-spatial Web Services
- Earth Geo-observation
- Geo-spatial fundamentals

We take here the opportunity to warmly thank all the members of the GEOProcessing 2016 technical program committee, as well as the numerous reviewers. The creation of such a high quality conference program would not have been possible without their involvement. We also kindly thank all the authors that dedicated much of their time and effort to contribute to

GEOProcessing 2016. We truly believe that, thanks to all these efforts, the final conference program consisted of top quality contributions.

Also, this event could not have been a reality without the support of many individuals, organizations and sponsors. We also gratefully thank the members of the GEOProcessing 2016 organizing committee for their help in handling the logistics and for their work that made this professional meeting a success.

We hope GEOProcessing 2016 was a successful international forum for the exchange of ideas and results between academia and industry and to promote further progress in the field of geographic information systems, applications, and services. We also hope that Venice, Italy, provided a pleasant environment during the conference and everyone saved some time to enjoy the unique charm of the city.

GEOProcessing Advisory Committee

Monica De Martino, Consiglio Nazionale delle Ricerche - Genova, Italy
Claus-Peter Rückemann, Leibniz Universität Hannover / Westfälische Wilhelms-Universität Münster / North-German Supercomputing Alliance (HLRN), Germany
Yerach Doytsher, Technion - Israel Institute of Technology, Haifa, Israel

GEOProcessing 2016

Committee

GEOProcessing Advisory Committee

Monica De Martino, Consiglio Nazionale delle Ricerche - Genova, Italy
Claus-Peter Rückemann, Leibniz Universität Hannover / Westfälische Wilhelms-Universität Münster / North-German Supercomputing Alliance (HLRN), Germany
Yerach Doytsher, Technion - Israel Institute of Technology, Haifa, Israel

GEOProcessing 2016 Technical Program Committee

Diana F. Adamatti, Universidade Federal do Rio Grande, Brazil
Ablimit Aji, Emory University, USA
Nuhcan Akçit, Middle East Technical University, Turkey
Zaher Al Aghbari, University of Sharjah, UAE
Mirko Albani, European Space Agency, Italy
Riccardo Albertoni, IMATI-CNR, Italy
Francesc Antón Castro, Denmark´s National Space Institute, Denmark
Thierry Badard, Université Laval - Québec, Canada
Petko Bakalov, Environmental Systems Research Institute, USA
Fabiano Baldo, Santa Catarina State University, Brazil
Fabian D. Barbato, ORT University - Montevideo, Uruguay
Thomas Barkowsky, University of Bremen, Germany
Michela Bertolotto, University College Dublin, Ireland
Reinaldo Bezerra Braga, Federal University of Ceará, Brazil
Budhendra L. Bhaduri, Oak Ridge National Laboratory, USA
Sandro Bimonte, Irstea | TSCF - Clermont Ferrand, France
Giuseppe Borruso, University of Trieste, Italy
Jean Brodeur, Natural Resources Canada / Government of Canada, Canada
David Brosset, Naval Academy Research Institute, France
Michael Cathcart, Electro-Optical Systems Laboratory / GTRI Georgia Institute of Technology, USA
Mete Celik, Erciyes University, Turkey
Yao-Yi Chiang, Spatial Sciences Institute - University of Southern California, USA
Chi-Yin Chow, City University of Hong Kong, Hong Kong
Christophe Claramunt Naval Academy Research Institute, France
Keith Clarke, University of California - Santa Barbara, USA
Konstantin Clemens, TU-Berlin, Germany
Eliseo Clementini, University of L'Aquila, Italy
Ana Cristina Costa, NOVA IMS Universidade Nova de Lisboa, Portugal

Bart Kuijpers, Hasselt University, Belgium
Rosa Lasaponara, CNR, Italy
Robert Laurini, INSA de Lyon - Villeurbanne, France
Ahmed Lbath, Université Grenoble Alpes, France
Dan Lee, Esri, USA
Lassi Lehto, National Land Survey of Finland, Finland
Fabio Luiz Leite Junior, UEPB - State University of Paraíba, Brazil
Jing Li, University of Denver, USA
Ki-Joune Li, Pusan National University, South Korea
Xun Li, Arizona State University, USA
Jugurta Lisboa, Federal University of Viçosa, Brazil
Qing Liu, CSIRO, Australia
Xuan Liu, IBM T.J. Watson Research Center - Yorktown Heights, USA
Zhi Liu, University of North Texas, USA
Victor Lobo, Portuguese Naval Academy / New University of Lisbon, Portugal
Cheng Long, Hong Kong University of Science and Technology, Hong Kong
Qifeng Lu, Sevatec Inc., USA
Miguel R. Luaces, University of A Coruña, Spain
Vincenzo (Enzo) Maltese, University of Trento, Italy
Jesus Marti Gavila, Universidad Politecnica de Valencia, Spain
Hervé Martin, Université Joseph Fourier - Grenoble, France
Bruno Martins, University of Lisbon | IST & INESC-ID, Portugal
Stephan Mäs, Technische Universität Dresden, Germany
Michael P. McGuire, Towson University, USA
Mark McKenney, Southern Illinois University Edwardsville, USA
Tomas Mildorf, University of West Bohemia - Pilsen, Czech Republic
Beniamino Murgante, University of Basilicata, Italy
Shawn D. Newsam, University of California - Merced, USA
Lena Noack, Royal Observatory of Belgium, Belgium
Alexey Noskov, Israel Institute of Technology (The Technion), Haifa, Israel
Daniel Orellana V., Universidad de Cuenca, Ecuador
Edison Camilo Ospina Álvarez, National University of Colombia - Medellín / QMC Telecom, Colombia
Okan Pala, North Carolina State University's Center for Geospatial Analytics, USA
Kostas Patroumpas, Athena Research Center, Greece
Peng Peng, Data Alibaba Recommendation Group, Alibaba Co., Ltd., China
Donna Peuquet, Pennsylvania State University, USA
Maurizio Pollino, ENEA - Italian National Agency for New Technologies - Rome, Italy
David Prosperi, Florida Atlantic University, USA
Sigrid Reiter, University of Liège, Belgium
Matthias Renz, Ludwig-Maximilians Universität München, Germany
Kai-Florian Richter, Department of Geography - University of Zurich, Switzerland
Armanda Rodrigues, NOVA LINCS - Universidade NOVA de Lisboa, Portugal
Henry Roig Llacer, Institute of Geosciences - University of Brasilia, Brazil

Sergio Rosim, National Institute for Space Research, Brazil
Claus-Peter Rückemann, Leibniz Universität Hannover / Westfälische Wilhelms-Universität Münster / North-German Supercomputing Alliance (HLRN), Germany
Eliyahu Safra, University of Haifa, Israel
Ayman Samy, KOC (Kuwait Oil Company) | GIS UNIT | Outsourced from Openware company (KUWAIT ESRI OFFICIAL DITRIBUTER), Kuwait
Markus Schneider, University of Florida, USA
Shashi Shekhar, University of Minnesota, USA
Spiros Skiadopoulos, University of Peloponnese - Tripoli, Hellas
Francesco Soldovieri, Istituto per il Rilevamento Elettromagnetico dell'Ambiente - Consiglio Nazionale delle Ricerche (CNR), Italy
Mudhakar Srivatsa, IBM T. J. Watson Research Center, USA
Lena Strömbäck, SMHI, Sweden
Kazutoshi Sumiya, University of Hyogo, Japan
Juergen Symanzik, Utah State University - Logan, USA
Ali Tahir, IGIS-NUST Islamabad, Pakistan
Naohisa Takahashi, Nagoya Institute of Technology, Japan
Ergin Tari, Istanbul Technical University, Turkey
Maristela Terto de Holanda, University of Brasilia, Brazil
Jean-Claude Thill, University of North Carolina at Charlotte, USA
Laura Toma, Bowdoin College, Brunswick, USA
Paul M. Torrens, University of Maryland - College Park, USA
Luigi Troiano, University of Sannio, Italy
Theodore Tsiligiridis, Agricultural University of Athens, Greece
E. Lynn Usery, U.S. Geological Survey - Rolla, USA
Taketoshi Ushiama, Kyushu University, Japan
Michael Vassilakopoulos, University of Thessaly, Greece
Iván Esteban Villalón Turrubiates, Universidad Jesuita de Guadalajara, México
Fusheng Wang, Stony Brook University, USA
Jue Wang, Washington University in St. Louis, USA
Iris Weber, Institut für Planetologie, Westfälische Wilhelms-Universität Münster, Germany
Nancy Wiegand, University of Wisconsin-Madison, USA
John P. Wilson, University of Southern California, USA
Eric B. Wolf, US Geological Survey - Boulder, USA
Ouri Wolfson, University of Illinois, USA
Raymond Wong, Hong Kong University of Science and Technology, Hong Kong
Mike Worboys, University of Maine - Orono, USA
Ningchuan Xiao, The Ohio State University - Columbus, USA
Kristina Yamamoto, US Geological Survey, USA
Weiping Yang, Esri, USA
Xiaojun Yang, Florida State University, USA
Zhangcai Yin, Wuhan University of Technology, China
Shohei Yokoyama, Shizuoka University, Japan
Jin Soung Yoo, Indiana University - Purdue University Fort Wayne, USA

Copyright Information

Evaluation of CASE Tools with UML Profile Support for Geographical Database Design

Thiago Bicalho Ferreira, Jugurta Lisboa-Filho
Departamento de Informática
Universidade Federal de Viçosa
Viçosa, Minas Gerais, Brazil
e-mail: thiagao.ti@gmail.com, jugura@ufv.br

Sergio Murilo Stempliuc
Faculdade Governador Ozanan Coelho (FAGOC)
Ubá, Minas Gerais, Brazil
e-mail: smstempliuc@gmail.com

***Abstract*— GeoProfile is a Unified Modeling Language (UML) profile developed for the conceptual modeling of geographical databases. It uses the entire UML infrastructure including Computer-Aided Software Engineering (CASE) tools. Additionally, the Model Driven Architecture (MDA) approach along with constraints specified in Object Constraint Language (OCL) can be used in CASE tools to transform models until the generation of Structured Query Language (SQL) source code. This paper describes the evaluation of a set of CASE tools with UML profile support based on specific requirements for the use of the MDA approach, OCL constraints and other elements to aid the conceptual modeling of geographic databases using the UML GeoProfile. Based on the results, geographical databases designers can choose the tool that best suits your project or use the evaluation methodology used here to evaluate other CASE tools.**

Keywords-UML Profile; CASE tools; MDA; OCL; Geographical Database.

I. INTRODUCTION

Given the complexity of spatial data, researchers have dedicated themselves for over the last twenty years to adapt original formalisms of the Entity-Relationship (ER) model and the Object-Oriented model aiming to allow the conceptual modeling of geographic databases [1][2].

These researches proposed several conceptual models such as OMT-G [3], MADS [4], GeoOOA [5], UML-GeoFrame [6], Perceptory [1], GEOUML [7], STGL Profile [8], and ChronoGeograph [9]. Moreover, several specific Computer-Aided Software Engineering (CASE) tools have been implemented for these models. The use of several conceptual models and tools then created issues such as the lack of a standard in geographical database (GDB) modeling and the lack of interoperability among the them. In face of such problems, references [10] have proposed a Unified Modeling Language (UML) profile called GeoProfile.

GeoProfile can use all of UML's infrastructure, which includes Object Constraint Language (OCL) to define integrity constraints and Model Driven Architecture (MDA) for the transformation between its different abstraction levels [10][11]. Moreover, one of the advantages of using a UML profile is that it can be used in different CASE tools. However, not all tools offer the same features, which difficult the GDB designer to choose one. Examples of CASE tools with UML support include Enterprise Architect, Papyrus, StarUML, Visual Paradigm, and IBM Rational Software Architect.

In order to compare these tools and in the context of this study, some characteristics or features were prioritized such as the support to the UML Profile definition, validation of OCL constraints, and application of the MDA approach. The key aspect is that the tools need to allow models to be created using UML GeoProfile, the transformation among the different levels established by the MDA architecture, the syntactic e semantic validation of spatial OCL constraints, and that the models should be implemented from scripts generated for a selected database management system.

This paper aims to describe the evaluation of a set of CASE tools considering important requirements from the conceptual project to the implementation of the geographical database.

The remaining of the paper is structured as follows. Section II briefly explains the representation of geographical data, the UML GeoProfile, the MDA approach and the syntax to specify OCL expressions. Section III presents a description of each CASE tool analyzed according to the goal of this study. Section IV shows the requirements, the methodology and the result of the tool evaluations. Section V presents the conclusions and future works.

II. GEOGRAPHICAL DATABASE MODELING CONCEPTS

This section presents a literature review identifying the main concepts that contribute to the conceptual GDB modeling.

A. Representing Geographical Information in Computers

The representation of geographical space in computers is a challenge faced by researchers. According to Longley et al. [12], the world is infinitely complex and computing systems are finite, thus, it is up to the designer to limit the amount of details to be captured from the environment mapped. The two main approaches on computing are the continuous (fields) and discrete (objects) representations. Another representation also employed is in the form of networks, which takes into account graph theory.

Figure 1 shows part of a city with a sports center and represents part of this city focusing on the roads and the stadium. The GDB of Figure 1(b) must be conceptually modeled containing all structures of interest in the system while leaving aside other information such as the type of vegetation, vacant plots, terrain, and other characteristics that may be abstracted from Figure 1(a).

In order to design the conceptual data schema, first the vector structures used to represent the boundaries of each geographic entity must be understood, which is normally specified through basic geometric shapes: point, line and polygon (area) [13]. Figure 1(b) presents the use of these three types of vector structures. For instance, the stadium may be spatially represented as a point or as a polygon (multiple spatial representation); the main east road, as a line; and the sports center, as a polygon.

Additionally, presenting the structures, Figure 1(b) illustrates the relationship among the vector objects, which shows the stadium "is within" the sports center, the sports center "touches" the road to the stadium, the main west road "is near" the sports center, but does not "touch" it.

Such relationships are known as topological relationships and have been discussed by [14] and [15] and used by [16].

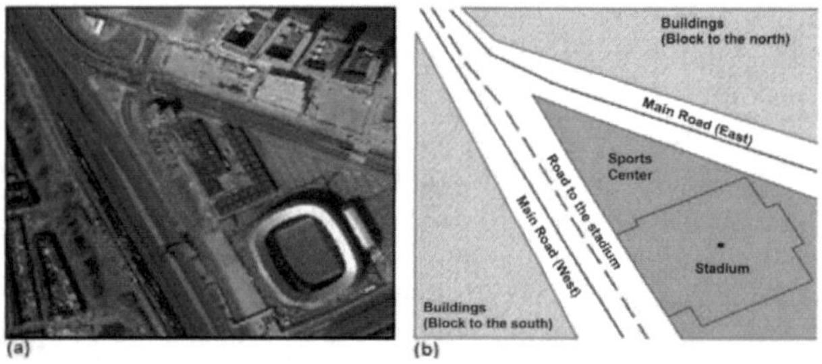

Figure 1. (a) Photograph of part of a city with a sports center between roads. (b) Spatial representation of this area. Source: Adapted from [17].

B. Model-Driven Architecture (MDA)

According to Kleppe et al. [18], MDA is a framework standardized by the [19] for the development of software employing a Model-Driven Development (MDD) view.

The MDA approach consists of three abstraction levels, namely, CIM, PIM and PSM. Computation-Independent Model (CIM) does not show details of the system's structure, but rather the environment in which the system will operate. Platform-Independent Model (PIM) is an independent model of any implementation technology containing the software requirements. Platform-Specific Model (PSM) specifies details about the platform in which it will be implemented. The artifacts produced by the MDA approach are formal models that can be processed by computers and, after undergoing transformations, will get to a final source-code step (top-down approach) or to high levels of abstraction (bottom-up approach). Figure 2 illustrates the action of transformation tools at MDA levels.

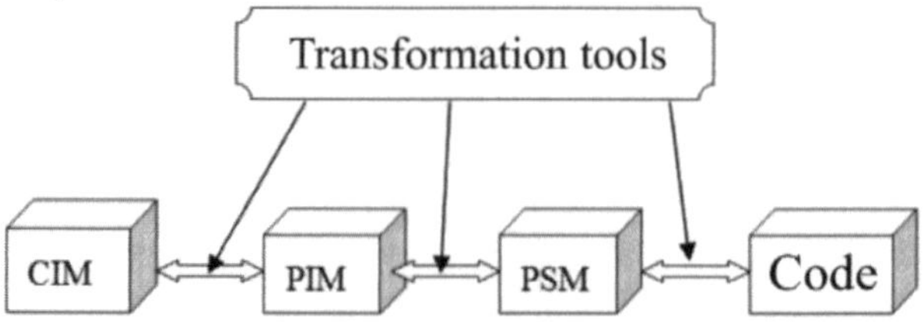

Figure 2. Use of transformation tools in the MDA approach. Source: Adapted from [18].

C. Object Constraint Language (OCL)

Conceptual modeling makes the problem easier to be understood through abstraction, thus enabling risk management and contributing to error correction early in the project, which minimizes the cost of maintenance [20]. However, Warmer and Kleppe [21] state that conceptual models may not be able to represent all requirements, resulting in problems to those who interpret them.

The OCL, adopted by OMG [22] since version 2.0, was defined as a formal language to complement the conceptual modeling using UML. Using OCL ambiguity-free integrity constraints can be created, which makes it possible to specify the data consistency wanted in the system at a high level of abstraction. Since it is a formal language, it can be processed by CASE tools until the source-code generation, which enables more powerful and satisfactory data consistency [21]. OCL is currently at version 2.4 [23].

The OCL expressions represents constraints that are needed in the system and not how they should be implemented. The evaluation of a constraint on the data always yields a Boolean value [21]. The syntax of a typical expression in OCL that represents a condition has the format presented by Code 1.

```
<context>
  inv:<expression>          (1)
```

Code 2 illustrates a hypothetical example of OCL constraint that specifies that a Brazilian municipality must be larger than 3,000 km² (note: The smallest Brazilian municipality, Santa Cruz de Minas, MG, is 3,565 km²). A detailed specification of the OCL can be found in [21] [23].

```
context Municipality
  inv:self.area > 3000      (2)
```

D. UML GeoProfile

In order to provide elements for specific domains without becoming excessively complex, UML has an extension mechanism called Profile. A UML Profile consists of: a metamodel; a set of stereotypes presented through texts in the form of <<text>> or through graphical icons called pictograms; tagged values; and constraints; all grouped in a stereotyped package called <<profile>>, thus formalizing the UML builder extension [24].

GeoProfile is a UML profile proposed for the geographical data modeling comprising the main characteristics of the existing models in the field [25]. GeoProfile is employed at the CIM and PIM levels of the MDA approach, using OCL constraints as a resource to validate the conceptual scheme generated by the designer [26].

The GeoProfile stereotypes are extensions of the Association and Class metaclasses. The stereotypes extended from the Class metaclass allow representing the geographic space in the discrete view (e.g., points, lined and polygons), in the continuous view (e.g., large cells and triangular networks), and through networks (nodes and arcs). The temporal aspects can also be represented with the stereotypes made up of tagged values that store instant and range values. The extended stereotypes of the Association metaclass allow representing topological relationships (e.g., touches and within) among the geographical stereotypes, and the temporal relationship (Temporal) among the temporal objects.

For the extended stereotypes of the Class metaclass, the abstracted stereotypes have been defined: <<Network>>, to group network stereotypes; <<GeoObject>>, to group the discrete view stereotypes; <<GeoField>>, to group the continuous view stereotypes; and <Arc>>, to group the <<UnidirectionalArc>> and <<BidirecionalArc>> stereotypes that represent the possible links between the nodes of a network.

III. CASE Tools Analyzed

The tools analyzed in this study were chosen according to the ease of access to the software and documentation. These tools are open source and commercial with some support to the UML profile and are well known by the software development community. The sub-sections below describe the results of the analysis made on the following CASE tools, exploring the resources they offer compared to the GeoProfile: Enterprise Architect (EA) version 12.0, Papyrus UML2 Modeler (Papyrus) version 1.12.3, StarUML–UML/MDA Plataform (StarUML) version 5.0.2.1570, Visual Paradigm for UML (VP) version 10.2 and IBM Rational Software Architect (RSA) version 9.0.

A. Enterprise Architect (EA)

Enterprise Architect (EA) [27] is a commercial CASE tool licensed by Sparx Systems that allows the visual creation of UML profiles and insertion with syntactic validation of OCL expressions. EA does not offer resources for semantic validation of OCL expressions.

Additionally, being a modeling tool, it acts as an MDA transformation tool, with its own language for transformation between the model levels. This language can be modified so that the users are able to reach the last MDA approach level, the source code [28]. Since the modeling in this paper refers to GDB, the last MDA step is the Data Definition Language (DDL) source code, which EA is able to generate.

The GeoProfile stereotypes in the EA tool can be represented graphically ⊡ or textually <<point>>. The tool also offers resources for multiple stereotype representation, e.g., depending on the scale, a city may be modeled as a point or a polygon <<point, polygon>>.

The advantage at using EA is that it does not allow the insertion of extended stereotypes of the Class metaclass in Association elements and vice versa. The problem is that it allows the use of abstract stereotypes in conceptual models, e.g. the abstract GeoProfile stereotypes: <<Arc>>, <<GeoField>>, <<GeoObject>>, <<Network>> and <<NetworkObj>>.

B. Papyrus UML2 Modeler

Papyrus UML2 Modeler [29] is an open-source tool based on the Eclipse environment and licensed by Eclipse (Eclipse Public License). It has a visual environment to insert UML profiles, thus providing support to insertion and syntactic validation of OCL constraints. However, it does not semantically validate these constraints.

Adding graphical icons to the stereotypes is possible. Thus, a class or association can be represented by stereotypes as follows: only text, only graphical icon, or graphical icon and text. The Papyrus tool allows multiple representation to be specified through stereotypes, but, in case the graphical representation is used, only the first stereotype used by the designer is presented.

Additionally, restricting the use of abstract GeoProfile stereotypes in conceptual models, in this CASE tool other GeoProfile stereotypes can only be used with correct UML elements, i.e., an extended stereotype of the Association metaclass cannot be used in a class defined by the Class metaclass.

The Papyrus tool does not support the MDA approach, the transformation language and DDL code generation.

C. StarUML

StarUML [30] is an open-source tool whose profile insertion is done through an Extensible Markup Language (XML) document. This tool does not support OCL and, despite being considered MDA, the features offered are incomplete. What it allows is transforming a model (PIM) into source code without going through the PSM. The source codes can be generated for the languages Java, C++ and C#. StarUML does not have a transformation language and the conceptual models produced from GeoProfile cannot be transformed into DDL source code.

Although multiple stereotype representation is not supported by the tool, the designer can choose between graphical and text representation, but only text is supported in associations. Therefore, the possible class stereotype representations are: textual, graphical, and textual and graphical. The tool can also restrict the use of abstract stereotypes at the same time that the others can be properly used with UML elements.

D. Visual Paradigm for UML (VP)

With an intuitive modeling environment, the commercial tool Visual Paradigm for UML [31] supports the visual creation of UML profiles. The stereotypes can be presented graphically or textually, with support for multiple representation with the graphical ones.

The tool does not allow the use of extended stereotypes of different metaclasses, as described in section A however, it does allow abstract GeoProfile stereotypes to be used during conceptual modeling.

The tool allows incomplete MDA approach, transforming PIM straight into source code. Nevertheless, it does not support DDL code generation from UML class diagrams, just only from those created through the ER model. Thus, the GeoProfile conceptual models cannot be transformed into DDL code.

This tool also does not support the syntactic and semantic validation of OCL constraints on conceptual models created from GeoProfile.

E. Rational Software Architect (RSA)

RSA [32] is a commercial CASE tool licensed by IBM that allows the visual creation of UML profiles. This tool supports the use of profiles and is designed to allow syntactic and semantic validation of OCL constraints applied to UML diagrams.

The representations by the stereotypes in an association or class may take place as follows: only textual stereotype, only graphical stereotype, and representation by the textual and graphical stereotypes. However, the multiple representation by the stereotypes can take place in two ways: All stereotypes applied to the class or association must be in textual format or the first stereotype applied takes on the graphical format and the others on textual format.

The tool does not allow inserting extended stereotypes of the Class metaclass in association elements and vice versa, and the stereotypes defined as abstract in GeoProfile cannot be used in the UML elements.

RSA has incomplete support to MDA since it does not natively allow DDL source-code generation. Although there is a transformation mechanism in which the origin, target, and some settings regarding the mapping in the transformation from model into source code can be determined, RSA does not have an MDA transformation language. Therefore, with RSA's native features and mechanisms, these transformations cannot be performed on models created from the GeoProfile.

IV. Results of the CASE Tools Comparison

This section initially presents a set of requirements the CASE tools must meet to support conceptual GDB modeling based on the GeoProfile. Next, it presents the method used in the evaluation, the results and the final classification of the CASE tools analyzed.

This method, originally proposed by Rosario and Santos Neto [33], was used in exploratory research involving software project management tools. This method was also applied by Câmara et al. [34] on comparison of development environments for systems of Volunteered Geographic Information (VGI).

A. Requirements Survey

Based on the literature and on the descriptions of each CASE tool, this paper proposes requirements to evaluate which tool has the greatest amount of features to support the GeoProfile use, aiming the transformation of data models at the different MDA levels and to specify integrity constraints at conceptual level using OCL. Table I lists these requirements.

B. Evaluation Method of CASE Tools

In the context of this study, the requirements were classified as follows:

- Requirements that are Essentials: Weight 3;
- Requirements that are Important: Weight 2;
- Requirements that are Desirable: Weight 1.

Additionally, to the weight attributed to requirements, a scale must be defined for how well the tools satisfy each one. They may not satisfy (NS), partially satisfy (PS), or satisfy (S) a requirement. Therefore, the following scales can be attributed:

- Does not satisfy the requirement: A scale with value 0 is attributed;
- Partially satisfies the requirement: A scale with value 1 is attributed;
- Satisfies the requirement: A scale with value 2 is attributed.

Based on this evaluation, the classification of each tool was calculated by adding up the products of the importance weight (W) and the satisfaction scale (S) for each requirement (n), represented by Formula (3). References [33] originally proposed this method.

$$X = \sum_{i=1}^{n} S_i . W_i \qquad (3)$$

TABLE I. REQUIREMENTS TO EVALUATE CASE TOOLS

	Requirement description
Rq 01	Correct attribution of GeoProfile stereotypes in the UML elements
Rq 02	Restriction to the use of abstract stereotypes in elements of the model
Rq 03	Support to syntactic validation of OCL constraints
Rq 04	Support to semantic validation of OCL constraints
Rq 05	Support to MDA transformations
Rq 06	Support to transformation language
Rq 07	Support to graphical exhibition of profile stereotypes
Rq 08	Support to multiple representation through stereotypes
Rq09	Support to visual profile creation
Rq 10	Support to DDL code generation
Rq 11	Open-source tool

C. Evaluation of the CASE Tools

In order to evaluate each CASE tool and its practical capacity regarding the theoretical functionalities predicted for a UML profile, particularly GeoProfile, the requirements presented in Table I were classified according to the following criteria:

- The requirements considered essential are those that support MDA;

- Requirements that aid in transformations between MDA models are considered important;
- Requirements that care for quality of the GDB models are considered important;
- Requirements that facilitate understanding and contribute to the adoption of the tool are considered desirable.

Table II presents the classification of the requirements regarding their importance level, which are *Essential*, *Important* or *Desirable*. Table III presents the way each CASE tool satisfies the requirements of Table I. At the end, the summary of the evaluation based on Formula (3) is presented using the data from Tables II and III.

TABLE II. CLASSIFICATION OF THE REQUIREMENTS BASED ON THE IMPORTANCE LEVEL.

Importance	Requirements
Essential	Rq05
Important	Rq 01, Rq 03, Rq 04, Rq 06, Rq 08, Rq 10
Desirable	Rq 02, Rq 07, Rq 09, Rq 11

Table III shows the level of satisfaction for each of the CASE tools analyzed, considering each of the 11 requirements. A CASE tool may or may not support a requirement, or provide partial support. For example, EA offers full support for Rq 01. The assigned scale for this level of satisfaction is 2. Meanwhile, Rq 01 was classified as "important" in Table II, therefore receiving weight 2. So when Formula (3) is applied, the sum of (scale x weight) is calculated for all requirements. Thus, the total sum for EA is 30. The same method was used for all the other tools.

An analysis of Table III shows that the Enterprise Architect tool was the one that best satisfied the requirements for the transformation of conceptual models so that the OCL constraints can be used in the tool. Since it has a customizable transformation language, the OCL constraints can be transformed into integrity constraints along with the SQL code generated in the last MDA level.

Another situation that can be observed in Table III is that the CASE tool RSA provides the best features to use the OCL constraints since it allows for both syntactic and semantic validations.

V. CONCLUSIONS AND FUTURE WORK

With this paper is possible to observe that the tools evaluated do not have features to meet all the theoretical needs of UML, mainly regarding the use of profiles, MDA and OCL. However, they all support conceptual GDB modeling using GeoProfile.

The results of the comparison show that at the time this paper was written the EA could be considered the best CASE tool regarding transformations at the different MDA levels of models created using the GeoProfile. The RSA can be considered the tool that best supports OCL constraints due to its semantic validation, which makes the conceptual models less prone to errors. Among the free-software tools, Papyrus stood out compared to StarUML for supporting the GeoProfile.

Based on the results in this paper, a designer intending to use GeoProfile can know which CASE tool currently best meets the needs of the GDB project. However, it is important to point out that all tools analyzed are being constantly improved, which can change the results of this comparison at any moment.

The method employed, originally proposed by Paranhos and Santos Neto [33], can be used for different comparisons so that designers can establish their own requirements and assign importance weights and satisfaction scales to each one.

As future research, studies are being done aiming to reach interoperability of conceptual geographical data models created from different conceptual metamodels specific for geographical databases, whose transformation base is the GeoProfile metamodel.

TABLE III. CLASSIFICATION OF THE CASE TOOLS

CASE	Enterprise Architect			Rational Software Architect			Visual Paradigm			Papyrus			StarUML		
Requirement	S	PS	NS	S	PS	NS	S	PS	NS	S	PS	NS	S	PS	NS
Rq 01	X			X			X			X			X		
Rq 02			X	X					X	X			X		
Rq 03	X			X					X	X					X
Rq 04			X	X					X			X			X
Rq 05	X				X			X				X		X	
Rq 06	X					X			X			X			X
Rq 07	X			X			X			X			X		
Rq 08	X			X			X			X					X
Rq 09	X			X			X			X					X
Rq 10	X					X	X					X			X
Rq 11			X			X			X	X			X		
Total	30			25			19			20			12		

ACKNOWLEDGEMENTS

Project partially funded by the Brazilians agencies FAPEMIG and CAPES. We also thank the support of Faculdade Governador Ozanan Coelho (FAGOC).

REFERENCES

[1] Y. Bédard, S. Larrivee, M. J. Proulx, and M. Nadeau, "Modeling geospatial databases with plug-ins for visual languages: a pragmatic approach and the impacts of 16 years of research and experimentations on perceptory". Lecture Notes in Computer Science 3289, 2004 pp. 17-30.

[2] F. Pinet, "Entity-relationship and object-oriented formalisms for modeling spatial environmental data," Environmental Modelling & Software, vol. 33, 2012, pp. 80-91.

[3] K A. V. Borges, C. A. Davis, and A. H. F. Laender, "OMT-G: An ObjectOriented Data Model for Geographic Applications", GeoInformatica, vol. 5, no. 3, 2001, pp. 221-260.

[4] C. Parent, S. Spaccapietra, and E. Simonyi, "Modeling and Multiple Perceptions," in S. Shekhar and H. Xion, Eds., Encyclopedia of GIS, Berlim: Springer-Verlag, 2008, pp. 682-690.

[5] G. Kösters, B. Pagel, and H. Six, "GIS-Application Development with GeoOOA," International Journal of Geographical Information Science, vol. 11, no. 4, 1997, pp. 307-335.

[6] J. Lisboa Filho and C. Iochpe "Modeling with a UML Profile," in S. Shekhar and H. Xiong, Eds. Encyclopedia of GIS, Berlim: Springer-Verlag, 2008, pp. 691-700.

[7] A. Belussi, M. Negri, and G. Pelagatti, "Geouml: a geographic conceptual model defined through specialization of iso tc211 standards". Proc. 10th EC GI & GIS Workshop, ESDI State of the Art, Warsaw, Poland, 2004, pp. 23-25.

[8] A. Miralles and T. Libourel, "Spatial database modeling with enriched model driven architecture," in S. Shekhar and H. Xion, Eds., Encyclopedia of GIS, Berlim: Springer-Verlag, 2008, pp. 700-705.

[9] D. Gubiani and A. Montanari, "ChronoGeoGraph: an expressive spatiotemporal conceptual model," Proc. of the Fifteenth Italian Symposium on Advanced Database Systems, 2007, pp. 160-171.

[10] G. B. Sampaio, F. R. Nalon, and J. Lisboa-Filho, "GeoProfile-UML Profile for Conceptual Modeling of Geographic Databases," Proc. Int. Conf. on Enterprise Information Systems (ICEIS), Funchal-Madeira, Portugal, 2013, pp. 409-412.

[11] F. R. Nalon, J. Lisboa-Filho, K. A. V. Borges, J. L. Braga, and M. V. A. Andrade, "Using MDA and a UML Profile integrated with International Standards to Model Geographic Databases," Proc. Brazilian Symposium on Geoinformatics (GeoInfo), 2010, pp. 146-157.

[12] P. A. Longley, M. F. Goodchild, D. J. Maguire, and D. W. Rhind, Geographic Information Science and Systems, Danvers: Wiley, 2010.

[13] G. Câmara, "Computational representation of spatial data", In: M. A. Casanova, G. Câmara, C. A. Davis Jr., L. Vinhas, and G. R. Queiroz (Org.), "Bancos de Dados Geográficos". Curitiba: EspaçoGeo, cap. 1, 2005, pp. 1-44. (in Portuguese)

[14] E. Clementini, P. Di Felice, and P. Oosterom, "A Small Set of Formal Topological Relationships Suitable for End-User Interaction," Proc. Int. Symposium on Advances in Spatial Databases, 1993, p. 277-295.

[15] M. J. Egenhofer and R. D. Franzosa, "Point-set topological spatial relations," International Journal of Geographic Information Systems, vol. 5, no. 2, 1991, pp. 161-174.

[16] A. A. A. Ribeiro, S. M. Stempliuc, and J. Lisboa-Filho, J., "Extending OCL to specify and validate integrity constraints in UML-GeoFrame conceptual data model," Proc. Int. Conf. on Enterprise Information Systems (ICEIS), Angers Loire Valley, France, 2013, pp. 329-336.

[17] R. Kothuri, A. Godfrind, and E. Beinat, Pro Oracle Spatial for Oracle Database 11g, USA: Apress, 2007.

[18] A. Kleppe, J. Warmer, and W. Bast, MDA Explained: The Model Driven Architecture: Practice and Promise, Boston: Addison-Wesley, 2nd ed., 2003.

[19] OMG., MDA Guide, OMG Document formal/2003-06-01 edition, Needham, MA, USA, Version 1.0.1, 2003.

[20] G. Booch, J. Rumbaugh, and I. Jacobson, UML: user guide, Elsevier. Rio de Janeiro, 2nd ed., 2005.

[21] J. Warmer and A. Kleppe, The Object Constraint Language: Getting Your Models Ready for MDA, Boston: Addison Wesley, 2nd ed., 2003.

[22] OMG., Object Constraint Language, OMG Document formal/2006-05-01 edition, Needham, MA, USA. Version 2.0, 2006.

[23] OMG., Object Constraint Language, OMG Document formal/2014-02-03 edition, Needham, MA, USA. Version 2.4, 2014.

[24] H. Eriksson, et al., "UML 2 Toolkit", Wiley Publishing. Indianapolis. 552p, 2004.

[25] J. Lisboa Filho, G. B. Sampaio, F. R. Nalon, and K. A. V. Borges, "A UML profile for conceptual modeling," Proc. Int. Workshop on Domain Engineering (DE@CAISE), 2010, pp. 18-31.

[26] F. R. Nalon, J. Lisboa-Filho, J. L. Braga, K. A. V. Borges, and M V. A. Andrade, "Applying the model driven architecture approach for geographic database design using a UML Profile and ISO standards," Journal of Information and Data Management, vol. 2, no. 2, 2011, pp. 171-180.

[27] Sparx Systems, Enterprise Architect 12.1. [Online]. Available from: http://www.sparxsystems.com.au/ 2016.04.13

[28] T. B. Ferreira, S. M. Stempliuc, and J. Lisboa-Filho, "Data Modeling with UML Profile GeoProfile and Transformations in MDA tool Enterprise Architect," Actas. Conferencia Ibérica de Sistemas y Tecnologías de Informacion (CISTI), Barcelona, AISTI | ISEGI, 2014, pp. 603-608.

[29] Eclipse Foundation, Papyrus Modeling Environment. [Online]. Available from: http://www.eclipse.org/papyrus/ 2016.04.13

[30] Star UML, StarUML 2: A sophisticated software modeler. [Online]. Available from: http://staruml.io/ 2016.04.13

[31] Visual Paradigm International, Visual Paradigm, [Online]. Available from: https://www.visual-paradigm.com/ 2016.04.13

[32] IBM, Rational Software Modeler. Rational Software Architect Family. [Online]. Available from: http://www-03.ibm.com/software/products/en/rational-software-architect-family 2016.04.13

[33] R. D. D. Paranhos and I. Santos Neto, Comparative study of change control tools in the software development process, Salvador: Universidade Católica do Salvador, 2009.

[34] J. H. S. Câmara, T. Almeida, D. R. Carvalho, et al., "A comparative analysis of development environments for voluntary geographical information Web systems," Proc. of Brazilian Symposium Geoinformatics (GEOINFO), 2014, pp. 130-141.

Impact of DEM Processing on the Geotechnical Instability Analysis of Waste Heaps in Wallonia

Odile Close
Faculty of Bioengineering
UCL, Université Catholique de Louvain
Louvain-la-Neuve, Belgium
e-mail: odile.close@gmail.com

Nathalie Stéphenne, Christophe Frippiat
Risques chroniques
ISSeP, Institut Scientifique de Service Public
Liège, Belgium
e-mail: n.stephenne@issep.be, ch.frippiat@issep.be

Abstract— **This paper evaluates the effects of Digital Elevation Model (DEM) processing on the geotechnical instability analysis of four waste heaps located in Western Wallonia. For this purpose, an infinite slope stability equation has been computed on each cell of two DEMs (LiDAR and ERRUISSOL, available on the geoportal of Wallonia). By processing raw LiDAR data with various interpolation techniques and spatial resolutions, we tested their influence on the slope stability analysis. In order to better understand the real geotechnical stability, several datasets have been used in the analysis (field observation, historical aerial photography and ortho-photos). Our results show that interpolation techniques and spatial resolution affect the DEM quality in regard to slope stability analysis. In particular, by removing striped patterns resulting from data acquisition, the Triangulation technique facilitates stability assessment. According to our findings, 10 m resolution is sufficient and adequate for stability analysis while 1 m resolution overestimates the risk of slope failure.**

Keywords: Digital Elevation Model, LiDAR, geotechnical risk, geospatial web services, terrain analysis

I. INTRODUCTION

A first inventory of Walloon facilities at risk submitted to European Union authorities identified geotechnical failure as being one of the major risks linked to coal mine waste heaps [1]. The inventory responded to the obligation imposed on Member States by Directive 2006/21/EC to identify the risks related to waste facilities.

Light Detection and Ranging (LiDAR) Digital Elevation Model (DEM) was used by [2] for the investigation of landslides risks. However, these authors identified some interpretation issues in the stability results derived from the LiDAR DEM provided by the regional authorities on their web geoportal [3]. Their study has quantified the risk of geotechnical failure by using a geotechnical factor of safety computed on a cell basis using the topography of the facility. The topography was extracted either from a regional-scale DEM with a spatial resolution of 10 m (ERRUISSOL model; [4]) or a new DEM dataset using LiDAR scanner which has been acquired by the Walloon Region. This dataset with 1 m resolution has been averaged to 10 m for direct comparison with the ERRUISSOL DEM. As the ERRUISSOL DEM leads to underestimation of geotechnical instabilities, it is possible through the LiDAR DEM to identify new waste heaps at risk.

LiDAR is a powerful system for producing a DEM on account of its ability to collect three-dimensional information very effectively over large areas [5]. However, there are many ways of processing a DEM that use different interpolation techniques and spatial resolution. Several authors [6]-[10] have shown that some interpolation methods are more appropriate than others in certain circumstances, the method chosen thus having the potential of affecting the quality of the DEMs produced [11]. For example, according to [14] Triangulated Irregular Networks (TINs) is the best interpolation algorithm for fluvial environment topics and [5] consider that the DEM generated by Binning is efficient for analysing the terrain features (communication, energy, agriculture, etc.).

This paper studies technical choices in terms of resolution and interpolation methods by processing raw LiDAR data rather than provided DEM and analysing their potential for the specific topic of landslide risks. Comparing processing methods with detailed field knowledge increases the applicability of these data in the Walloon decision making process.

The paper is structured as follows. Section II presents the studied area and the context of the study. Section III describes the different datasets. Section IV explains the methodology used. Finally, Section V presents our conclusions.

II. CASE STUDY

The present study evaluates the slope stability of four waste heaps located in western Wallonia (Heribus, 14-17 et Siege Social, Crachet 7-12 and Saint-Placide). The study area is located in the Borinage Region (Figure 1). In the past century, the economy of the Region was based on coal exploitation. However, this activity has been closed for a long time (the last operating coal mine in Wallonia closed in 1984). Nowadays, waste heaps and abandoned buildings are the only surviving traces of this period.

According to Directive 2006/21/EC, Member States have to establish an inventory of closed mining waste facilities potentially posing a serious threat to human health or environment. In Wallonia, two main risks associated with mining waste deposits have been identified: the risk of geotechnical failure and the risk of spontaneous combustion [1].

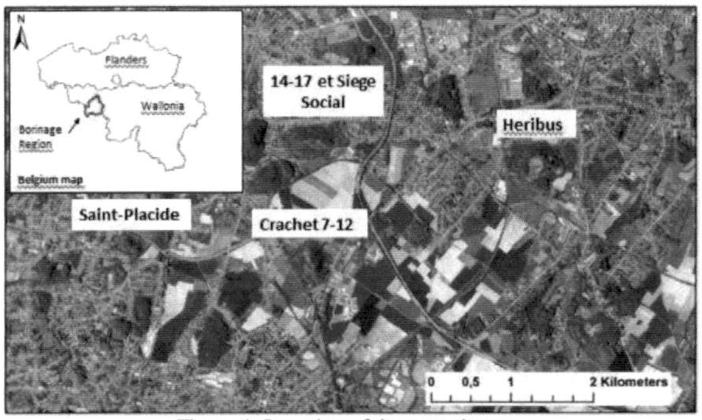

Figure 1. Location of the waste heaps

The inventory has been made in a Geographical Information System (GIS) using existing datasets provided either by European institutions (European Environmental Agency, EEA) or regional authorities (Service Public de Wallonie, SPW). GIS inputs include topographic data (location of settlements, surface waters, terrain, etc.), census figures, protected areas, land use/land cover surfaces, delineation of groundwater bodies (according to the Water Framework Directive), and site data that are specific to the waste facilities under consideration (location, contents, geometry, etc.). Table 1 illustrates some of the criteria used to define the level of risk for each Walloon coal tips. The four waste heaps object of this study were all classified in category 5, which means that they involve at least one specific risk and that there is at least one potential target located in the immediate vicinity of the closed waste facility referenced on the web site [12].

TABLE 1 CHARACTERISTICS OF THE FOUR WASTE HEAPS AND CLASSIFICATION

Risk Assessment criteria	**Heribus**	**14-17 et Siege Social**	**Crachet 7-12**	**Saint-Placide**
Height > 20 m	yes (75.2 m)	yes (78.3 m)	yes (60.2 m)	yes (35.1 m)
Slope > 1/12	no	no	no	no
Volume > 70,000 m³	yes (5.2 10^6 m³)	yes (4.1 10^6 m³)	yes (3.4 10^6 m³)	yes (1.5 10^6 m³)
Main foundation slope > 33%	no	yes	yes	yes
Materials exposed to the wind	no	no	no	no
Waste facility uncovered	no	no	no	no
Target within 1 km	yes (watercourse and population)	yes (population)	yes (population)	yes (population)
Natura 2000 site within 1 km	yes	yes	yes	no
Risk classification	5	5	5	5

III. DATA

The quantitative inventory established to answer the Directive permits a pre-selection of the facilities that demand further risk analysis. The four heaps selected in this paper need this detailed analysis which refers to geotechnical modelling, field observations and visual interpretation of ancillary data such as orthos-photos and historical photos (table 2). These data support the understanding of the geotechnical risk but this paper has a specific technical focus on the comparison of both DEM data sources, LiDAR and ERRUISSOL.

A. *Preliminary information*

The preliminary knowledge about the waste heaps was gathered by analyzing ortho-photos and historical aerial photography. The ortho-photos are from different periods (2006-2007, 2009-2010, 2012-2013) and are available on the geoportal of the Walloon Region [13]. The historical aerial photographs are provided by the SPW on two dates: 1954 and 1969.

TABLE 2 PRELIMINARY AND FIELD INFORMATION OF THE FOUR WASTE HEAPS

	Aspects	**Assessment**
Preliminary information	Vegetation	Ortho-photos analysis (2006-2007, 2009-2010, 2012-2013) (http://geoportail.wallonie.be)
	Erosion	Ortho-photos analysis (2006-2007, 2009-2010, 2012-2013), DEM visualisation (http://geoportail.wallonie.be)
	Evidence of landslide	Aerial photography (1954, 1969)
	Evidence of spontaneous combustion	Literature
Field information	Vegetation	Presence of trees, grass
	Erosion	Presence of gully
	Evidence of landslide	Scarp, absence of vegetation, inclined vegetation
	Evidence of spontaneous combustion	Presence of burnt coal and fumaroles

B. *Field observations*

A field campaign was conducted on the four waste heaps to consider the following four aspects: (i) the presence of vegetation, (ii) the presence of gully erosion, (iii) traces of landslide and (iv) traces of spontaneous combustion. Vegetation cover is an important factor to consider when studying the susceptibility of slope failure because plant roots contribute to maintaining slope stability by increasing soil cohesion [15]. The second factor observed, gully erosion, is an erosional process whereby drainage lines are generated by ephemeral streams. This erosional landform may affect the soil stability by increasing the slope gradient [16]. The third factor examined, previous landslides, is significant because it indicates that the coal tip was not stable in the past. The absence of vegetation may indicate that a landslide has occurred. Finally, spontaneous combustion can lead to soil instability in two ways. Firstly, it may induce the formation of cavities underneath the surface and be responsible for the apparition of a discontinuity plane between burning coal zones and unburnt coal zones that may lead to the occurrence of a landslide [17]. Secondly, the surface of a burning waste heap can reach a temperature superior to 100°C. These burning zones scatter the grass vegetation and prevent the growing of trees [18].

All the waste heaps examined exhibit evidences of past and/or current spontaneous combustion. Fumaroles have even been observed on 14-17 et Siege Social. The trace of a landslide has been observed in Heribus but was not noticed on the other coal tips. Heribus seems to be the most inclined to slope failures since it has already undergone two sliding events in the past (1992 and 1994) [17] and because of its land cover, burning zones and history. All waste heaps, except Saint-Placide, show traces of gully erosion. The vegetation cover was generally abundant except on the zones where spontaneous combustion has been observed and the zones on which landslides have occurred (Heribus).

C. *LiDAR data*

The data acquisition was performed between 2013 and 2014 over the Walloon Region with a Riegl Litemapper 6800i system. The LiDAR system operated a pulse repetition rate of 150 kHz. The flight altitude was between 1200 and 1500 meters and the point density was about 0.8 point/m². Point data is post-proceeded by analysing the laser time range, the laser scan angle (60°), the Global Positioning System (GPS) position and the Inertial Measurement Unit (IMU) [19]. LiDAR returns were classified as 'ground', 'vegetation', 'building', 'water' and 'unclassified' by the data provider [2]. The coordinate system used here was Belgian Lambert 72.

D. *ERRUISSOL data*

The ERRUISSOL data are the result of the integration of several datasets: (i) elevation points derived from orthophotos interpretation of the Projet Informatique de Cartographie Continue (PICC) at a scale of 1/1,000 with a distance of 50 m between points, (ii) points from the Digital Terrain Model (DTM) based on aerial photos at a scale of 1/10,000 with a distance of 20 m between points and, (iii) Digital Terrain Model (DTM) from local LiDAR flights on the watershed with a resolution of 1 point/m². The ERRUISSOL DEM has been produced in 2003 [2][4].

VI. METHOD

This paper addresses technical questions raised by [2] in the processing of raw LiDAR data. These authors point out that the slope ground surface map of the LiDAR DEM present zones with striped patterns that do not appear in the ERRUISSOL model (Figure 2). Image 1 shows ERRUISSOL data (10 m) and image 2 LiDAR data (1 m). Striped patterns appear on the high resolution image. These striped patterns indicate a succession of high and low slope. Their signification has not been elucidated in previous study and deserves our attention. To evaluate the impact of this technical issue on geotechnical assessment, the infinite slope stability model (section VI A) has been used on the basis of several DEM interpolation techniques (section VI B). The assessment of the susceptibility of slope failures is quantified by using a factor of safety that is computed on each pixel of a DEM with ArcGIS ©ESRI.

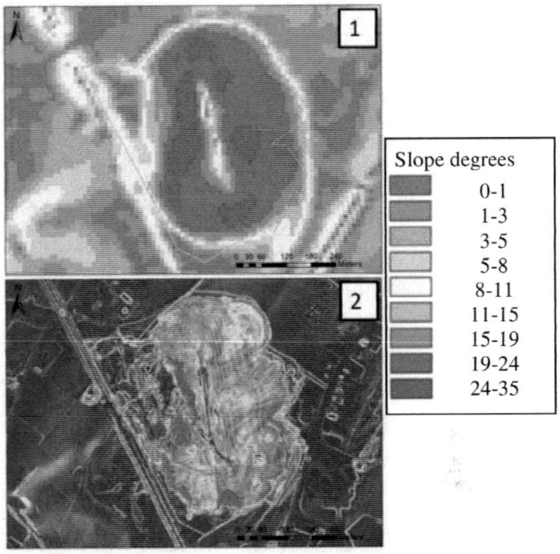

Figure 2 Slope ground surface (degrees) for crachet waste heap.

A. *Infinite slope stability model*

Slope stability analysis is commonplace in soil mechanic, engineering geology and geomorphology. Usually, to express the stability of a slope, a safety factor, F, is used. This factor is defined as the ratio of a maximum admissible load and the load value actually applied on the slope [2]. In this study, we refer to the infinite slope stability equation from [20] for dry conditions in the soil, adapted by [15]:

$$F = \tan\phi' / \tan\beta \qquad (1)$$

Where ϕ' is the effective soil friction angle (degrees) and β is the slope angle (degrees). The infinite slope model relies on the assumptions that the failure plane is mainly planar and parallel to the topographic surface and that the cohesion of the materials can be neglected [21].

According to several authors [21] [22], there is no risk of slope failure as long as the effective soil friction angle is greater than the slope angle value ($F > 1$). When $F = 1$, this indicates a state of limit equilibrium and when $F < 1$, this indicates a slope failure. Waste heap materials are generally composed of shale rock debris from 2 mm to 20 cm. Thus, the effective soil friction angle adopted here, following the paper of [2], is 35°. The factor of safety has been calculated for each pixel of the different DEMs produced.

B. *The interpolation techniques and spatial resolution*

Several factors may affect the quality of a DEM. First, are the factors related to the data collection, which mostly depend on the technology used and may have an impact on the DEM's quality. LiDAR data collection consists of several overlapping parallel strips acquired during flight plans. Most errors in LiDAR-derived DEM can be attributed to systematic and random errors by both laser scanner system and the GPS/IMU during data acquisition procedures [22]. These errors produce discrepancies in overlapping regions between neighboring strips [23]. The striped patterns observed on the waste heaps are thus probably related to an improper calibration of both systems [24]. Second, the density of points can be a source of errors. Indeed, a low-point density may result in an over-estimation of the height. Finally, the interpolation techniques used can influence the quality of a DEM [11].

In this study, we will not investigate collected data nor points of density because of missing information about data collection. Rather, we will concentrate on evaluating the performance of the two main techniques for generating DEMs with ArcGIS©ESRI: Binning and Triangulation.

- Binning technique

The principle of the Binning interpolation technique is to examine the elevation values that fall within a cell to determine the final value. In case of empty cells, natural neighbor works with Voronoi and Delaunay diagrams to find the closest point to input points and applies weights to them based on proportionate areas to interpolate a value [26]. With this technique, the interpolate elevations are guaranteed to be within the range of the sample used [27].

- Triangulation technique

The Triangulation interpolation method derives the cell value with a Triangulated Irregular Networks (TIN) based approach. TIN models and DEMs are two different ways of representing Earth surface in a digital data structure. Whereas DEMs rely on a regular grid surface representing the height of the terrain, TINs use irregular gridded models. A characteristic underlying the definition of the TIN is that it provides more points in rough irregular areas than in flat ones. The principle consists in connecting sampling points by lines to form triangles of irregular size and shape. The triangles are represented by planes, thereby permitting a more continuous representation of the terrain surface [28]. Moreover, TINs have shown their reliability for discontinuous shape (such as ridges) and breaks of slope [14], [29].

V. RESULTS

A. *Integration of several data*

Combination of field observations, ortho-photos, historical aerial photographs and DEMs support the assessment of the slope stability of the four coal tips with the example of Heribus (Figure 3).

The vertical cross-section analysis (G) calculated on the 1 m LiDAR data processed with Binning technique discloses a break of slope illustrated by an arrow. This zone is related to the slope failure zone in (D) and (E). This instability does not appear on the ERRUISSOL data with 10 m resolution (F). When looking at the most recent ortho-photo of 2012-2013 (C), this zone is hidden by vegetation but a discontinuous topography can clearly be identified on the historical photography (B). The aerial photography from 1969 permits to discerns the ground heterogeneity, imperceptible nowadays due to the presence of dense vegetation.

B. *The interpolation techniques and striped patterns*

The factor of safety is calculated on the basis of different interpolation techniques and spatial resolutions as illustrated for the Saint-Placide waste heap (Figure 4). With a 1 m resolution DEM, the striped patterns are present on the Binning technique (C) but not on the Triangulation technique (B). This discrepancy is due to the fact that whereas the Binning techniques interpolates on regularly gridded surface, the Triangulation technique interpolates elevation values on triangular surfaces of different size. Thus, the height differences between the overlapping areas resulting from the acquisition of the data are erased by means of the elevation value recorded in those triangles. Consequently, the problematic height discrepancy may be suppressed through the means of elevation values executed on an irregular surface. However, both interpolation techniques converge when the spatial resolution is 10 m and there are no striped patterns (E and F). The factor of safety for the ERRUISSOL DEM differs on some spots (D) from the other DEMs or field observations. The localisation of the instability areas does not coincide spatially on the three figures D (ERRUISSOL) in comparison to E and F (LiDAR data aggregated to 10 m). The reasons for these disparities are linked to the integration of inputs with various resolution and precision. However, this dataset was the only available for the Walloon inventory [1].

VI. DISCUSSION AND CONCLUSION

This paper studies technical choices in terms of resolution and interpolation techniques when processing raw LiDAR data. Two interpolation techniques from ArcGIS©ESRI have been tested to analyze their potential for the topic of geotechnical instability analysis. This method has been applied to four waste heaps located in Western Wallonia but could also be applied to other study sites in Belgium (Liège, Charleroi and De Kempen coalfield) or even in the heavily mined Nord-Pas-de-Calais in France.

The different sources of data are interestingly combined in this paper for describing, understanding and quantifying the risk of geotechnical failure. Field observations did not lead to the conclusion that there is a great risk of landslide on the four waste heaps because of the presence of dense vegetation and the absence of sliding events in the past. However, the factor of safety for LiDAR DEM with 1 m resolution presented numerous zones indicating a slope failure. Yet, the integration of all data permits a moderation of these results. There is no risk to have a serious impact on environment and/or human health since all the coal tips are covered by vegetation, have remained stable for many years and are not located in a residential area. However, should the authorities plan to change the affectation of the heap, a new risk analysis would be necessary.

Computing a factor of safety based on an infinite slope stability model assuming a 1 m grid size yields inadequate results. The resolution of 1 m does not seem to be consistent with the model assumptions. First, it is difficult to consider a slope as infinite when a high resolution of 1 m is used. Second, neglecting the grain cohesion appears to be inappropriate when there is dense vegetation on the waste heaps.

The utilization of raw LiDAR data improves the understanding of the impact of processing on the safety factor. Indeed, figures 2 and 3 show that different processing steps can influence the factor of safety. The Triangulation technique has demonstrated its ability to remove the striped patterns by establishing triangles which average the value of several cells, by contrast with the Binning technique, which uses a regular grid to interpolate elevation values. Because it smooths the cell value, the Triangulation technique is more appropriate to assess the slope stability here, but it could be less so under different circumstances.

While both LiDAR and ERRUISSOL data have the resolution proposed by the Protocol of the Directive, LiDAR data improve the stability assessment by comparison with ERRUISSOL. Indeed, whereas the LiDAR data have revealed that Heribus presents a risk of slope failure, this had not been detected by the ERRUISSOL data. Conversely, the ERRUISSOL data have found a risk of geotechnical instability where the LiDAR data did not discover anything.

ACKNOWLEDGMENT

This study has been done during a three months internship in the Institut Scientifique de Service Public (ISSeP). This work is related to a research subvention from the SPW, DGO3. LiDAR data have been acquired by the SPW, General secretariat. We thank the administration for providing this dataset and for their support.

REFERENCES

[1] C. Frippiat, N. Stephenne, M. Veschkens, and D. Pacyna 2013. Adaptation of the European Union risk assessment protocol for the pre-inventory of Walloon mining waste deposits. In: Mine closure 2013, pp. 495-507, edited by M. Tibbett, A. B. Fourie, and C. Digby (Australian Centre for Geomechnaics, Perth) 642p.

[2] N. Stephenne, C. Frippiat, M. Veschkens, M. Salmon, and D. Pacyna 2014. Use of a Lidar High Resolution Model for Risk Stability Analysis. EARSel eProceedings, Special Issue: 34th EARSel Symposium, 24-19.

[3] http://geoportail.wallonie.be/WalOnMap [accessed 2016-04-03]

[4] P. Demarcin, A. Degré, A. Smoos and, S. Dautrebande 2009. Projet ERRUISSOL, Cartographie numérique des zones à risque de ruissellement et d'érosion des sols en Région Wallonne. Rapport Final de convention DGO3-FUSAGx (Unité d'hydrologie et hydraulique agricole, Faculté universitaire des Sciences agronomiques de Gembloux, Gembloux, Belgium) 55p.

[5] N. Polat, M. Uysal and, A. S. Toprak 2015. An investigation of DEM generation process based on LIDAR data filtering, decimation, and interpolation methods for an urban area. Measurement 75, 50-56.

[6] D. Weber, and E. Englund 1992. Evaluation and comparison of spatial interpolators II. Mathematical Geology 24, 381-391.

[7] D. Weber, and E. Englund 1994. Evaluation and comparison of spatial interpolators II. Mathematical Geology 26, 589-603.

[8] A. Carrara, G. Bitelli, and R. Carla 1997. Comparison of techniques for generating digital terrain models from contours lines. International Journal of Geographical Information Science 11, 451-473.

[9] S. M. Robeson 1997. Spherical methods for spatial interpolation: review and evaluation. Catrography and Geographic Information Systems 24, 3-20.

[10] G. R. Hancock 2006. The impact of different gridding methods on catchment geomorphology and soil erosion over long timescale using a landscape evolution model. Earth Surface Processes and Landforms 31, 1035-1050.

[11] V. Chaplot et al. 2006. Accuracy of interpolation techniques for the derivation of digital elevation models in relation to landform types and data density. Geomorphology 77, 126-141.

[12] http://geologie.wallonie.be/soussol/exploitations/ssol_expl_dechets/ssol_dechets_risques) [accessed 2016-04-03]

[13] http://geoportail.wallonie.be/WalOnMap/ [accessed 2016-04-03]

[14] G. L. Heritage, D. J. Milan, A. R. G. Large, and I. C. Fuller 2009. Influence of survey strategy and interpolation model on DEM quality. Geomorphology 112, 334–344.

[15] G. B. Chirico, M. Borga, P. Tarolli, R. Rigon, and F. Preti 2013. Role of vegetation on slope stability under transient unsaturated conditions. Four Decades of Progress in Monitoring and Modeling of Processes in the Soil-PlantAtmosphere System: Applications and Challenges. Procedia Environmental Sciences 19, 932 – 941.

[16] J. Poesen, J. Nachtergaele, G. Verstraeten, and C. Valentin 2003. Gully erosion and environmental change: importance and research needs. Catena 50, 91–133.

[17] A. Monjoie, and C. Schroeder 2001. Instabilités de versants de terrils en relation avec l'autocombustion des schistes et charbons résiduels. Revue française de géotechnique N°95-96, 91-102.

[18] J. Nyssen, and D. Vermeersch 2010. Slope aspect affects geomorphic dynamics of coal mining spoil heaps in Belgium. Geomorphology 123, 109-121.

[19] SPW 2014. Rapport de production des données du relief de la Wallonie. Jambes, 5p.

[20] N. Lu, and J. Godt 2008. Infinite slope stability under steady unsaturated seepage conditions. Water Resources Ressearch 44, 1-13.

[21] M. Mergili, I. Marchesini, M. Rossi, F. Guzzetti, and W. Fellin 2014. Spatially distributed three-dimensional slope stability modelling in a raster GIS. Geomorphology 206, 178–195.

[22] J. A. White, and D. I. Singham 2012. Slope stability assessment using stochastic rainfall simulation. International Conference on Computational Science, ICCS 2012. Procedia Computer Science 9, 699 – 706.

[23] G. Vosselman, and H. G. Maas 2001. Adjustement and filtering of raw laser altimetry data. OEEPE Workshop Airborne Laser System and Interferometric SAR forDetailed Digital Elevation Models. Vol. 40, 62-72.

[24] M. Favalli, A. Fornaciai, M. T. Pareschi 2009. LiDAR strip adjustement: Application to volcanic areas. Geomorphology 111, 123-135.

[25] A. F. Habib et al. 2008. LiDAR strips adjustement using conjuguate linear features in overlapping strips. The International Archives of the Photogrammetry, Remote

Sensing and Spatial Information Sciences. Vol. XXXVII. Part B1. Beijing 2008, 385-390.

[26] R. Sibson 1981. A brief description of natural neighbour interpolation, in:V. Barnet (Ed.), Interpreting Multivariate Data, Wiley, Chichester, 21–36.

[27] P. V. Arun 2013. A comparative analysis of different DEM interpolation methods. The Egyptian Journal of Remote Sensing and Space Sciences 16, 133-139.

[28] T. Ali, and A. Mehrabian 2009. A novel computational paradigm for creating a Triangular Irregular Network (TIN) from LiDAR data. Nonlinear Analysis 71, e624-e629.

[29] I. D. Moore, R. B. Grayson, A. R. Ladson 1991. Digital terrain modelling: a review of hydrological, geomorphological, and biological applications. Hydrological Processes 5, 3-30.

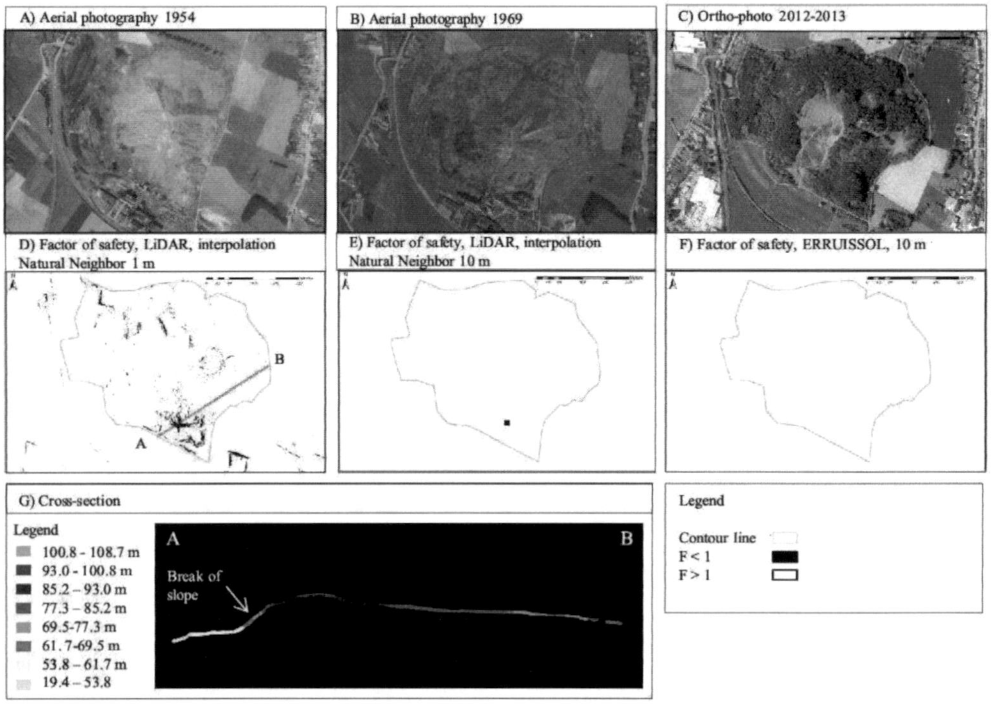

Figure 3. Integration of several datasets on the Heribus waste heap with two aerial photographs from 1954 (A) and 1969 (B), ortho-photo from 2012-2013 (C), a map showing the safety factor based on LiDAR data from natural neighbor Binning techniques with 1 m (D) and 10 m (E), a map displaying the safety factor based on ERRUISSOL data with 10 m (F) and a vertical cross-section (G) on a zone delimited on (D).

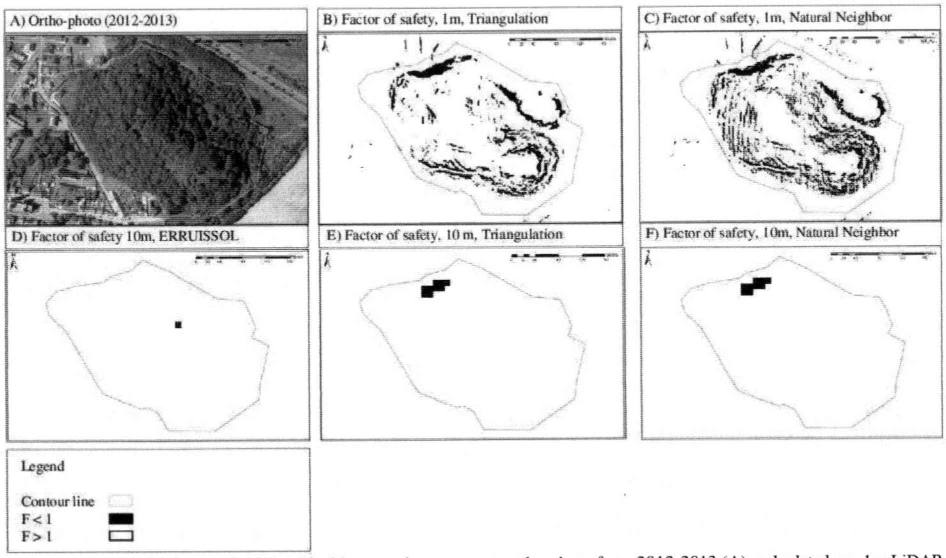

Figure 4. Factor of safety on the Saint-Placide waste heap seen on ortho-photo from 2012-2013 (A), calculated on 1m LiDAR data with the Triangulation technique (B) on 1m LiDAR data with natural neighbor Binning technique (C) on 10m on ERRUISSol dataset (D), on 1m LiDAR data aggregated to 10m with the Triangulation technique (E) on 1m LiDAR data aggregated to 10m with the natural neighbor Binning technique (F).

Uncovering User Profiles in Location-Based Social Networks

Soha A. Mohamed and Alia I. Abdelmoty

School of Computer Science & Informatics
Cardiff University
Wales, UK
Email: {AlySA, AbdelmotyAI}@cardiff.ac.uk

***Abstract*—With the current trend of embedding location services within social networks, an ever growing amount of users' spatio-temporal tracks are being collected that can be used to generate user profiles to reflect users' interests in places. User-contributed annotations of place, as well as other place properties, add a layer of important semantics that if considered, can result in more refined representations of user profiles. In this paper, semantic information is summarised as tags for places and a folksonomy data model is used to represent spatial and semantic relationships between users, places and tags. The model allows simple co-occurrence methods and similarity measures to be applied to build different views of personalized user profiles. Basic profiles capture direct user interactions, while enriched profiles offer an extended view of user's association with places and tags that takes into account relationships in the folksonomy. The main contribution of this work is the demonstration of how the different data dimensions captured on location-based social networks can be combined to represent useful views of user profiles.**

***Keywords*–GeoFolksonomy;User Profiles;Location-based Social Networks;.**

I. INTRODUCTION

This work focusses on Location-Based Social Networks (LBSN) that collect information on users' interests in physical places in the real world. By "switching on" location on devices, we are giving away information on our whereabouts, our daily routines, activities, experiences, and interests. Thus, in comparison to other personal information, location data are possibly the most crucial type of data of relevance to privacy, as it pulls together our virtual and physical existences and thus raises critical questions about privacy in both worlds. This work introduces methods for constructing user profiles that consider the different dimensions of the data captured from users on LBSN. These profiles, when made transparent to users of the network, should empower their sense of awareness and control of their data.

So far, previous works have studied data produced from LBSN from the point of view of enhancing the services provided by these networks, namely, for point of interest (POI) recommendations. There, the question of concern is to find places of interest to a user based on their history of visits to other places and their general interaction with the social network. Most works relied mainly on the spatial dimension of user data [1], with some works more recently exploring the relevance of the social and content data dimensions on these networks [2]. However, data dimensions are normally treated separately, or their outputs are combined in fused models.

In this paper, both semantic and spatial interactions of users are used to project distinct and complementary views of personalised user profiles. Thus, user's annotations on places they visit are compiled in semantic profiles, while collective user annotations on places are used to create specific profiles for places that encapsulate user's experiences in the place. Place profiles, in turn, are used to construct personalised user profiles. In comparison to previous works in the area of recommendations, LBSN data are treated as folksonomies of users, places and tags. User annotations in the form of tips, their interaction with places, in the form of check-ins, as well as general place properties, namely, place categories and tags, are analysed concurrently to extract relations between the three elements of the folksonomy. Simple co-occurrence methods and similarity measures are used to compute direct and enriched user profiles.

Thus, the proposed approach provides users with the ability to project different views of their profiles, using their direct interactions with the social network or extended with a holistic view of other users' interaction with the network in different regions of geographic space. Previous works attempting a similar approach used matrix factorization techniques to handle the multiple data dimensions, but did not consider the use of the range of content data as used in this paper. Sample realistic data from Foursquare are used to demonstrate the approach and evaluation results show its potential value. In particular, it is shown that enriched user profiles offer potentially more accurate views, than direct profiles, of user's spatial as well as semantic preferences. Hence, these should be considered when designing tools for enabling user awareness on these networks.

The rest of the paper is organised as follows. Section II provides an overview of related works. In Section III, a geo-folksonomy data model for LBSN is introduced and in Section IV different types of user profiles are defined. In Section V, the experiment used to evaluate the approach is described and its results are presented and discussed. The paper concludes in Section VI with an overview of future work.

II. RELATED WORK

Works on modelling user data in LBSN mainly consider two problems; a) place (or point of interest) recommendation, and b) user similarity calculation. Different types of data are used by different approaches, namely, geographic content, social content as well as textual annotations made by users. Also, different methods are used in analysing the data, for example, distance estimations for geographic data modelling and topic modelling for annotation data analysis.

In the area of POI recommendation, works range from generic approaches that uses the popularity of places [3] to recommendation methods that are based on user's individual preferences [4]. A useful survey of these approaches can be found in [5].

Based on check-in data gathered through Foursquare, Noulas and Mascolo [6] exploit factors such as the transition between types of places, mobility between venues and spatial-temporal characteristics of user check-in patterns to build a supervised model for predicting a user's next check-in. Ye,Lui and Lee [4] investigated the geographical influence with a power-law distribution. The hypothesis is that users tend to visit places within short distances of one another. Other works considered other distance distribution models [7]. Gao, Tang and Liu [8] considered a joint model of geo-social correlations for personalized POI recommendation, where the probability of a user checking in to a new POI is described as a function of correlations between user's friends and non-friends close to, and distant from a region of interest. Liu, Xiong and Papadimitriou [9] approached the problem of POI recommendations by proposing a geographical probabilistic factor model that combines the modeling of geographical preference and user mobility. Geographical influence is captured through the identification of latent regions of activity for all users of the LBSN reflecting activity areas for the entire population and mapping the individual user mobility over those regions. Their model is enhanced by assuming a Poisson distribution for the check-in count which better represents the skewed data (users visiting some places one time, while other places 100s of times). Whilst providing some useful insights for modelling the spatial dimension of the data, the above works do not consider the semantic dimension of the data.

Correlations between geographical distance and social connections were noted in [10] [2]. Techniques of personalized POI recommendation with geographical influence and social connections mainly study these two elements separately, and then combine their output together within a fused model. Social influence is usually modeled through friend-based collaborative filtering [11] [4] [12] with the assumption that a user tends to be friends with other users who are geographically close to him, or would want to visit similar places to those visited by his friends. Ying, Lu, Kuo, and Tseng [13] proposed to combine the social factor with individual preferences and location popularity within a regression-tree model to recommend POIs. The social factor corresponds to similar users; users with common check-ins to the user in question. In this paper, we also use this factor when extending user profiles to represent places of interest within the region of user activity.

More recently, the importance of content information for POI recommendation was recognised. Two types of content can be considered, attributes of places and user-contributed annotations. Place categories are normally used as an indication of user activity, thus a user visiting a French restaurant would be considered as interested in French food, etc. User annotations in the form of tips and comments are analysed collectively to extract general topics to characterise places or to extract collective sentiment indications about the place. Examples of works that considered place categories are [14] [15] [16] [17]. In [14] [15], Latent Dirichlet Allocation (LDA) model was used to represent places as a probability distribution over topics collected from tags and categories or comments made in a place and similarly aggregate all tips from places a user has visited to model a user's interest. Aggregation was necessary as terms associated with a single POI are usually short, incomplete and ambiguous. [16] on the other hand modelled topics from tweets and reviews from Twitter and Yelp, and assumed that the relations between user interests and location are derived from the topic distributions for both users and locations. In [17], a probabilistic approach is proposed that utilize geographical, social and categorical correlations among users and places to recommend new POIs from historical check-in data of all users. In this paper, we also model user's association to place through the place's relation to tags, but add the influence of other users relations in the place to the equation. Aiming at improving the effectiveness of location recommendation, Yang, Zhang, Yu and Wang [18] proposed a hybrid user POI preference model by combining the preference extracted from check-ins and text-based tips which were processed using sentiment analysis techniques. Sentiment analysis is an interesting type of semantics which we do not consider this work, but can be incorporated in future work.

Studying user similarity from LBSN data is useful, as information available about users, their locations and activities are considered to be sparse. User similarities can be exploited to predict types of activities and places preferred by a user based on those of users with similar preferences. So far, most works on user similarity mainly focused on structured, e.g., geographic coordinates, or semi-structured, e.g., tags and place categories, data. Recently, Lee and Chung [19] presented a method for determining user similarity based on LBSN data. While the authors made use of check-in information, they concentrated on the hierarchy of location categories supplied by Foursquare in conjunction with the frequency of check-ins to determine a measure of similarity. Mckenzie, Adams, and Janowicz [15] suggest exploring unstructured user-contributed data, namely tips provided by users. A topic modeling approach is used to represent users' interests in places. Venues (places in Foursquare) are described as a mixture of a given number of topics and topic signatures are computed as a distribution across venues. User similarity can then be measured by computing a dissimilarity metric between users' topic distribution. Their method of modelling venues is interesting, but it limits the representation of user profiles, where profiles are based on generated topics derived from collective user annotation on places. Thus, individualised association of users with the place is somewhat ignored. In contrast to the above approach, our model does not assume constraints on the number of topics represented by the tags, but combines the individual's association with both tags and place in the creation of user profiles.

III. GEO-FOLKSONOMY MODEL

The location-based social networking platform, Foursquare, was used as our source of data. It holds a large number of crowdsourced venues ($>$ 65 million places) from a user population estimated recently to around 55 million users. As the application defines it, a venue is a user-contributed "physical location, such as a place of business or personal residence.". Foursquare allows users to check in to a specific venue, sharing their location with friends, as well as other online social networks, such as Facebook or Twitter. Built with a gamification strategy, users are rewarded for checking in to locations with badges, in-game points, and discounts from advertisers. This game-play encourages users to revisit the application, compete against their friends and contribute check-ins, photos and tips. Tips consist of user input on a specific venue, normally describing a recommendation, experience or activity performed in the place.

In this work, we use a folksonomy data model to represent

user-place relationships and derive tag assignments from users' actions of check-ins and annotation of venues. In particular, tags are assigned to venues in our data model in two scenarios as follows.

1) A user's check-in results in the assignment of place categories associated with the place as tags annotated by this user. Thus, a check-in by user u in place r with the categories (represented as keywords) x, y and z, will be considered as an assertion of the form $(u, r, (x, y, z))$. This in turn will be transformed to a set of triples $\{(u, r, x), (u, r, y), (u, r, z)\}$ in the folksonomy.
2) A user's tip in the place also results in the assignment of place categories as tags, in addition to the set of keywords extracted from the tip. Thus, in the above example, a tip by u in r with the keywords $(t_1, \cdots, t_n)$, will be considered as an assertion of the form $(u, r, (x, y, z, t_1, \cdots, t_n))$, and is in turn transformed to individual triples between the user, place and tags in the folksonomy.

The process of extracting keywords from tips is done by tokenizing the tip into a set of words (terms) on white space and punctuation. Then we remove all words with non-latin characters and stop words. The output is a set of single words (term vector). Furthermore, we use Wordnet syntactic category and logical groupings for classifying the extracted terms. For example, Wordnet 'noun.act' category is used to filter action verbs and nouns to describe a user- or place- associated activity (ex. swimming, buying or eating).

The data capturing process results in the creation of a *geo-folksonomy*, which can be defined as a quadruple $\mathbb{F} := (U, T, R, Y)$, where U, T, R are finite sets of instances of users, tags and places respectively, and Y defines a relation, the tag assignment, between these sets, that is, $Y \subseteq U \times T \times R$, [20] [21].

A geo-folksonomy can be transformed into a tripartite undirected graph, which is denoted as folksonomy graph $\mathbb{G}_\mathbb{F}$. A geo-Folksonomy Graph $\mathbb{G}_\mathbb{F} = (V_\mathbb{F}, E_\mathbb{F})$ is an undirected weighted tripartite graph that models a given folksonomy $\mathbb{F}$, where: $V_\mathbb{F} = U \cup T \cup R$ is the set of nodes, $E_\mathbb{F} = \{\{u, t\}, \{t, r\}, \{u, r\} | (u, t, r) \in Y\}\}$ is the set of edges, and a weight w is associated with each edge $e \in E_\mathbb{F}$.

The weight associated with an edge $\{u, t\}, \{t, r\}$ and $\{u, r\}$ corresponds to the co-occurrence frequency of the corresponding nodes within the set of tag assignments Y. For example, $w(t, r) = |\{u \in U : (u, t, r) \in Y\}|$ corresponds to the number of users that assigned tag t to place r.

Figure 1 depicts the overall process of user profile creation. The process starts with data collection of check-ins and tip data from Foursquare, that are then processed to extract users, places and tags and their associated properties. The modelling stage includes the definition of relationships between the three entities and the application of folksonomy co-occurence methods to extract the different types of profiles. Place and tag similarity calculations are used to further extend the basic profiles to build different views of enriched user profiles.

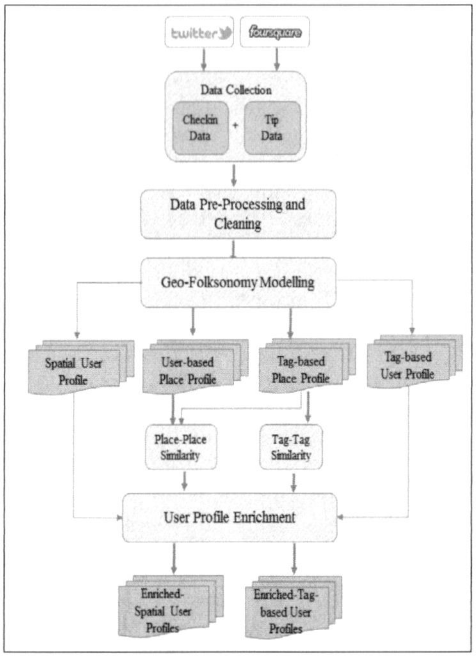

Figure 1. The framework of our system

IV. User Modeling Strategies

We propose an approach to modelling users in LBSN that represents a user's spatial, semantic and combined spatio-semantic association with place. A spatial user profile represents the user's interest in places, while a tag-based profile describes his association with concepts associated with places in the folksonomy model. A spatio-semantic profile describes the user specific interest in certain concepts associated with places in his profile. A user profile is built in stages. Starting with a basic profile that utilises direct check-in and annotation histories, a user profile is then extended by computing the relationship between places and concepts derived from the collective behaviour of other users in the dataset. A basic profile represents actual interactions with places, while the extended profile describes "recommended" associations given overall interactions between users, places and concepts in the dataset. We are able to model such interactions separately in the extended profile by controlling the similarity function used to create the profile. For example, we can focus on modelling the types of places visited by the user or take into account visit behaviour of other users whose profiles overlap with the user, as discussed below.

A. Basic User Profiles

Definition 1: **Spatial User Profile** A spatial user profile $P_R(u)$ of a user u is deduced from the set of places that u visited or annotated directly.

$$P_R(u) = \{(r, w(u, r)) | (u, t, r) \in Y, \\ w(u, r) = |\{t \in T : (u, t, r) \in Y\}|\} \quad (1)$$

$w(u, r)$ is the number of tag assignments, where user u assigned some tag t to place r through the action of checking-in or annotation. Hence, the weight assigned to a place simply corresponds to the frequency of the user reference to the place either by checking in or by leaving a tip.

We further normalise the weights so that the sum of the weights assigned to the places in the spatial profile is equal to 1. We use $\overline{P_R}$ to explicitly refer to the spatial profile where the sum of all weights is equal to 1, with $\overline{w}(u, r) = \frac{|\{t \in T:(u,t,r) \in Y\}|}{\sum_{i=1}^{n} \sum_{j=1}^{m} |\{t_i \in T:(u,t_i,r_j) \in Y\}|}$, where n and m are the total number of tags and resources, respectively. More simply,

$\overline{w}(u,r) = \frac{N(u,r)}{N_T(u)}$, where $N(u,r)$ is the number of tags used by u for resource r, while $N_T(u)$ is the total number of tags used by u for all places.

Correspondingly, we define the tag-based profile of a user; $P_T(u)$ as follows.

Definition 2: **Semantic User Profile** A semantic user profile $P_T(u)$ of a user u is deduced from the set of tag assignments linked with u.

$$\begin{aligned} P_T(u) =& \{(t, w(u,t)) | (u,t,r) \in Y, \\ & w(u,t) = |\{r \in R : (u,t,r) \in Y\}|\} \end{aligned} \quad (2)$$

$w(u,t)$ is the number of tag assignments where user u assigned tag t to some place through the action of checking-in or annotation.

$\overline{P_T}$ refers to the semantic profile where the sum of all weights is equal to 1, with $\overline{w}(u,t) = \frac{N(u,t)}{N_R(u)}$, where $N(u,t)$ is the number of resources annotated by u with t and $N_R(u)$ is the total number of resources annotated by u.

Furthermore, we define a spatio-semantic profile of a user $P_{RT}(u)$, that is a personalised association between user, place and tag.

Definition 3: **Spatio-Semantic User Profile** Let $\mathbb{F}_u = (T_u, R_u, I_u)$ of a given user $u \in U$ be the restriction of $\mathbb{F}$ to u, such that, T_u and R_u are finite sets of tags and places respectively, that are referenced from tag assignments performed by u, and I_u defines a relation between these sets: $I_u := \{(t,r) \in T_u \times R_u | (u,t,r) \in Y\}$.

A spatio-semantic user profile $P_{RT}(u)$ of a user u is deduced from the set of tag assignments made for place r by u.

$$\begin{aligned} P_{RT}(u) =& \{([r,t], w_u([r,t])) | (t,r) \in I_u, \\ & w_u([r,t]) = |\{t \in T_u : (t,r) \in I_u\}|\} \end{aligned} \quad (3)$$

where $w([r,t])$ is how often user u assigned tag t to place r.

$\overline{P_{RT}}$ is the spatio-semantic profile where the sum of all weights is equal to 1, with $w_u([r,t]) = \frac{N(u,[r,t])}{N_{RT}(u)}$, where $N(u,[r,t])$ is the number of times u annotate r with t, and $N_{RT}(u)$ is the total number of tags assigned by u for r. (Note that tag assignment by users for a place comes from both the explicit action of annotation as well as implicit action of checking-in as represented in the geo-folksonomy model).

B. Place and Tag Profiles

So far, the basic user profile provides only a limited view of the user association with places and concepts derived directly from captured data. Basic profiles reduce the dimensionality of the folksonomy space by considering only 2 dimensions at a time; user-place and user-tag, leading to a loss of correlation information between all three elements. Users profiles can be extended to represent possible latent relationships in the data. Thus a user profile can be used to present places (respectively tags) similar to those in the basic profile, where similarity between places (respectively tags) is measured through the collective actions of other users of check-ins and annotations.

To compute tag-tag similarity, profiles for tags are first defined through the places they are used to annotate. Thus, a *place-based tag profile* ($P_R(t)$) of a tag t is a weighted list of places r that are annotated by t. That is, $w(r,t)$ is determined by the number of users' check-ins and tips that resulted in assigning t to r in the geo-folksonomy. Similarity between tags is defined as the cosine similarity between their place-based tag profiles as follows.

$$CosSim(t_1, t_2) = \frac{|P_R(t_1) \cap P_R(t_2)|}{\sqrt{|P_R(t_1)|.|P_R(t_2)|}} \quad (4)$$

On the other hand, similarity between places is defined by measuring the similarity of their tag-based and user-based profiles. Let $P_T(r)$ and $P_U(r)$ be the tag-based place profile and user-based place profile for place r (defined in a similar manner to user profiles above). Conceptually, a tag-based place profile is a description of the place by the tags assigned to it and a user-based place profile is an account of users' visits to the place.

Cosine similarity between tag-based place profiles ($CosSim_{tag}(r_1, r_2)$) and between user-based place profiles ($CosSim_{user}(r_1, r_2)$) construct a tag-oriented ranking and user-oriented ranking, respectively. These similarity rankings can be aggregated using the so-called Borda method [22] to compute a generalised similarity score between two places as shown in Equation 5

$$PlaceSim(r_1, r_2) = \gamma * CosSim_{tag}(r_1, r_2) + (1-\gamma) * CosSim_{user}(r_1, r_2) \quad (5)$$

where $0 \leq \gamma \leq 1$ is a parameter that determines the balance of importance given to similarity scores from $P_T(r)$ and $P_U(r)$. Conceptually, similarity between two places is a function of the overlap between their tag assignments only (for $\gamma = 0$), a measure of their common visitors only (for $\gamma = 1$), or both (for γ between 0 and 1).

C. Enriched User Profiles

We extend the basic user profiles by the information extracted from the computation of tag and place similarity above. The enriched user profiles will therefore present a modified view of how users are associated with places that reflect collective user behaviour on the LBSN.

Definition 4: **Enriched Spatial User Profile** An enriched spatial user profile $\acute{P}_R(u)$ of a user u is an extension of the basic profile by places with the highest degree of similarity to places in $\overline{P_R(u)}$. Let R_u be the set of all places in $\overline{P_R(u)}$ and w_i is the weight associated with place i in the profile.

$$\begin{aligned} &\acute{P}_R(u) = \{< r_i, w_i > | \\ &w_i = \left\{ \begin{array}{ll} w_i & ,\text{if } r_i \in R_u \\ w_i * Max(PlaceSim(r_i, r_j)) & ,\forall (r_i \in \{R - R_u\} \wedge r_j \in R_u) \end{array} \right\} \end{aligned} \quad (6)$$

We compute the maximum similarity of the 10 most similar places in the dataset for every place in the basic user profile, and use the highest similarity score as the weight for the new place in the enriched user profile. The process of building the enriched spatial profile from place similarity with $\gamma = 1$ is shown in Figure 2.

Definition 5: **Enriched Tag-based User Profile** An enriched tag-based user profile $\acute{P}_T(u)$ of a user u is an extension of the basic profile by tags with the highest degree of similarity to tags in $\overline{P_T(u)}$. Let T_u be the set of all tags in $\overline{P_T(u)}$ and w_i is the weight associated with tag i in the profile.

$$\begin{aligned} &\acute{P}_T(u) = \{< t_i, w_i > | \\ &w_i = \left\{ \begin{array}{ll} w_i & ,if t_i \in T_u \\ w_i * Max(Sim(t_i, t_j)) & ,\forall (t_i \in \{T - T_u\} \wedge T_j \in T_u) \end{array} \right\} \end{aligned} \quad (7)$$

```
procedure SpatialEnrichment(P_R(u),γ)
    for all place r_i in Spatial-Profile P_R(u) do
        Compute PlaceSim(r_1, r_2) from Equation 5.
        Find top-10 similar places r_j to each r_i in P_R(u)
        for each < r_j, sim > in top similar places do
            w_j = w_i * sim
            add < r_j, w_j > to P_R(u)
        end for
    end for
    return Ṕ_R(u)
end procedure
```

Figure 2. Algorithm for building the enriched user profile with $\gamma = 1$.

A similar algorithm to that of enriching place profiles is used for choosing the tags and weights.

Definition 6: **Enriched Spatio-Semantic User Profile**

An enriched spatio-semantic user profile $\acute{P}_{RT}(u)$ of a user u is an extension of the basic profile by tags and places with the highest degree of similarity to tags in $\overline{P_{RT}(u)}$. Let T_u be the set of all tags in $\overline{P_T(u)}$, R_u be the set of all places in $\overline{P_R(u)}$ and w_{ij} is the weight associated with tag i and place j in the profile.

$$\acute{P}_{RT}(u) = < [r_i, t_j], w_u(r_i, t_j) > | w_u(r_i, t_j) = \left\{ \begin{array}{ll} w_u(r_i, t_j) & , \text{if } r_i \in R_u and t_j \in T_u \\ w_u(r_i, t_j) * Max(PlaceSim(r_i, r_k)) & , t_j \in P_T(r_k) \wedge r_k \in \{R - R_u\} \\ 0 & otherwise \end{array} \right\} \quad (8)$$

The spatio-semantic profile is extended with the most similar places to the user profile and these are assigned a weight computed using the place similarity value for all tags in their place-tag profiles and 0 for tags that are not in their profile. Thus the user simply inherits relationships with all the tags and their associated weights from basic places that are deemed similar to those in his profile.

1) User Profile Example: Here an example is given of a sample user profile created from the dataset used in this work. 'user349' checked in 600 different venues, with associated 400 venue categories. Note that one venue can have more than one venue category. Figure 3 shows the top 20 tags in his semantic user profile. Figure 4 shows filtered tags from his profile representing human activity (approximately 5% of all tags), as derived by mapping to Wordnet noun.act category.

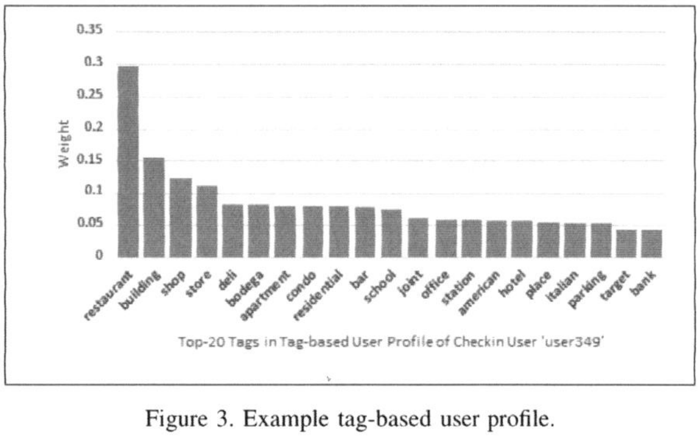

Figure 3. Example tag-based user profile.

Figures 5 and 6 show the spatial profile and the enriched spatial profiles for user 'user349', respectively. $\gamma = 0.5$ was used in the place similarity equation of the enriched profile. The size of the dots in the figures represents the weight of the place in the profile.

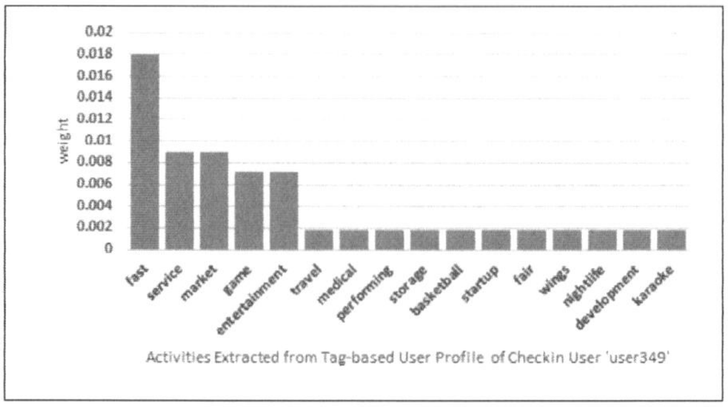

Figure 4. Sample of tags representing activities in a semantic user profile.

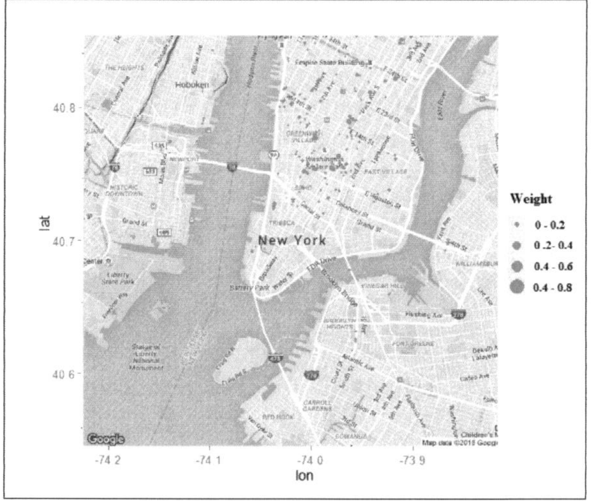

Figure 5. Spatial user profile for user 'user349'.

V. Experiments and Results

A. Datasets

Approximately (10 months) of check-in data in New York city were collected from Foursquare between April 2012 and February 2013 [23]. This data consists of 227,428 anonymized user check-ins, with venue ids, venue category, longitude and latitude of venues and time stamps of check-ins. The data was then used to recursively extract venue-related tips (tip id, text and time stamp), and subsequently all venues for users related

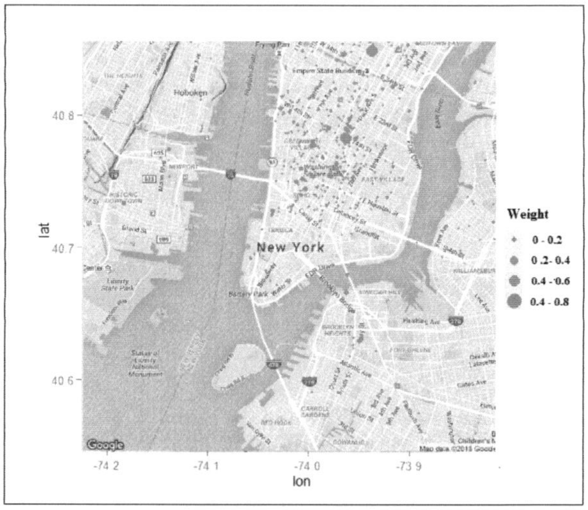

Figure 6. Enriched spatial user profile for user 'user349' with $\gamma = 0.5$.

procedure SPATIO-SEMANTIC TOP-K RECOMMENDER(γ,TopK)
  for each u_i **do**
    SpatialEnrichment($P_R(u_i), \gamma$)
  end for
  for all u_i, u_j **do**
    Fetch profiles $P_R(u_i), P_R(u_j)$
    Compute CosSim(u_i, u_j) .
  end for
  for each U_i **do**
    Fetch most similar user u_j
    Sort $< r_i, w_i >$ of $P_R(u_j)$
    Recommend TopK r_i that are not in $P_R(u_i)$
  end for
  return TopK $< r_i, w_i >$
end procedure

Figure 7. Spatio-semantic Top-K recommendation algorithm

to the tips collected. 604,924 tips were collected for 167,786 users in 36,940 venues. Time stamps of the tip data range from January 2009 to June 2015.

Experiments were carried out using a sample of 20 users with a high frequency of check-ins and co-location rate (10 users with an average of 601 check-ins) and tips (10 users with and average of 95 tips). Table I shows summary statistics of the sample dataset used.

TABLE I. EXPERIMENT DATASET

Number of Venues	2,041
Total number of Checkins	4,212
Total Number of Tips	942
Total Number of Tags	3,357
Number of users	20
Total Number categories	317
Total Number of Relationships	17,955
Average Checkins/user	601
Average tag/user	363

B. Experiment Setup

The evaluation experiment aims to measure the impact of using the full range of content captured on LBSN when building user profiles in comparison to using only partial views based on check-in information. The experiment takes the form of place (and tag) top-N recommendation problem using the different constructed user profiles based on the users profiles cosine similarities and seeks to establish how well the profiles reflect the user spatial and semantic characteristics when using the LBSN. The algorithm used for computing the top-N recommendations using spatial profiles is shown in figure 7.

We use recall@N, precision@N and F1@N as our success measures, where N is the predefined number of places (or tags) to be recommended. Recall measures the ratio of correct recommendations to the number of true places (or tags) of a test check-in or tip record, whereas precision measures the ratio of correct to false recommendations made. Recall and precision are given by

$$recall = \frac{TP}{TP + FN}$$
$$precision = \frac{TP}{TP + FP}$$

where true positives (TP) is the number of correct place (or tags) recommended, false positive (FP) is the number of wrong recommendations and false negatives (FN) is the number of true place (or tags) which were not recommended. F1 is a combination of recall and precision and is given by

$$F1 = \frac{2 * precision * recall}{precision + recall}$$

The values of TP, FP, FN are determined by randomly splitting the users into two sets; the training set and the testing set. Multi-fold cross-validation was used to ensure a fair partitioning between test data and training data. Data were split 90% for training and 10 % for testing, and the process was repeated 5 times to create 5 folds and the mean of the performance was reported.

C. Evaluation of Spatial Profiles

Results for the enriched user profiles using the proposed top-N recommendation method are presented. Different versions of the enriched spatial profiles, using different place similarity measures were created, a) using $\gamma = 0$ (to represent place-tag similarity only), b) using $\gamma = 1$, (to represent place-user similarity only), and c) using $\gamma = 0.5$ for an aggregated view of both effects. Hence, result sets are shown for the following user profiles. 1. Enriched-Spatial(Tag) 2. Enriched-Spatial(User) 3. Enriched-Spatial(All).

We compare the results of the top-N recommendation using the three different profiles with traditional Item-based Collaborative Filtering (IBCF) [24] and User-based collaborative Filtering (UCBF) [25] approaches, applied against the basic spatial user profile for recommending top-1, 2, 3, 4, 5, 10, 20, 30, 40, 50. Figures 8 and 9 and 10 show the precision, recall and F1-measure for all all approaches. As is shown in the figures, enriched user profiles demonstrate significantly better performance in comparison to the traditional approaches. In particular, the F1 measure for the combined profile (Spatial + All) outperforms the UBCF approach by 10% on average and the IBCF approach by 12% on average.

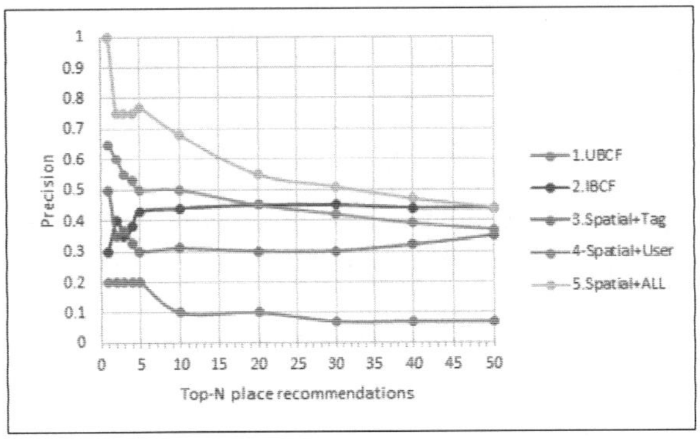

Figure 8. Precision values for the top-N place recommendations.

D. Evaluation of Semantic profiles

A similar experiment was carried out to evaluate the semantic user profiles. Again, the results were compared to the UBCF and IBCF approaches. Figures 11, 12 and 13 show the results of the top-10, 20, 30, 40, and 50 tag recommendations using the different methods.

As shown in Figure 11, the enriched semantic profile demonstrates significant improvements with respect to both the

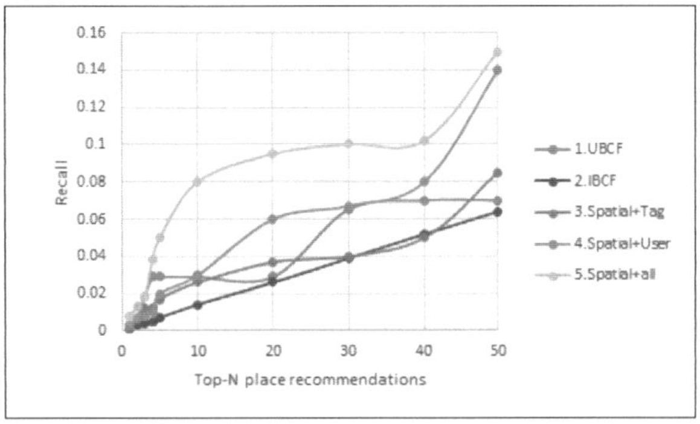

Figure 9. Recall values for the top-N place recommendations

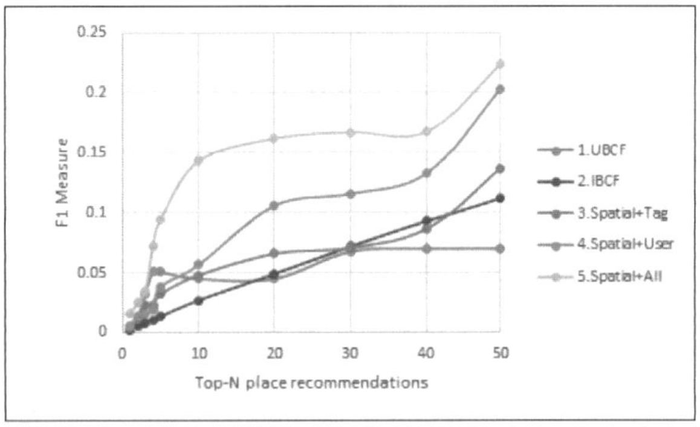

Figure 10. F1 measure values for the top-N place recommendations

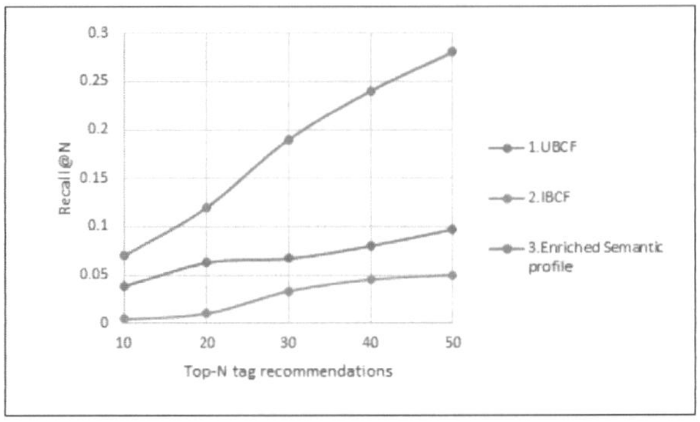

Figure 12. Recall values for the top-N tag recommendations.

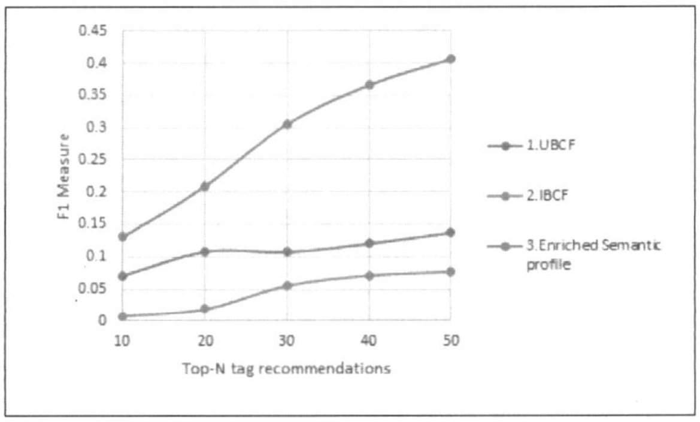

Figure 13. F1 measure values for the top-N tag recommendations.

UBCF and IBCF approaches. Results demonstrates the quality of the enriched semantic user profiles, and thus confirm their utility for more accurate representations of user profiles. .

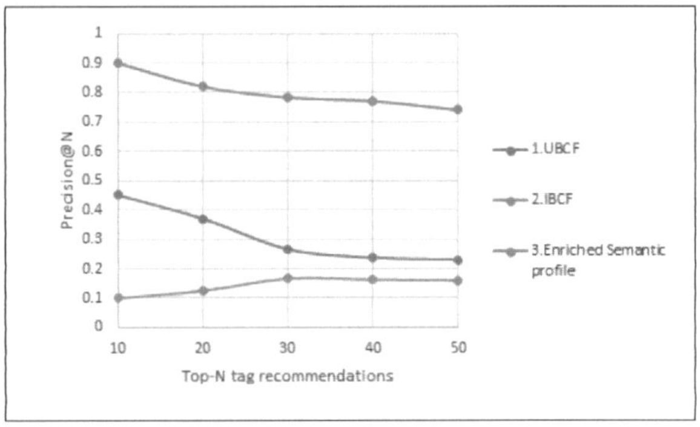

Figure 11. Precision values for the top-N tag recommendations.

VI. Conclusions

This paper considers the problem of user profiling on location-based social networks. Both the spatial (where) and the semantic (what) dimensions of user and place data are used to construct different views of a user's profile. A place is considered to be associated with a set of tags or labels that describe its associated place types, as well as summarise the users' annotations in the place. A folksonomy data model and analysis methods are used to represent and manipulate the data to construct user profiles and place profiles. It is shown how user profiles can be extended from a basic model that describes user's direct links with a place, to an enriched profiles describing richers views of place data on the social network. The model is flexible and can be adjusted to focus on the spatial and semantic dimensions separately or in combination. Results demonstrate that the proposed methods produce user profiles that are more representative of user's spatial and semantic preferences. To our knowledge, no other works have proposed similar treatments of the problem before. Future work will consider a larger number of users as well as the effect of user check-in behaviour on the results. The temporal dimension of the data adds another layer of complexity and is also the subject of future work.

VII. Acknowledgment

Soha is fully supported by a grant from the Egyptian Cultural Affairs and Missions Sector, and their support is duly acknowledged.

References

[1] Q. Li, Y. Zheng, X. Xie, Y. Chen, W. Liu, and W.-Y. Ma, "Mining user similarity based on location history," in Proceedings of the 16th ACM SIGSPATIAL international conference on Advances in geographic information systems. ACM, 2008, pp. 34–42.

[2] H. Gao, J. Tang, and H. Liu, "Exploring social-historical ties on location-based social networks." in ICWSM, 2012, pp. 114–121.

[3] X. Cao, G. Cong, and C. S. Jensen, "Mining significant semantic locations from gps data," Proceedings of the VLDB Endowment, vol. 3, no. 1-2, 2010, pp. 1009–1020.

[4] M. Ye, X. Liu, and W.-C. Lee, "Exploring social influence for recommendation-a probabilistic generative model approach," arXiv preprint arXiv:1109.0758, 2011.

[5] J. Bao, Y. Zheng, D. Wilkie, and M. Mokbel, "Recommendations in location-based social networks: a survey," GeoInformatica, vol. 19, no. 3, 2015, pp. 525–565.

[6] A. Noulas and C. Mascolo, "Exploiting foursquare and cellular data to infer user activity in urban environments," in 14th International Conference on Mobile Data Management, vol. 1. IEEE, 2013, pp. 167–176.

[7] J.-D. Zhang and C.-Y. Chow, "igslr: personalized geo-social location recommendation: a kernel density estimation approach," in Proceedings of the 21st ACM SIGSPATIAL International Conference on Advances in Geographic Information Systems. ACM, 2013, pp. 334–343.

[8] H. Gao, J. Tang, and H. Liu, "gscorr: modeling geo-social correlations for new check-ins on location-based social networks," in Proceedings of the 21st ACM international conference on Information and knowledge management. ACM, 2012, pp. 1582–1586.

[9] B. Liu, H. Xiong, S. Papadimitriou, Y. Fu, and Z. Yao, "A general geographical probabilistic factor model for point of interest recommendation," Knowledge and Data Engineering, IEEE Transactions on, vol. 27, no. 5, 2015, pp. 1167–1179.

[10] E. Cho, S. A. Myers, and J. Leskovec, "Friendship and mobility: user movement in location-based social networks," in Proceedings of the 17th ACM SIGKDD international conference on Knowledge discovery and data mining. ACM, 2011, pp. 1082–1090.

[11] D. Zhou, B. Wang, S. M. Rahimi, and X. Wang, "A study of recommending locations on location-based social network by collaborative filtering," in Advances in Artificial Intelligence. Springer, 2012, pp. 255–266.

[12] H. Wang, M. Terrovitis, and N. Mamoulis, "Location recommendation in location-based social networks using user check-in data," in Proceedings of the 21st ACM SIGSPATIAL International Conference on Advances in Geographic Information Systems. ACM, 2013, pp. 374–383.

[13] J. J.-C. Ying, E. H.-C. Lu, W.-N. Kuo, and V. S. Tseng, "Urban point-of-interest recommendation by mining user check-in behaviors," in Proceedings of the ACM SIGKDD International Workshop on Urban Computing. ACM, 2012, pp. 63–70.

[14] B. Liu, Y. Fu, Z. Yao, and H. Xiong, "Learning geographical preferences for point-of-interest recommendation," in Proceedings of the 19th ACM SIGKDD international conference on Knowledge discovery and data mining. ACM, 2013, pp. 1043–1051.

[15] G. McKenzie, B. Adams, and K. Janowicz, "A thematic approach to user similarity built on geosocial check-ins," in Geographic Information Science at the Heart of Europe. Springer, 2013, pp. 39–53.

[16] B. Hu and M. Ester, "Spatial topic modeling in online social media for location recommendation," in Proceedings of the 7th ACM conference on Recommender systems. ACM, 2013, pp. 25–32.

[17] J.-D. Zhang and C.-Y. Chow, "Geosoca: Exploiting geographical, social and categorical correlations for point-of-interest recommendations," in Proceedings of the 38th International ACM SIGIR Conference on Research and Development in Information Retrieval. ACM, 2015, pp. 443–452.

[18] D. Yang, D. Zhang, Z. Yu, and Z. Wang, "A sentiment-enhanced personalized location recommendation system," in Proceedings of the 24th ACM Conference on Hypertext and Social Media. ACM, 2013, pp. 119–128.

[19] M.-J. Lee and C.-W. Chung, "A user similarity calculation based on the location for social network services," in Database Systems for Advanced Applications. Springer, 2011, pp. 38–52.

[20] F. Abel, "Contextualization, User Modeling and Personalization in the Social Web," PhD Thesis, Gottfried Wilhelm Leibniz University Hannover, April 2011.

[21] A. Hotho, R. Jäschke, C. Schmitz, and G. Stumme, "Information retrieval in folksonomies: Search and ranking," in Semantic web: research and applications, proceedings. Springer, 2006, pp. 411–426.

[22] C. Dwork, R. Kumar, M. Naor, and D. Sivakumar, "Rank aggregation methods for the web," in Proceedings of the 10th international conference on World Wide Web. ACM, 2001, pp. 613–622.

[23] D. Yang, D. Zhang, V. W. Zheng, and Z. Yu, "Modeling user activity preference by leveraging user spatial temporal characteristics in lbsns," Systems, Man, and Cybernetics: Systems, IEEE Transactions on, vol. 45, no. 1, 2015, pp. 129–142.

[24] B. Sarwar, G. Karypis, J. Konstan, and J. Riedl, "Item-based collaborative filtering recommendation algorithms," in Proceedings of the 10th international conference on World Wide Web. ACM, 2001, pp. 285–295.

[25] M. J. Pazzani, "A framework for collaborative, content-based and demographic filtering," Artificial Intelligence Review, vol. 13, no. 5-6, 1999, pp. 393–408.

Modelling Urban Expansion: A Multiple Urban-Densities Approach

Ahmed Mustafa, Ismaïl Saadi, Mario Cools, Jacques Teller
Local Environment Management & Analysis (LEMA), ArGEnCo
University of Liège
Liège, Belgium
e-mail: a.mustafa@ulg.ac.be, ismail.saadi@ulg.ac.be, mario.cools@ulg.ac.be, jacques.teller@ulg.ac.be

***Abstract*— Most existing spatio-temporal urban expansion models consider urban land-use as a binary process, through the identification of urban versus non-urban areas. The main aim of this study is to analyze and model the expansion of multiple urban densities in Wallonia, Belgium. To this end, this study employs a multinomial logistic regression model that enables to visualize the consequence of different urban densities expansion. Cadastral datasets of years 2000 and 2010 are used to set four urban classes (non-urban, low-density, medium-density and high-density urban). Besides, several socio-economic, geographic and political driving forces dealing with urban development were operationalized to create maps of urban expansion probability for each urban density class. These probability maps are then utilized to predict future urban expansions for years 2020 and 2030. The model is validated using relative operating characteristic method for different urban classes. Our results suggest that different urban densities expansions are mainly linked to zoning status, neighboring areas that are urban and accessibility. Most importantly, this study highlights that the contribution of different driving forces to urban expansion process varies along with urban density.**

Keywords- multinomial logistic regression; urban expansion; urban densities; driving forces.

I. INTRODUCTION

Rapid urbanization is one of the crucial global issues affecting the physical features of the Earth. As a consequence, a series of urban expansion modelling approaches has been proposed. Most existing urban expansion models are based on a regular grid composed of square cells of dimension between 30x30m to 300x300m [1]–[6]. Typically, these models address urban expansion as a binary process, through the identification of urban versus non-urban land-uses. Most urban cells at these dimensions comprise a mix of different land-uses. For instance, a cell classified as urban land-use may covered by 60% built-up surface and 40% open-space surface. This causes an erroneous estimation of urban expansion pattern. This paper proposes an urban expansion model that enables modelling three urban classes: low-density urban, medium-density urban and high-density urban. Cadastral datasets (CAD) are used to set urban densities. The MLR is employed to model future urban expansion in Wallonia, Belgium. Frist, urban land-use maps are prepared for years 2000 and 2010 based on CAD data. Next, the MLR model is applied to correlate the observed urban expansion pattern for different urban densities with a number of indicators related to distances, topological, neighborhood, socioeconomic factors and land-use policies. Finally, the MLR's outcomes will be utilized to model urban expansion scenarios for years 2020 and 2030 based on linear extrapolation of observed urban expansion between 2000 and 2010. Relative operating characteristic (ROC) method validates the MLR's outcomes.

This paper focuses on assessing the change from non-urban land-use (reference class) into one of urban density classes. The paper is organized as follows. Section II introduces the case study area, urban expansion model and data. Section III wraps up the results and discussions. Section IV concludes the paper findings.

II. MATERIAL AND METHODS

A. *Study area*

Wallonia (south Belgium) is landlocked, accounts for 55% of the territory of Belgium with a total area of 16,844 km². It comprises five provinces: Hainaut, Liège, Luxembourg, Namur, and Walloon Brabant. The main urban areas are Charleroi, Liège, Mons and Namur. They are all characterized by a historical city-center, around which the urban development expanded. The total population in 2010 was 3,498,384 inhabitants that makes up a third of Belgium population (Fig. 1).

B. *Outline of the model*

The analysis presented here consists of two main parts: (I) estimating probability maps of three urban classes (low, medium and high-density urban) versus non-urban class and (II) develop future urbanization scenarios for years 2020 and 2030.

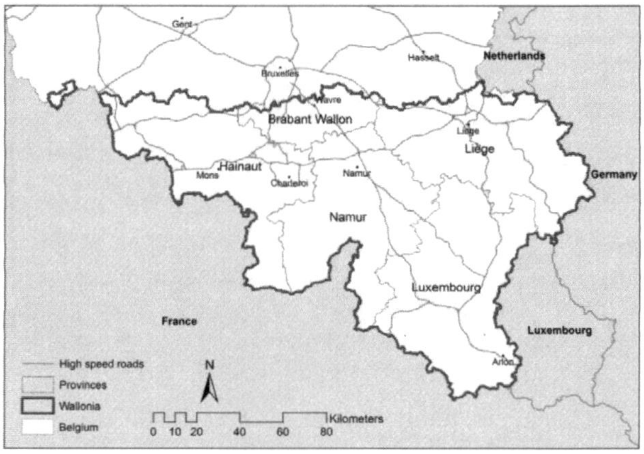

Figure 1. Study area.

The dependent variables (changes from non-urban to one of urban density classes) for the MLR model is defined using CAD. CAD is a vector dataset representing buildings in two dimensions as polygons. Each building comes with different attributes from which the construction date is the most important attribute for our study. Using the construction date, two urban land-use raster-grids were generated for 2000 and 2010 years. First, CAD vector data were rasterized at a very fine cell dimension 2x2m. The rasterized cells were then aggregated to obtain a 50x50m raster-grid. Thus, each aggregated cell has a density value that exhibits the number of rasterized 2x2m cells. The magnitude of density value is then used to represent four urban classes: (class0) non-urban, (class1) low-density urban, (class2) medium-density urban and (class3) high-density urban. Geometrical interval classification method is used to set thresholds that define urban land-use classes (table I). This classification scheme works fairly well on continuous data.

TABLE I. URBAN CLASSES DENSITY RANGE IN NUMBER OF 2X2M CELLS (% OF 50X50M CELL AREA).

Class	*Minimum*	*Maximum*
Class0 (non-urban)	0	25 (4)
Class1 (low-density)	25 (4)	56 (9.12)
Class2 (medium-density)	57 (9.12)	174 (28)
Class3 (high-density)	175 (28)	625 (100)

The independent variables for the MLR model (X), urbanization driving forces, are selected based on expert knowledge of our study area as well as on literature review [6][7]. Table II summarizes the complete list of the selected urbanization driving forces. A zoning map (land-use policy) was developed by discerning the zones where urban development is not permitted and the zones that are designated for urban based on the regional development plan. All data used in this study is represented at 50x50m raster-grid square. The independent variables are measured in different units and therefore we standardized all continuous X. If some of X comparatively measure the same phenomena, then strong collinearities will cause the erroneous estimation of the MLR's parameters. Consequently, a multicollinearity test was examined in the initial stage using the variance inflation factors (VIF). Montgomery and Runger (2003) recommended that the VIFs should not exceed 4. The VIF test results for all X suggest that the variables digital elevation model (DEM) and slope measure the same phenomena and that is also represented between population density and employment rate. In a refining stage, the DEM and employment rate variables have been suppressed. The VIF values for the refining stage implies that all X variables included in this stage show a very low degree of multicollinearity and therefore are introduced in the MLR. Both dependent and independent variables may exhibit spatial autocorrelation, which may have biased the results of the regression analysis [9]. These issue can be addressed through a data sampling approach [5][7]. For the model calibration, 45000 cells were randomly selected. Cells that were urban in 2000 were not included in the samples.

TABLE II. LIST OF SELECTED URBANIZATION DRIVING FORCES.

Driver	*Name*	*Unit*
X1	DEM	Meter
X2	Slope	Percent rise
X3	Distance to Road1 (high-speed roads)	Meter
X4	Distance to Road2	Meter
X5	Distance to Road3	Meter
X6	Distance to Road4 (local roads)	Meter
X7	Distance to railway stations	Meter
X8	Distance to high-populous cities	Meter
X9	Distance to medium-populous cities	Meter
X10	Number of class1 cells within a 5x5 window	Number
X11	Number of class2 cells within a 5x5 window	Number
X12	Number of class3 cells within a 5x5 window	Number
X13	Population density	inh/km²
X14	Employment rate	Percent
X15	Zoning	Binary (0 non-urban, 1 urban)

The general form of the MLR can be represented as, given K_0 (non-urban class) as the reference class:

$$\log\left(\frac{P(Y=k_1)}{P(Y=k_0)}\right) = \alpha_{k_1} + \beta_{k_1 1}X_1 + \beta_{k_1 2}X_2 + \ldots + \beta_{k_1 n}X_n$$

$$\ldots \tag{1}$$

$$\log\left(\frac{P(Y=k_n)}{P(Y=k_0)}\right) = \alpha_{k_n} + \beta_{k_n 1}X_1 + \beta_{k_n 2}X_2 + \ldots + \beta_{k_n n}X_n$$

where $\log\left(\frac{P(Y=k_n)}{P(Y=k_0)}\right)$ is the natural logarithm of class k_n against the reference class, α is the intercept, β is the regression coefficients of class k_n. The probabilities of each class can be calculated with the following formula:

$$P(Y=k_0) = \frac{1}{1+e^{\log(k_1)}+\ldots+e^{\log(k_n)}}$$

$$\ldots \tag{2}$$

$$P(Y=k_n) = \frac{e^{\log(k_n)}}{1+e^{\log(k_1)}+\ldots+e^{\log(k_n)}}$$

III. RESULTS AND DISCUSSIONS

Fig. 2 shows different urban density classes of the observed 2000 urban land-use. High-density urban lands are concentrated in the existing urban centers. Medium-density lands tend to be located around cities in suburbs and low-density lands tend to be found in rural and remote locations.

The MLR's outcomes are probability of urbanization maps for each class based on vectors of regression coefficients β and intercepts α. Table III gives the MLR's results. All explanatory variables are statistically significant on one or more urban classes.

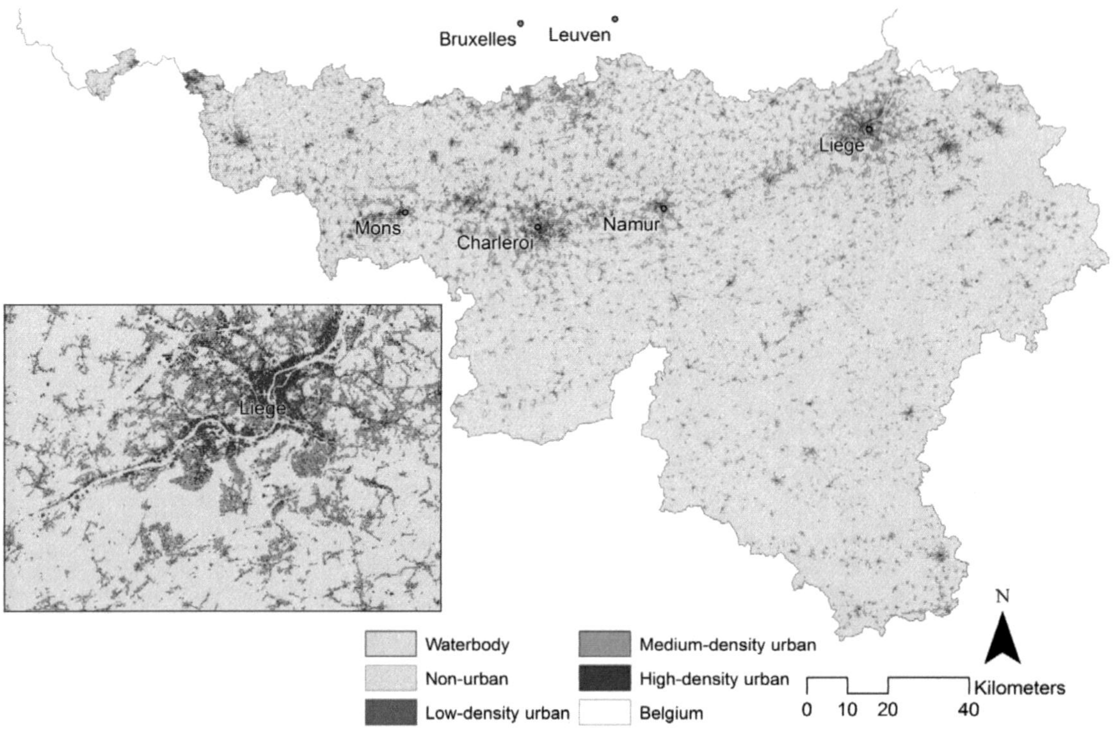

Figure 2. Urban density classes of 2000.

TABLE III. THE COEFFICIENTS (B) OF THE MLR MODEL (CLASS0 IS THE REFERENCE CLASS).

Driver	*Class1*	*Class2*	*Class3*
Intercept	-2.869	-2.979	-3.381
X1	N.I.	N.I.	N.I.
X2	0.036	-0.148*	-0.141*
X3	-0.038	-0.125*	-0.534*
X4	-0.004	-0.112*	-0.350*
X5	-0.137*	-0.160*	-0.238*
X6	-0.356*	-0.306*	-0.160*
X7	0.097*	0.001	-0.174*
X8	0.006	0.053*	0.122*
X9	-0.054*	-0.104*	-0.069*
X10	0.391*	0.305*	-0.024
X11	0.153*	0.151*	0.036*
X12	0.091*	0.214*	0.460*
X13	0.000	0.147*	0.138*
X14	N.I.	N.I.	N.I.
X15	3.317*	2.930*	2.692*

* Indicate significance at P <= 0.05 level
N.I. not included

The impact of different drivers varies along with urban density. Urban expansion of all urban density classes are extremely correlated with zoning status (X15). Distances to Road1 and Road2 (X3 and X4) have a noticeable impact on the development of high density projects (class3). The impact of distance to Road4 (X6), is generally decreasing with increasing urban densities.

The ROC-values of the probability maps of 2000-2010 are 0.94, 0.93 and 0.88 for classes 1, 2 and 3 respectively. That means the probability maps can be used for reliable predictions of the future urban expansion patterns.

The assessed probability maps for the period 2000–2010 have been used to generate spatially-explicit urbanization scenarios for 2020 and 2030 by (I) quantifying the necessary area for future expansion for each urban density class and (II) selecting the cells with the highest values from urbanization probability maps for each class until the required areas are met. This generates urban expansion map for each 2020 and 2030. Next, the expansion maps are combined with the 2010 actual map to produce the urban distribution map. Waterbodies, that are defined using zoning plan, are introduced as a constrained.

The future necessary areas for each density class are calculated on the basis of a linear extrapolation of the actual urban expansion between 2000 and 2010. The urban area is expected to increase, given the actual 2010 urban area, to 3.6% in 2020 and to 10.1% in 2030. The percentage of each urban density class expansion to the total expansion between 2000 and 2010 were about 56% low-density, 35% medium-density and 9% high-density lands. These percentages are then used to estimate the required urban lands in 2020 and 2030 for each urban density class (Fig. 3).

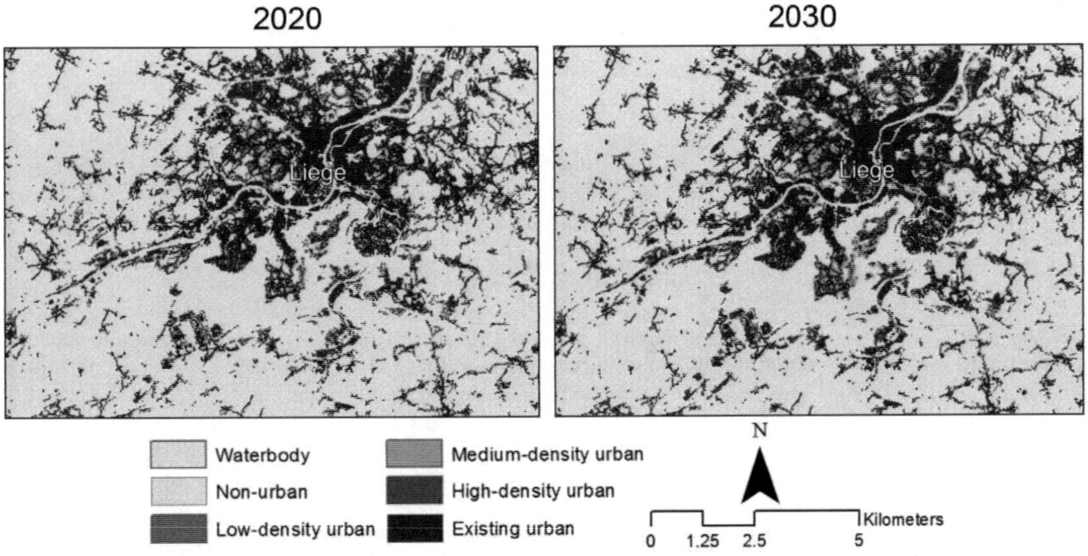

Figure 3. Developed scenario for the Liège metropolitan area.

IV. CONCLUSIONS

This paper employs the MLR model to examine drivers of urban expansion in Wallonia (Belgium) and to forecast near-future urban expansion using Belgian cadastral data (CAD). Four classes, non-urban, low-density, medium-density and high-density urban, are defined as dependent variables for the MLR. Urban expansion for each urban class versus non-urban class is predicted for 2020 and 2030 based on 2000 and 2010. Several variables are selected and introduced in the MLR model as independent variables. It was found that all the independent variables have impacts on urban expansion in Wallonia, but their relative importance are varied with density. However, it can be concluded that policies, number of existing urban lands within neighborhood and accessibility are the most important determinants of urban expansion process. A validation of the MNL showed that the model's outcomes allows to predict future urban expansion patterns with a relatively high explanatory power.

Based on liner extrapolation of urban expansion between 2000 and 2010, expansion scenarios are proposed to simulate 2020 and 2030 urban patterns. This study's findings would help decision makers and urban planners in enhancing understanding of urban expansion in Wallonia. Most importantly, it can serve as input to hydrological modelling.

Finally, future extension of this research will be dedicated to analyze the urbanization process within existing medium and low-density urban areas instead of only studying the change from non-urban to one of urban density classes.

ACKNOWLEDGMENT

The research was funded through the ARC grant for Concerted Research Actions and through the Special Fund for Research, both financed by the Wallonia-Brussels Federation.

REFERENCES

[1] D. Guan, et al., "Modeling urban land use change by the integration of cellular automaton and Markov model," Ecol. Model., vol. 222, no. 20–22, Oct. 2011, pp. 3761–3772.

[2] Z. Hu and C. P. Lo, "Modeling urban growth in Atlanta using logistic regression," Comput. Environ. Urban Syst., vol. 31, no. 6, Nov. 2007, pp. 667–688.

[3] C. A. Jantz, S. J. Goetz, and M. K. Shelley, "Using the Sleuth Urban Growth Model to Simulate the Impacts of Future Policy Scenarios on Urban Land Use in the Baltimore-Washington Metropolitan Area," Environ. Plan. B Plan. Des., vol. 31, no. 2, 2003, pp. 251–271.

[4] X. Liu, X. Li, X. Shi, S. Wu, and T. Liu, "Simulating complex urban development using kernel-based non-linear cellular automata," Ecol. Model., vol. 211, no. 1–2, Feb. 2008, pp. 169–181.

[5] A. Rienow and R. Goetzke, "Supporting SLEUTH – Enhancing a cellular automaton with support vector machines for urban growth modeling," Comput. Environ. Urban Syst., vol. 49, Jan. 2015, pp. 66–81.

[6] A. Mustafa, I. Saadi, M. Cools, and J. Teller, "Measuring the Effect of Stochastic Perturbation Component in Cellular Automata Urban Growth Model," Procedia Environ. Sci., vol. 22, 2014, pp. 156–168.

[7] H. Cammerer, A. H. Thieken, and P. H. Verburg, "Spatio-temporal dynamics in the flood exposure due to land use changes in the Alpine Lech Valley in Tyrol (Austria)," Nat. Hazards, vol. 68, no. 3, Sep. 2013, pp. 1243–1270.

[8] D. C. Montgomery and G. C. Runger, Applied Statistics and Probability for Engineers, Fourth. New York: John Wiley & Sons, 2003.

[9] K. P. Overmars, G. H. J. de Koning, and A. Veldkamp, "Spatial autocorrelation in multi-scale land use models," Ecol. Model., vol. 164, no. 2–3, Jun. 2003, pp. 257–270.

Spatial Regression in Health: Modelling Spatial Neighbourhood of High Risk Population

Stefania Bertazzon
Department of Geography
University of Calgary
Calgary, Alberta, Canada
e-mail: bertazzs@ucalgary.ca

Abstract— **Many health conditions affect certain individuals more than others: for example, adults over 65 years of age are more affected by cardiovascular disease than younger individuals. Therefore, the spatial pattern of the disease incidence can be modelled more effectively through the residential pattern of higher risk groups. The method is demonstrated through a spatial regression of the association of cardiac catheterization and socioeconomic determinants in Calgary (Canada). Over a 5-year interval, 45% of catheterizations are performed on seniors, that constitute 9% of the population. Seniors' residential location is therefore used as an auxiliary process to model the spatial weights of the regression model. This spatial model leads to a more realistic neighbourhood configuration, yielding more reliable regression estimates. Based on the residential location of the population at greater risk, the model presents low sensitivity to variations in the supporting geographic units. The use of a relevant auxiliary process is general and applicable to a range of conditions; it constitutes a promising alternative to the direct estimation of spatial parameters on the primary process. Overall, the spatial weights matrix based on at risk population shall increase the reliability of spatially autoregressive multivariate epidemiological models.**

Keywords- health geography; spatial regression analysis; spatial correlation; cardiovascular condition; cardiac catheterization; Seniors; risk population; residential location.

I. INTRODUCTION

Geographic information science (GIS) has been increasingly employed in population health research due to its ability to analyze interactions of health determinants in space [12], [26]. This has furthered the integration between geography and health sciences, promoting the development of more effective spatial analytical methods [13], whose reliable results can be translated into policy decisions [9], [27].

Most geographical phenomena, e.g., disease prevalence and population distribution, exhibit variations across space and self-similarity over short distances. These properties, known as spatial dependence and non-stationarity [24], are known to hamper the reliability of analytical models, by increasing the uncertainty of the estimated parameters [2]. In their presence, analytical models may lead to ineffective, or even harmful, health policy decisions. Spatial analytical methods offer a valid response to this problem; however, their ability to improve the model reliability [2], [3] depends on the representation of spatial interactions embedded in the model.

Health and its determinants interact in space [17]; hence, these interactions can be modelled by multivariate regression [2]. While local analytical methods [15] are concerned with spatial non-stationarities, spatial autoregressive methods [2] address the uncertainty stemming from spatial dependence. The specification of a spatially autoregressive model requires the definition of a neighbourhood of spatial units: the more accurate the neighbour definition, the more reliable the model estimates. Ideally, an accurate neighbourhood definition rests on a deep knowledge of the spatial process involved; more often one must estimate spatial dependencies using statistical methods, which are typically applied to the dependent variable [3], [8], [2]. In such situations, we propose the application of those statistical methods to another spatial process, which is related to the dependent, and which is better understood, if not within the health sciences, within geography or urban studies. The latter process effectively serves as an auxiliary process, in that it is used to estimate the spatial parameters that will provide a more realistic neighbourhood representation, enhancing the reliability of the regression model.

Here, a multivariate spatial regression model [2] analyses the socioeconomic determinants of cardiac catheterization cases. For a 5-year period in the study region, a large proportion of catheterizations affect seniors, i.e., individuals aged 65 years and older: over 45% of catheterizations are performed on seniors, where seniors account for 9% of the total population (12.6% of adults). While only 0.01% of adults under 65 receive catheterizations, almost 7% of seniors do. Visual observation suggests that catheterization cases are spatially associated with seniors' residential location (refer to Figure 1). *Seniors* is therefore proposed as the auxiliary process.

The association between older age and cardiovascular disease is well known and has received much attention in the health literature [29], [28], [30]. Their spatial association has received less attention, although this relationship has been examined at the neighbourhood level [16], [18], [5], [25]. The city of Calgary was chosen as an interesting study area, where seniors' residential location presents a clustered distribution, facilitating the identification of spatial associations.

In the following, Section II provides the context of this study; Section III describes the methods employed; Section

IV outlines the results; Section V provides a discussion, and Section VI draws the main conclusions of the study.

II. Background

One of the leading causes of death in the developed world, cardiovascular disease is known to be associated with a number of risk factors, including age and gender, limited physical activity, smoking, and diet [18], [5]. Often these factors correlate with demographic and socioeconomic characteristics, such as age, occupation, and income, which can be measured by census variables [10], [31], [7], [4].

Calgary is one of the largest Canadian cities. Located in the foothills east of the Rocky Mountains, it covers a large and regular geographic area; its population is relatively young, affluent, and highly educated [40]. Due to its economy and history, its population presents a pseudo-concentric distribution, where age and socioeconomic status decrease as distance from the city center increases [36].

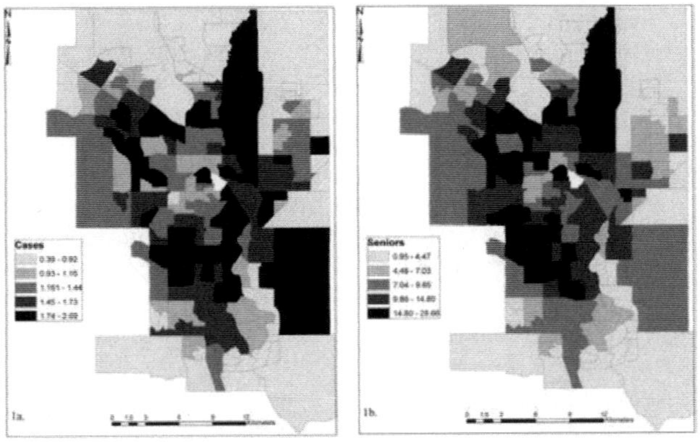

Figure 1. Distribution of Catheterization Cases and Seniors in Calgary.

The clinical data were provided by the Alberta Provincial Project for Outcome Assessment in Coronary Heart Disease (APPROACH) initiative [22], a clinical registry begun in 1995 with the collection of cardiac catheterization data. Cardiac catheterization is a procedure performed on individuals with cardiovascular conditions [22]. The province of Alberta has a publicly funded health care system, therefore there are no financial costs associated with the procedure. We acknowledge the limitations of catheterization in representing cardiovascular disease; however, more appropriate variables, e. g., hospitalizations for acute coronary syndrome (ACS), were not systematically collected at the postal code spatial aggregation level until much more recently. Analyzing the latter, more recent data would have been problematic, due to changes in the 2011 census [39]. However, a comparison of catheterization and ACS over a two-year period of overlapping data (2006–2007) suggests that the proposed method shall be transferable to ACS data as soon as all the clinical data and census variables become available.

For this study, we considered only patients aged 20 years and older, residing in Calgary, who had one or more catheterizations between 1999 and 2003 (for multiple procedures, only the first one was retained). Patient residential addresses, at the postal code level, were spatially aggregated to match census spatial units, using postal code conversion files (PCCF +) [43]. Demographic and socioeconomic variables were drawn from the 2001 census of Canada. To match the clinical records, census variables were trimmed to represent the population aged 20 years or older. Clinical records and census variables were normalized for each spatial unit, i. e., divided by the total pertinent population and multiplied by 1,000 [34]. This variable, representing cardiac catheterization cases, is named "*Cases*" in Table 1. For the regression analysis, clinical records were age- and sex-standardized [42]: the standardized variable is named "*Standardized Cases*". The variable "*Seniors*" (Table 1) emphasizes the difference between workforce and retirees as the sum of all age groups aged 65 years and older.

All the analyses are conducted on census tracts, which are relatively small and stable spatial units, with population between 2,500 and 8,000 residents, located in a census metropolitan area [39]. For the 2001 census, the spatial database contains 181 census tracts, and the clinical database contains 11,430 catheterization cases over the 5-year interval (Figure 1).

III. Methods

Cross-correlations and spatial correlation analyses assess the association of catheterization cases and seniors' residential location. Spatial correlation, measured by bivariate Moran's I, extends this comparison to neighbouring census tracts. Spatial autocorrelation, measured by Moran's I, assesses the self-similarity of each variable over neighbouring census tracts. Moran's I ranges from −1 for negative spatial autocorrelation, to +1 for positive spatial autocorrelation, with 0 indicating spatial randomness [19]. The calculation of Moran's I requires the definition of a spatial weights matrix, W (discussed below).

Spatial autoregression aims at enhancing the reliability of estimates in the presence of spatial dependencies (1).

$$Y = X\beta + \rho WY + \varepsilon \quad (1)$$

Spatial autoregression also requires a spatial weights matrix, W, which selects the spatial units deemed spatially dependent [20], and an autoregressive parameter, ρ (rho), is estimated. This study uses a simultaneous autoregressive specification and maximum likelihood estimation [2]. Following conventional practice, the regression is computed on the age- and sex-standardized dependent variable. Indirect standardization [42] employs age and sex groups, therefore inflating the correlation between the dependent and those demographic variables. For this reason, the regression model does not include demographic variables, even though the exclusion some variables, and particularly of *Seniors*, has a large impact on the model's goodness of fit. The use of *Seniors* as an auxiliary process mitigates this impact.

The spatial weights matrix is viewed as a tool to enhance model reliability. In the absence of a spatial specification, the model reliability is decreased by the presence of spatial autocorrelation in the regression residuals, which inflates the variance associated with the regression parameters [2]. Therefore, the spatial weights matrix is designed to best capture the spatial autocorrelation in the dependent variable, so that most of the spatial autocorrelation can be accounted for by the model, leaving insignificant spatial autocorrelation in the residuals. The basic form of a spatial weights matrix is a binary structure, where a threshold distance, or a number of nearest neighbours, selects the neighbouring spatial units where the variable is expected to exhibit spatial autocorrelation. Often a weight is added, in order to model distance decay effects [20]. This matrix involves the specification of three parameters: distance threshold, distance metric, and distance decay function. By dynamically adjusting these three parameters, the spatial weights matrix can yield different values of the spatial autocorrelation index. Of the three parameters, distance exerts the greatest influence, by determining how many spatial units are deemed spatially autocorrelated. There are several methods to define this parameter for areal units, such as census tracts [20]. Here, we use a distance threshold based on nearest neighbours, for the following reasons: census tracts are not necessarily meaningful for the spatial pattern of cardiovascular disease; they tend to be small and pseudo-rectangular in the city center, but in the outskirts they tend to be larger and less regular (refer to Figure 1). The latter feature forms a pattern of spatial units, liable to confound the pattern of the variables recorded in those units. To reduce this confounding effect, the number of nearest neighbours is preferred. To further de-emphasize the geometry of census tracts, we consider the distance between their centroids.

The second parameter is the distance metric. Among many distance metrics used in geography, the most common is the Euclidean metric: the straight line distance measurement between two points, 'as the crow flies'. In many North American cities, connectivity occurs over a pseudo-rectangular road pattern, better modelled by the Manhattan distance, which measures distance between points along a rectangular path with right angle turns. Connectivity over a complex or mixed network can be more accurately represented by metrics of the class known as Minkowski distance [38], which yields patterns intermediate between straight line and right angle. It is described by (2), of which Euclidean and Manhattan distances are special cases, where the key parameter, p, can take any value between 1 (Manhattan) and 2 (Euclidean).

$$d_{ij} = \left[(x_i - x_j)^p + (y_i - y_j)^p \right]^{1/p} \tag{2}$$

The p value can be estimated to best approximate empirical distance or travel time. Within this class of metrics, an appropriate choice can refine the selection of spatial units, producing buffer shapes close to the physical, pseudo-rectangular connectivity pattern of the census tracts.

The third parameter is the distance decay function, which weights the interaction among spatial units by their distance. Interaction tends to decrease as the distance between units increases: a number of functions have been developed to model this relationship [14], [35]. Commonly the distance decay function is calibrated by a weight, often another variable, which normalizes the relationship [8].

The interaction of these parameters in the spatial weights matrix affects the estimated spatial autocorrelation and hence the reliability of the regression estimates. In the proposed method, the three parameters are calibrated on the spatial autocorrelation of the auxiliary process, as opposed to the dependent variable, or primary process.

Seniors' residential location is understood better than the distribution of *Cases*, and can be explained by socio-economic and urban traits. Because of their high spatial association, the spatial autocorrelation of *Seniors* should be more meaningful than that of *Cases*; therefore, the method is expected to increase the reliability of the regression estimates, along with their interpretability.

Statistical analyses were conducted in TIBCO Spotfire S+ 8.2. Maps were obtained in ESRI ArcGIS 10.

IV. RESULTS

Spatial autocorrelation and cross-correlation analysis, initially run on the parameters (k = 3, p = 2), yields two important results (Table 1): *Cases* exhibits significant but moderate spatial autocorrelation (I = 0.36), whereas *Seniors* exhibits a much higher value (I = 0.54); the two variables exhibit high cross-correlation (0.60) and spatial correlation (0.35). Together, these results confirm that clustering of *Cases* tends to occur in association with the clustering of seniors' residential location.

TABLE I. CORRELATIONS AND SPATIAL CORRELATIONS

	Clinical records		Demographhic variables		Economic variables	Education		Family status
	Cardiac cath. cases	Age & sex std. cases	Age 65 and over	Age 55 to 64	Family median income	Secondary or lower education	Non-univ. post-sec. degree	2 parents with children
	Cases	Std.Cases	Seniors	Age.55-64	Fam.Income	Secondary	Trades	Families
Cases	0.36 **	0.75 **	0.66 **	0.50 **	-0.08 ns	0.25 **	-0.22 **	-0.25 **
Std.Cases	0.34 **	0.53 **	0.87 **	0.71 **	0.17 *	-0.09 ns	-0.42 **	-0.21 **
Seniors	0.35 **	0.49 **	0.60 **	0.31 **	-0.03 ns	-0.09 ns	-0.35 **	-0.47 **
Age.55-64	0.22 **	0.27 **	0.09 ns	0.50 **	0.29 **	-0.01 ns	-0.26 **	0.15 ns
Fam.Income	-0.06 ns	0.00 ns	-0.08 ns	0.10 ns	0.50 **	-0.68 **	-0.33 **	0.56 **
Secondary	0.09 ns	-0.12 ns	-0.17 *	0.04 ns	-0.42 **	0.75 **	0.25 **	-0.09 ns
Trades	-0.03 ns	-0.11 ns	-0.18 *	0.02 ns	0.02 ns	0.17 *	0.42 **	0.04 ns
Families	-0.13 ns	-0.19 *	-0.42 **	-0.05 ns	0.35 **	0.04 ns	0.24 **	0.74 **

Diagonal: univariate Moran's I. Upper half: Pearson's correlation . Lower half: bivariate Moran's I.

The correlation analysis also identifies additional traits consistent with this result. Significant and negative spatial cross-correlation between *Seniors* and *Families* indicates a highly clustered if not dichotomous spatial pattern, where neighbourhoods dominated by younger families alternate with neighbourhoods mostly occupied by seniors. The high correlation of *Income* with *Families* and with *Education* suggests that income exerts a strong but indirect influence on the spatial clustering, so that the observed residential pattern appears associated with economic factors, in addition to demographic ones. Nonetheless, age is confirmed as the variable that exhibits the most distinct spatial pattern.

Following these results, we analyze the response of the spatial autocorrelation index to variations in the spatial contiguity parameters for the variables *Cases*, *Standardized Cases*, and *Seniors*, as summarized in Figure 2. For all the parameter combinations, *Seniors* exhibits the highest spatial autocorrelation values, whereas the values of *Cases* are constantly significant but relatively low. Age- and sex-standardized *Cases* exhibits greater spatial autocorrelation than the non-standardized variable. The number of nearest neighbours (k) has a greater impact on the spatial autocorrelation value, whereas the distance metric only provides minor adjustments. Since the analysis shows that the spatial autocorrelation is best expressed by relatively small neighbourhoods, defined by low k values, distance decay weighting will not be discussed.

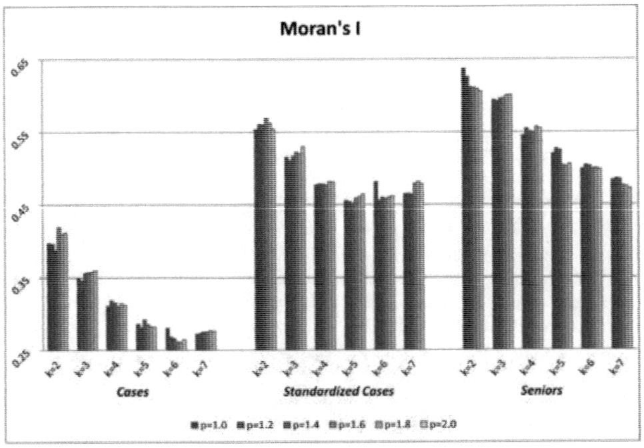

Figure 2. Moran's I as function of distance threshold and distance metric.

The spatial autocorrelation of *Cases* declines steeply as the number of nearest neighbours increases, and only parameter combinations of $k = 2$ and p values between 1.6 and 2 effectively capture the spatial autocorrelation in the variable. These combinations define very small neighbourhoods, as distance is short and metrics close to the straight line produce the shortest distance between centroids. For *Standardized Cases*, the trend is similar to *Cases*, and the most visible effect of the standardization is the increased value of the spatial autocorrelation. The spatial autocorrelation of *Seniors* exhibits less variation than *Cases* in response to variations in both parameters.

The regression summarized in Table 2 is based on the spatial contiguity parameters derived from the auxiliary process. It models the association between the standardized dependent variable and the pool of socioeconomic variables. Significant explanatory variables are: low education attainments, family median income, and, -with negative coefficient- family status and technical education, the latter possibly a proxy for young, low income groups. The model does not include the variable that is most highly correlated with the dependent, i.e., *Seniors*, due to the confounding effect of the indirect standardization; hence, the goodness of fit is relatively low, e.g., the value of Anselin's [1] pseudo-R^2 is 0.19.

TABLE II. SPATIAL AUTOREGRESSIVE MODEL

Stdandardized Cases (k = 3, p = 2)				
	β value	Std.Error	t value	Pr(>|t|)
Intercept	14.03	3.77	3.73	0.00
Secondary	0.03	0.00	6.01	0.00
Family median income	0.14	0.03	5.28	0.00
Trades	-0.02	0.01	-4.04	0.00
Families	-0.01	0.00	-3.40	0.00
Log. likelihoo	Pseudo R^2	Res. Std. Error	Rho	Residual Moran
-679.70	0.19	2.92	0.32	-0.03

In its linear specification, the model exhibits significant residual spatial autocorrelation; hence, a spatially autoregressive specification is presented, which exhibits non-significant residual spatial autocorrelation and a significant autoregressive coefficient, rho. The calibration of the spatial contiguity parameters on the auxiliary process, *Seniors*, is a way of representing this variable in the model. Arguably, this method enhances the significance of the rho parameter, attaining a more effective reduction of the residual spatial autocorrelation and more reliable regression estimates.

V. DISCUSSION

The use of an auxiliary process as an alternative to the use of the primary process for the definition of the spatial contiguity parameters leads to an improved neighbourhood definition, enhancing the reliability of the spatial regression model estimates. In the application discussed here, the resulting neighbourhoods are larger; moreover, the analytical results are less sensitive to variations in the neighbourhood size, defined by the contiguity parameters. These two results are important and related, both deriving from the choice of a particular auxiliary process. The tiny neighbourhoods defined by the primary process model clusters of *Cases*, which are isolated, as shown by their low spatial autocorrelation. Conversely, the larger neighbourhoods calibrated on the auxiliary process effectively model the distribution of the highest-risk population, *Seniors*. Therefore, modelling *Cases* based on the spatial distribution of *Seniors* provides not only a more reliable, but also a more interpretable model. Overall, the greater analytical stability obtained through of the use of the auxiliary process is an important result, which can potentially reduce the impact of the modifiable aereal unit problem (MAUP) [41], [32]. As a future research direction, the method discussed here shall be tested on different spatial units, e.g., communities or dissemination areas, where previous analyses [6] implemented directly on the primary process have suggested a large impact of the MAUP.

The distribution of seniors in Calgary has been studied within several disciplines, and it is conceptually understood as influenced by economic cycles and age of community, among other factors [36]. As an additional line of enquiry, the spatial structure of the process shall be analyzed in light of that literature. Conversely, other crucial aspects, such as range of spatial interaction, shall be confirmed by qualitative analyses [23], [37]. An integration of these two lines of enquiry is expected to substantially improve understanding and representation of the spatial pattern of the high-risk population in its interaction with the incidence of cardiac disease [33].

For the application presented here, the k parameter chosen through the auxiliary process is only marginally larger than the one selected for the primary process, and the difference between the resulting neighbourhoods is larger in conjunction with the distance metric. Hence, the impact on the reliability of the regression estimates may be moderate if measured simply by variance indicators; however, the spatial contiguity parameters also affect model inference, impacting model selection procedures. Comparing several regression specifications, the exclusion of *Seniors* as a predictor is constantly accompanied by increased residual spatial autocorrelation, suggesting that *Seniors* is the process associated with the observed spatial autocorrelation.

One important extension of the current analysis will be its application on acute coronary syndrome (ACS). While the proposed method presents many advantages with respect to modelling seniors, it shifts the analytical focus away from younger adults, where the prevalence is very low (0.01%), and its spatial modelling remains challenging.

A number of health conditions are associated with specific demographic segments, and in all those cases, the use of an appropriate auxiliary process can improve the analytical results. Testing is underway on ACS, congenital birth defects and child obesity. Further lines of enquiry shall include analyses of different contiguity configurations, such as threshold distance vs. nearest neighbours, distance metrics beyond the Minkowski range, and assessment of distance decay functions on larger neighbourhoods.

VI. CONCLUSION

A multivariate regression model estimates the association between cardiac catheterization and socioeconomic factors. In the presence of spatial dependencies, the use of a spatially autoregressive model increases the reliability of the model estimates. Such reliability can be further improved by an appropriate definition of the spatial contiguity parameters. Of the catheterizations recorded over five years in the study area, almost half are performed on individuals aged 65 years or older. This association is well known and understood, and it suggests a strong association between seniors' residential location and the spatial pattern of catheterization cases. Therefore, *Seniors* is identified as an auxiliary process for the calibration of the spatial contiguity parameters of the model. The method enhances the reliability of the regression estimates and the model selection, rendering the auxiliary-based model more efficient, interpretable, and stable over variations in the supporting spatial units.

ACKNOWLEDGMENT

I acknowledge the Natural Science and Engineering Research Council of Canada and the GEOIDE Network of Centers of Excellence for funding this research. I also thank APPROACH initiative researchers for data and support. I am grateful to Olesya Elikan and all the students who helped with bibliographic research, database maintenance, and visualization.

REFERENCES

[1] L. Anselin, SpaceStat tutorial. Regional Research Institute, West Virginia University, Morgantown, West Virginia, 1993.

[2] L. Anselin, Spatial Econometrics: Methods and Models. Kluwer Academic Publisher, New York, 1988.

[3] L. Anselin, A. K. Bera, R. Florax and M. J. Yoon, "Simple diagnostic tests for spatial dependence", Regional Science and Urban Economics 26, 1996, 77-104.

[4] T. Augustin, T. A. Glass, B. D. James and B. S. Schwartz, "Neighbourhood psychosocial hazards and cardiovascular disease: The Baltimore Memory Study". American Journal of Public Health 98, 2008, 1664–1670.

[5] A. J. Bagnall, S. G. Goodman, K. A. A. Fox, R. T. Yan, J. M. Gore, A. N. Cheema, T. Huynh, D. Chauret, D. H. Fitchett, A. Langer and A. T. Yan, "Influence of Age on Use of Cardiac Catheterization and Associated Outcomes in Patients With Non-ST-Elevation Acute Coronary Syndromes". American Journal of Cardiology 103, 2009, 1530-1536.

[6] S. Bertazzon, S. Olson and M. Knudtson, "A spatial analysis of the demographic and socioeconomic variables associated with cardiovascular disease in Calgary (Canada)". Applied Spatial Analysis and Policy 3, 2010, 1-23.

[7] B. Chaix, M. Rosvall and J. Merlo, "Neighborhood socioeconomic deprivation and residential instability: Effects on incidence of ischemic heart disease and survival after myocardial infarction". Epidemiology 18, 2007, 104–111.

[8] N. Cressie, Statistics for Spatial Data. Wiley, New York, 1993.

[9] M. Cutchin, "The need for the ''new health geography'' in epidemiologic studies of environment and health". Health & Place 13, 2007, 725–742.

[10] A. V. Diez Roux, S. Merkin, D. Arnett, L. Chambless, M. Massing, F. Nieto, P. Sorlie, M. Szklo, H. Tyroler and R. Watson, "Neighborhood of residence and incidence of coronary heart disease". New England Journal of Medicine 345, 2001 99-106.

[11] S. Dray, "A New Perspective about Moran's Coefficient: Spatial Autocorrelation as a Linear Regression Problem". Geographical Analysis 43, 2001, 127–141

[12] T. Dummer, "Health geography: supporting public health policy and planning". Canadian Medical Association Journal 178, 2008, 1177–1180.

[13] P. Elliott and D. Wartenberg, "Spatial epidemiology: Current approaches and future challenges". Environmental Health Perspectives 12, 2004, 998–1006.

[14] A. Fotheringham, "Spatial structure and distance-decay parameters". Annals of the Association of American Geographers 71, 1981, 425-436.

[15] A. Fotheringham, C. Brundson and M. Charlton, "Geographically weighted regression: A natural evolution of the expansion method for spatial data analysis". Environment & Planning A 30, 1998, 1905–1927.

[16] A. Gatrell, G. Lancaster, A. Chapple, S. Horsley and M. Smith, "Variations in use of tertiary cardiac services in part of North-West England". Health & Place. 8, 2002, 147-153.

[17] A. Gatrell, J. Popay and C. Thomas, 2004. "Mapping the determinants of health inequalities in social space: can Bourdieu help us?" Health & Place. 10, 2004, 245-257.

[18] Y. Gerber, S. Jacobsen, R. Frye, S. Weston, J. Killian and V. Roger, "Secular Trends in Deaths From Cardiovascular Diseases: A 25-Year Community Study". Circulation 113, 2006, 2285-2292.

[19] A. Getis, "A history of the concept of spatial autocorrelation: A geographer's perspective". Geographical Analysis 40, 2008, 297–309.

[20] A. Getis and J. Aldstadt, "Constructing the Spatial Weights Matrix Using a Local Statistic". Geographical Analysis 36, 2004, 90-104.

[21] A. Getis and K. Ord, "The analysis of spatial association by use of distance statistics". Geographical Analysis 24, 1992, 189–206.

[22] W. A.Ghali and M. L. Knudtson, "Overview of the Alberta provincial project for outcome assessment in coronary heart disease". Canadian Journal of Cardiology 16, 2000, 1225–1230.

[23] A. Grimes, D. C. Maré and M. Morten, "Defining Areas and Linking Geographical Data: an Example from New Zealand". Population, Space and Place 16, 2010, 165–170.

[24] R. Haining, The Special Nature of Spatial Data. The SAGE handbook of spatial analysis.In: A.S. Fotheringham and P. A. Rogerson (Eds.) SAGE, London, 2009.

[25] E. J. Holowaty, T. A. Norwood, S. Wanigaratne, J. J. Abellan and L. Beale, "Feasibility and utility of mapping disease risk at the neighbourhood level within a Canadian public health unit: an ecological study". International Journal of Health Geographics 9,21, 2010.

[26] A. Iftimi, F. Montes, A. M. Santiyán and F. Martínez-Ruiz, 'Space–time airborne disease mapping applied to detect specific behaviour of varicella in Valencia, Spain', *Spatial and Spatio-temporal Epidemiology,* 14–15, 2015, 33-44.

[27] V. Jürgens, S. Ess, M. Schwenkglenks, T. Cerny, and P. Vounatsou, 'Using lung cancer mortality to indirectly approximate smoking patterns in space', *Spatial and Spatio-temporal Epidemiology,* 14–15, 2015, 23-31.

[28] G. A. Kaplan and J. E. Keil, "Socioeconomic factors and cardiovascular disease: a review of the literature". Circulation 88, 1993, 1973–1998.

[29] E. G. Lakatta, Age-associated cardiovascular changes in health: impact on cardiovascular disease in older persons". Heart Failure Reviews 7, 2002, 29-49.

[30] D. Manuel, M. Leung and K. Nguyen, "Burden of cardiovascular disease in Canada". Canadian Journal of Cardiology 19, 2003, 997–1004.

[31] J. McKay and G. A. Mensah, The Atlas of Heart Disease and Stroke. Geneva: World Health Organization, 2005.

[32] M.-P. Parenteau and M. C. Sawada, "The modifiable areal unit problem (MAUP) in the relationship between exposure to NO2 and respiratory health". International Journal of Health Geographics, 2011, 10:58.

[33] B. Preston and M. W. Wilson, 'Practicing GIS as Mixed Method: Affordances and Limitations in an Urban Gardening Study', *Annals of the Association of American Geographers,* 104 (3), 2014, 510-29.

[34] S. Preston, P. Heuveline and M. Guillot, Demography: Measuring and Modeling Population Processes. Blackwell Publishing, Oxford, 2000.

[35] A. Rattner and B. A. Portnov, "Distance decay function in criminal behaviour: a case of Israel". Annals of Regional Science 41, 2007, 673–688.

[36] B. Sandalack and A. Nicolai, The Calgary Project: Urban form/urban life. University of Calgary Press, Calgary, 2006.

[37] S. M. Santos, D. Chor and G. L. Werneck, "Demarcation of local neighborhoods to study relations between contextual factors and health". International Journal of Health Geographics 9, 34, 2010.

[38] R. Shahid, S. Bertazzon, M. Knudtson and W. Ghali, "Comparison of distance measures in spatial analytical modeling for health service planning". BMC Health Services Research 9:200, 200.

[39] Statistics Canada, 2002. Cartographic Boundary Files: 2001 Census. Reference Guide. Catalogue no. 92F0171GIE. Minister of Industry, Ottawa.

[40] Statistics Canada. The Canadian Population in 2011: Population Counts and Growth. http://www12.statcan.gc.ca/census-recensement/2011/as-sa/98-310-x/98-310-x2011001-eng.cfm. Last accesses on 3/15/2016.

[41] D. J. Unwin, "GIS, spatial analysis and spatial statistics". Progress in Human Geography 20, 1996, 540-441.

[42] L. Waller and C. Gotway, Applied Spatial Statistics for Public Health Data.: John Wiley & Sons, Hobeken, 2004.

[43] R. Wilkins and S. Khan, 2010. PCCF + Version 5G User's Guide. Catalogue no. 82F0086-XDB. Health Statistics Division, Statistics Canada, Ottawa. http://data.library.utoronto.ca/datapub/codebooks/cstdli/pccf_health/pccf5h/MSWORD.PCCF5H.pdf. Last accessed on 3/15/2016.

On improving data quality and topology in vector spatial data

Nina Solomakhina*,†, Thomas Hubauer* and Silvio Becher*
* Siemens AG, Munich, Bavaria, 81739, Germany
Email: fname.lname@siemens.com
† École Polytechnique Fédérale de Lausanne, 1015 Lausanne, Switzerland
Email: fname.lname@epfl.ch

Abstract—**Data quality is an important issue for a spatial data, especially for topological relations between geographical features. Errors and inconsistencies found in Geographical Information System (GIS) data often misrepresent topological structure of the dataset and, therefore, geoprocessing and spatial analysis (e.g., network analysis) do not yield reliable results. The focus of this paper is to identify and correct topological errors in vector spatial data, in network data in particular. We present the method for identifying and correcting dangling line in datasets aiming to reconstruct incorrect topological relations between lines and other features. We tested the proposed approach on the real-world energy network data.**

Keywords–spatial data; networks; topology; topology errors

I. Introduction

Spatial data is very diverse. It comes in different formats and types: satellite or aerial photographs, hand drawn or printed maps, raster or vector graphics files and other numerous formats. Before performing further data mining and exploring procedures, it is necessary to assess whether data is suitable for their application, e.g., whether satellite and photo images require noise removal procedures or tables and Geographical Information System (GIS) files require duplicate removal procedures.

Digitized GIS data has two major formats - raster and vector graphics. Broadly speaking, raster graphics typically uses a grid of colored pixels to build the image, whereas vector graphics uses points, lines and simple geometric shapes. Vector graphics data is quite widespread and is represented by numerous formats, such as shapefiles (one of the most popular spatial data formats), GML (XML-like grammar developed by the Open Geospatial Consortium (OGC)), KML/KMZ (extension of XML for spatial data developed by Google) and others. The prevalence of vector formats for spatial data can easily be explained by its numerous advantages, including compatibility with relational databases and possibilities to easily scale and combine vector layers or update data. For instance, in comparison with raster data, vector format allows more efficient encoding of a topology and hence offers more analysis capabilities for networks, such as roads, rivers, rails and energy networks.

However, data seldom comes clean and accurate, and this statement holds for geospatial data as well. It might be inaccurate or outdated, and, as consequence, the topological structure of vector data can be corrupted leading to incorrect encoding of geographical features. In this paper, we discuss on topological errors in vector data and, in particular, on one of the most frequent problems in the network data: incorrect connections to line features. Further, we demonstrate and evaluate proposed methods on a real-world geographical data.

The rest of the paper is organized as follows: the next section introduces data quality and topology for vector spatial data. Overview of related work is provided in Section III. In Section IV, we propose a method for correcting topological errors in data and further in Section V we evaluate proposed method on our use case data. Section VI concludes the paper.

II. Data quality and topology in GIS vector data

In order to keep spatial data as accurate and complete as possible, a set of general quality criteria was defined [1]. These criteria are called *the elements of spatial data quality*:

1) Lineage - the history of the dataset, i.e., how was this data derived and how was the data transformed and processed;
2) Positional accuracy - a measure of accuracy of absolute and a relative positions of geographic features in the dataset;
3) Attribute accuracy - a measure of accuracy of quantitative and qualitative attributes of geographical features;
4) Completeness - a measure of whether all geographical features and their attributes were included in the set and, if otherwise, selection criteria which attributes were omitted;
5) Logical consistency - compliance with the structure of data model, absence of apparent contradictions in data;
6) Semantic accuracy - correct encoding of geographical features, i.e., the difference between geographical features in a given data set and in reality;
7) Temporal information - validity period for a given data set, dates of its observation and any updates performed;

Poor-quality data does not conform to one or several elements of quality. For example, irresponsible documentation affects lineage and temporal information quality; map transformations and generalizations cause attribute and semantic inaccuracies. Other typical sources for deficiencies in data quality elements include data collection, data conversions and transfer between different formats and coordinate systems.

Insufficient data quality is especially critical for vector data since its topology can be disturbed. *A geospatial topology* enforces rules concerning relationships between geospatial features representing real-world objects. These rules are called *topology rules* [2]. They are formulated using spatial predicates such as `Contains`, `Covers`, `Disjoint`, `Intersects`, and others. The geospatial topology determines and preserves relationships between geographical features. For instance, in road or telecommunication datasets topology is what makes a set of lines to be a network. It is essential for spatial data analysis, e.g., for querying or routing. Different types of topologies are distinguished, depending on the feature classes presenting in the dataset, for example, the arc-node topology defines relations between lines. Similarly, the polygon topology determines relations between polygons [3]. According to the type of the topology and data model requirements appropriate topology rules are defined. For example, both buildings and climatic zones can be encoded by multipolygons, but those representing buildings are allowed to have gaps between them, whereas multipolygons representing climatic zones are forbidden to have void areas between them. Errors in data may lead to violations of these topology rules, incorrect definitions of relationships between features, and, therefore, failure to meet data quality criteria. Such errors are called *topological errors*.

As it was mentioned above, vector data is especially suitable for networks, such as roads, electricity grids and others. Similarly, for each network dataset there are corresponding topology rules defined by a data model, requirements and further characteristics of data. However, some topological errors are typical for all kinds of networks, including: dangling lines, i.e., not precise connection of lines to the other features. These errors occur quite frequently in network data breaking its topology and, as one of the consequences, corrupting results of data analysis. In this paper, we concentrate on dangling lines and propose techniques for connecting them to the ending points correctly.

III. RELATED WORK

There was a lot of research on spatial data quality since 1990s, when the geographic information science took its roots. Also for vector spatial data there exist various methods of detection and correction of topological errors.

In general, all features shall be checked for violating defined quality criteria, topology rules or any other restrictions set by the data model. There are two main possibilities to conduct spatial data quality assessment and improvement in practice: using GIS or Computer Aided Design (CAD) software. CAD platforms provide an environment supplying graphic operators and algorithms for data processing, such as checking intersections, creating features, etc. Authors of [4], [5], [6], [7] and other works present systems that detect and correct wide range of topological errors operating objects in CAD environment. Some modules also treat positional inaccuracies and logical inconsistencies, such as identifier duplications [4], and semantic inaccuracies, such as self-intersections of features [7] .

In the first place, GIS is a system for storing and displaying geospatial information. However, present-day GIS software often offers some analysis functionalities, including checking validity of the topology in data. GRASS GIS [8], QGIS [9], ArcGIS [10] and other similar software packages find and fix errors in two- and three-dimensional data. For example, ArcGIS allows to choose from 28 topology rules and detects features that violate these rules [11]. It is important to mention though, that in GIS software checking validity functionality is often aimed rather for faster rendering and simplification then for an efficient data analysis.

One of the most important concepts underlying topological error detection and correction is a *tolerance gap* (also called

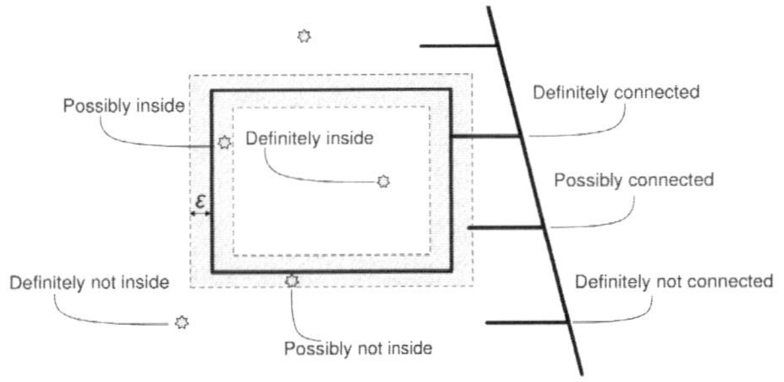

Figure 1. Epsilon-bounded tolerance gaps.

an error band, *a search radius*, or *an epsilon-bounded error region*) which is defined as an area around the feature expanded by epsilon in all directions from the boundary of the feature [12]. Its purpose is to access the cartographic error in feature relations. For example, for each polygon in a dataset with a polygon-point topology epsilon-bounded tolerance gap allows to separate surrounding features in four groups (see Figure 1): (i) definitely lying inside of the polygon, (ii) possibly lying inside of the polygon, (iii) possibly not lying inside of the polygon, (iv) definitely not lying inside of the polygon. Similarly, in case of the polygon-line topology, tolerance gap allows to find possibly connected lines (see also Figure 1). The value of epsilon shall be application dependent and mark out as many of doubtful features as possible. Tolerance gaps are widely used for detection and correction of dangling lines, slivering polygons and other topological errors [12], [3], [13], [5]. Additionally, error bands are used in probabilistic and fuzzy logic approaches [14], [15], [16] for spatial topology, where error band is introduced as an uncertainty in the boundary of features.

However, epsilon-bounded tolerance gaps suggest to connect dangles to the closest feature around. It might be incorrect choice leading to semantic inaccuracies, in case the object, encoded by dangling line, is connected to the other object rather than closest one. Purpose of our work was to correct numerous dangling features in real-world data and to rebuild network topology as accurate as possible. Existing systems mostly use the epsilon-bounded tolerance gap method for dangling lines and, therefore, produced semantic inaccuracies in the dataset. In order to avoid these inaccuracies, we propose a novel method for correcting dangling lines. In this method we suggest to respect the network structure, distinguish features that are already connected to the network from the features that are not yet connected. This method can be especially relevant for road, utility, telecommunication and other network. We also introduce error band in our method as an aid for correcting dangles. In further sections we detail our method for detection and correction topological errors in vector data.

IV. DATA QUALITY IMPROVEMENT

Vector data tends to dominate in network and other applications, where it is important to analyze relations between features such as connectivity and adjacency. However, it might not be possible to yield reliable analysis results due to poor data quality affecting the topological structure of the dataset.

One of the common data quality discrepancies for vector spatial data are positional inaccuracies of features. Positional

Figure 2. Dangling lines: a) undershoot, b) overshoot, c) correcting dangle using a tolerance gap

accuracy shows whether the geographical position of a feature corresponds to the real-world position of the object it represents. While constructing the dataset and transforming between formats and coordinate systems, geographical characteristics of the feature may be affected. Transformations that lead to positional inaccuracies might not only comprise transformations between coordinate systems, but also map generalizations and transformations of attribute format, e.g., rounding coordinate values.

As it was mentioned in Section II, there are two types of positional accuracy: (i) absolute, that defines an absolute geographical position of a feature, (ii) relative, that defines a position of a feature with respect to the other features. Dangling lines are an example of a relative positional inaccuracy. Line is called dangling, if its beginning or ending point does not agree with any other features. Typically, line features are encoded as a pair of points, multiline features - as a sequence of points. The first point is called *the beginning of the line* and the last point is *the ending of the line*. However, for our approach it is not significant, whether the beginning or the ending of a dangling line is not connected properly. Therefore, throughout the next sections we say *endpoint of a line* without specifying whether it is the first or the last point in the sequence of coordinates encoding the line.

Dangling lines are also often called undershoots or overshoots indicating on the type of displacement of the feature. Sometimes this topology error may be cleared by introducing tolerance gaps. Figures 2a, 2b illustrates examples of an undershoot and an overshoot correspondingly for a line-polygon topology. Figure 2c illustrates the process of restoring the connection by introducing a tolerance gap around polygon A, checking the containment relationship between the tolerance gap of the polygon and the endpoint of the line v, which allows to conclude that line v is possibly connected to polygon A, and finally building a corrected line v'.

In Figure 3 we schematically illustrate several possible situations when simply introducing tolerance gaps is not enough. Positional and relative inaccuracies in different layers may superpose and lead to a situation similar to the one shown in Figure 3a, when the endpoint of the conduit is not reached by the tolerance gap with the defined ε. In this case, we suggest using a tolerance gap for the endpoint of a line and a stepwise increment of ε as shown in Figure 4a. However, increment of ε shall be limited in order not to produce false connections. Figures 3b and 3c show cases, when the line v_2 is possibly connected to several polygons. We suggest two further actions to remove the uncertainty:

1) Build a line with the same slope and offset as the initial line or as the corresponding segment of an initial multiline. Among all candidates choose the feature that lies on the line and is the closest to the endpoint of the dangling line. According to this technique, Figures 3b and 4b shows the connection of the dangle v_2 to the polygon B, since the continuation of a line segment v_2 intersects it.
2) Among all candidates, filter out those features that are already connected to other lines. This remark, however, depends on topology rules specific to the data, but nevertheless is true for many common types of networks. According to the technique shown in previous point, in Figures 3c and 4c polygon A is the endpoint of the dangle. However, filtering out polygons A and C that are both connected to other lines, we choose polygon B out of all candidates.

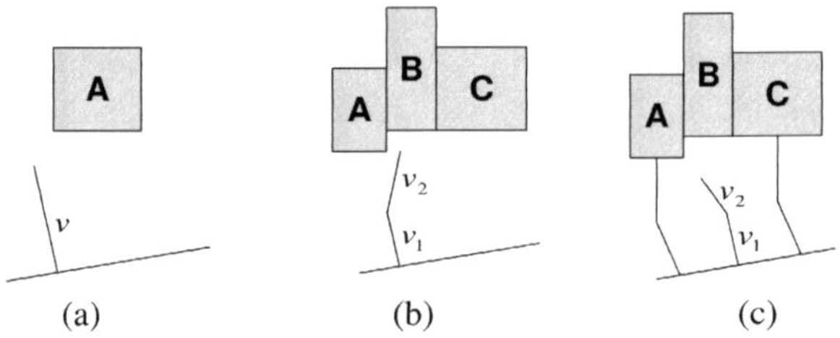

Figure 3. Other possible occurrences of dangling lines.

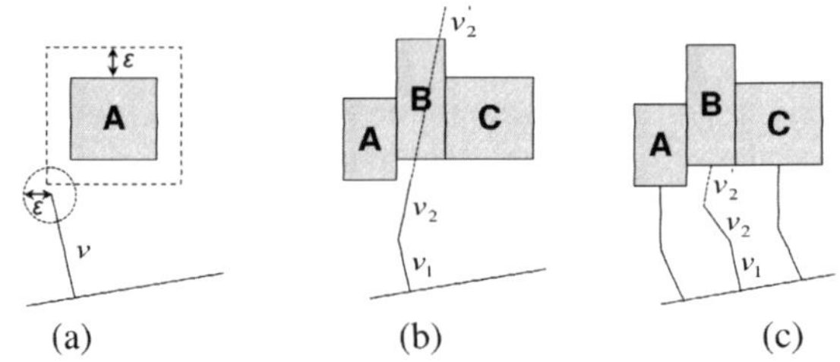

Figure 4. Correction of dangling lines for cases introduced in Figure 3.

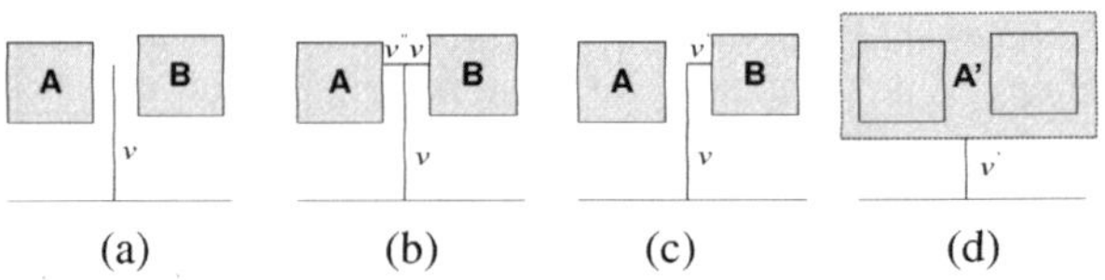

Figure 5. Different possibilities of treating dangling line in case it is unclear to which feature it is connected to.

Candidates for a correct connection of a dangle usually are determined by an exhaustive search, as in [4] and other works. However, it might be computationally intensive in case of a large dataset. We suggest to reduce search of the candidates by a part of a dataset, i.e., analogously with an epsilon value defining the width of tolerance gaps, we suggest to define an Ω value determining a search area around the dangling feature, i.e., as an envelope around the dangle expanded by Ω in all directions.

One of the indecisive cases that can not be solved using technique suggested above is shown in Figure 5a. In this case, both candidate polygons A and B are not connected to another lines, equidistant from the dangling line v and do not lie on the line of the dangle. The best solution in this case is to consult a domain expert who knows the data and can exactly point out the correct relationship between features. Otherwise, there are the following possibilities how to treat this problem: (i) leave it as it is (see Figure 5a), (ii) connect both polygons (see Figure 5c), (iii) choose (ot guess) which polygon to connect (on Figure 5a polygon B is chosen), or (iv) merge two polygons into a new bigger polygon A' connected to the endpoint (see Figure 5d).

Each of these methods has its pros and cons. For example, the methods (ii) and (iii) are suitable if there are few such cases, but if there are a lot of them, it creates an abundance of synthetically introduced connections and, as a consequence, might result in geospatial topology which contains a lot of semantic and positional inaccuracies, and as a consequence has nothing to do with real-world topology. The latter method is not suitable, when these polygons represent different objects in real world and have completely different characteristics. The semantic accuracy of the dataset is affected by merging two polygons into one. On the other hand, ignoring dangling features might lead to oversimplification and logical inconsistencies, for example, when we ignore a dangling line that is supposed to connect buildings to an electricity substation, and there are no other substations in the network delivering electricity to consumers. The choice of a particular method depends on a field of application, type of data, task at hand and quality criteria for a particular dataset.

1: **for** $i = 0$ **to** n **do**
2: **if** $\nexists f \in F$ s.t. beginning of $l_i \in L$ connected to f **and** $\nexists l \in L, l \neq l_i$ s.t. beginning of $l_i \in L$ connected to l **then**
3: $L' \leftarrow l_i$
4: **else if** $\nexists f \in F$ s.t. ending of $l_i \in L$ connected to f **and** $\nexists l \in L, l \neq l_i$ s.t. ending of $l_i \in L$ connected to l **then**
5: $L' \leftarrow l_i$
6: **end if**
7: **end for**
8: **for** $i = 0$ **to** m **do**
9: **if** $\exists l \in L$ such that $f_i \in F$ connected to l **then**
10: $F' \leftarrow f_i$
11: **end if**
12: **end for**

Figure 6. Inspecting lines and features in dataset, building L', F' sets .

We suggest algorithms shown in Figures 6 and 7 summarizing techniques we introduced above. We use the algorithm in Figure 6 to process all data and to determine spatial relationships between features. Let $\{L\}_{i=1}^{n}$ be a set of line features and $\{F\}_{i=1}^{m}$ be a set of point and polygon features.

1: **for all** $l' \in L'$ **do**
2: **for all** $f \in F$ s.t. f is contained in $E(l', \Omega)$ **do**
3: **if** ending of l' is contained in $e(f, \varepsilon)$ **then**
4: $F(l') \leftarrow f$
5: **end if**
6: **end for**
7: **if** $|F(l')| = 1$ **then**
8: **return** connection l' to f
9: **else if** $|F(l')| > 1$ **and** $|F(l') \setminus F'| = 1$ **then**
10: $g = F(l') \setminus F'$
11: **return** connection l' to g
12: **else if** $|F(l')| > 1$ **then**
13: **for all** f in $F(l')$ **do**
14: **if** extension of l' crosses f **and** $f \notin F'$ **then**
15: **return** connection l' to f
16: **else if** extension of l' crosses f **and** $|F(l') \setminus F'| = \emptyset$ **then**
17: **return** connection l' to f
18: **else if** $f \notin F'$ **and** f is the closest to l' **then**
19: **return** connection l' to f
20: **end if**
21: **end for**
22: **else if** $F(l') = \emptyset$ **then**
23: **repeat**
24: increase ε
25: **until** $F(l') \neq \emptyset$ **or** $\varepsilon < threshold$
26: **end if**
27: **end for**

Figure 7. Detecting and correcting dangling lines in a vector spatial dataset.

The algorithm builds sets $L' \subseteq L$ of dangling lines and $F' \subseteq F$ of features having connection to lines. The first *for* loop iterates over all n line features in the dataset; l_i denotes the current line. During the loop the *if* conditionals on lines 2 and 4 check, whether there are no feature f from the set of all features F and no line l from the set of lines L that are connected to a beginning or to an ending of l_i. If it is the case for at least one endpoint of the line l_i, it is added to a set of dangling lines L' as shown on lines 3 and 6 of the algorithm. The second *for* loop iterates over set F of features and checks whether there exists line l that is connected to this feature. If yes, feature f_i is added to a set F' of connected features. Thus, the set F' of features connected to lines and the set $F \setminus F'$ of unconnected features are built. These two procedures are separated in the pseudo code in Figure 6 for an easier understanding where do sets L' and F' come from. However, set F' can be built during the first *for*-loop.

Algorithm listed in Figure 7 performs data cleaning. It consists of one *for* loop that iterates over a set L' of dangling lines and attempts to connect it to a feature. Firstly, we introduce values Ω, ε that defines a size of an envelope $E(l', \Omega)$ around any dangling line l' and an epsilon-bounded tolerance gap $e(f, \varepsilon)$ around any feature f. In a nested *for* loop we iterate over features from $E(l', \Omega)$ and building a set $F(l')$ of features that are possibly connected to dangle l' using epsilon-bounded tolerance gap method. The cardinality of the set $F(l')$ determines the next actions. If $F(l')$ has only one element, then a connection between this element and the endpoint of l' is restored (see *if* conditional on lines 7-9). In case $F(l')$ is not empty and has more than one member, we

TABLE I. Networks size in one of the districts in Geneva canton in Switzerland

	Lines	Points and Polygons
Electricity	524	338
District heating	164	120
Water treatment	821	263
Buildings	-	407
Σ	1509	1128

apply techniques elaborated above. In particular, we use a *for* loop on lines 13-21 to search for the nearest feature that is not the member of the set F' and crossed by the extension line with the same slope and offset as the line l'. If all candidate features are not the members of the set F', we connect l' to the feature that is the nearest in crossed by the extension line. Finally, if $F(l')$ is empty, we increase tolerance value ε and repeat the search for candidates, unless ε can not be increased anymore (see *if* conditional on lines 22-26). Note, that in Figure 7 we process lines with dangling endings, in case of dangling beginning of a line procedures are similar.

V. CASE STUDY: URBAN ENERGY NETWORKS

A. Data cleaning

We applied data cleaning techniques introduced above to urban energy networks data. In this section, we describe this data and demonstrate the result of application of geoprocessing procedures.

The authors of this paper work in the European project "CI-NERGY" [1], which aims to develop urban decision making and operational optimization software tools to minimize non-renewable energy use in cities. In particular, the authors' expertise lies in the area of analysis of energy networks. Spatial data plays a crucial role since it provides a topology of the network, precise geographical positions of network equipment and consumers as well as connections between them. We perform routing, breadth-first, depth-first and other algorithms on the spatial data. Therefore data of sufficient quality is especially crucial to gain as precise layout of the network as possible and to produce meaningful results of analysis. One of the case studies in the project is the canton of Geneva, located in the south-western corner of Switzerland. Geneva energy networks data was provided by SIG Geneva[2].

We evaluated our methods on one of the districts in canton Geneva, Table I provides a short overview of its network size. The provided spatial data is stored in ESRI shapefile format and consists of a building layer and network layers, typically three layers per network. Buildings are represented as polygon and multipolygon features. Points, polygons and multipolygons depict installations and other equipment in networks, and lines and polylines depict conduits and pipes connecting those installations and buildings. However, the data included some positional inconsistencies caused by data losses and errors during conversion from internally used format to commonly used ESRI shapefiles. In particular, it concerned network conduits layers representing connection of the buildings and other objects to networks, which resulted in dangling features, not precisely connected to networks. Based on this information,

[1]The CINERGY, Smart cities with sustainable energy systems Marie Curie Initial Training Network (ITN) project: http://ci-nergy.eu/About.html

[2]SIG: Swiss supplier of local energy services http://http://www.sig-ge.ch/

TABLE II. Comparison of number of buildings connected to the network

	Electricity	DH	Water
Uncleaned data	137	4	37
Proposed technique	139	37	93
Clients in the database	140	37	101

we concluded that positional accuracy of buildings is not disturbed and dangling features and imprecise connections are caused by inconsistencies in network conduits spatial data rather than in building layers. Moreover, different network layers suffered from different displacement of features. Such, electricity network was the least affected and in most cases missing connections could be restored using tolerance gap technique with ε value not increasing 5-10 meters, whereas for the water network large ε values were needed in order to find features possibly connected to dangling lines. In dense districts of the city it resulted in large search sets and a need to choose which feature to connect.

We implemented data cleaning and graph construction methods in Java. We aimed to create a module that would be independent from existing GIS or CAD software and could detect and process dangles in datasets with line-line, line-point and line-polygon relationships. We used GeoTools Java library [22] for manipulation with shapefiles and geometries.

Figure 8 illustrates the results of application of data cleaning procedures to the real-world data. Electricity lines are shown as green lines, water pipes - black, district heating pipes - red. We used QGIS software for visualization of shapefiles [9].

B. Evaluation

Apart from the geospatial data we have also received aggregated consumption data from our city partners in Geneva. That allows us to evaluate our approach. We apply path search algorithm on the networks before and after data cleaning and compare, which buildings are connected to the network and which are listed in the client database. Dangling lines result in the absence of the path between buildings, that are connected to the network. In Table II we compare results of search of connected buildings before and after data cleaning. Electricity grid data had a sufficient quality, being almost completely connected and containing only a couple of dangling lines. On the contrary, district heating and water networks had poor quality and in most cases lines representing pipes did not connect to the polygons representing buildings. For electricity and water, method returned a very good result without false negatives and false positives, connecting correct clients to the network in the spatial data. In case of the district heating our procedure resulted in a complete connectedness of the network and correction of each dangling line. For the water network unfortunately it was not the case, as it contains multiple occurrences of a situation shown in Figure 5a. Therefore, out of 101 building that shall be connected to the water network only 37 were correctly connected in the initial dataset and 93 were correctly connected using procedures described in this paper.

VI. CONCLUSION

In this paper, we considered data quality in vector data, and, in particular, data quality inconsistencies corrupting geospatial

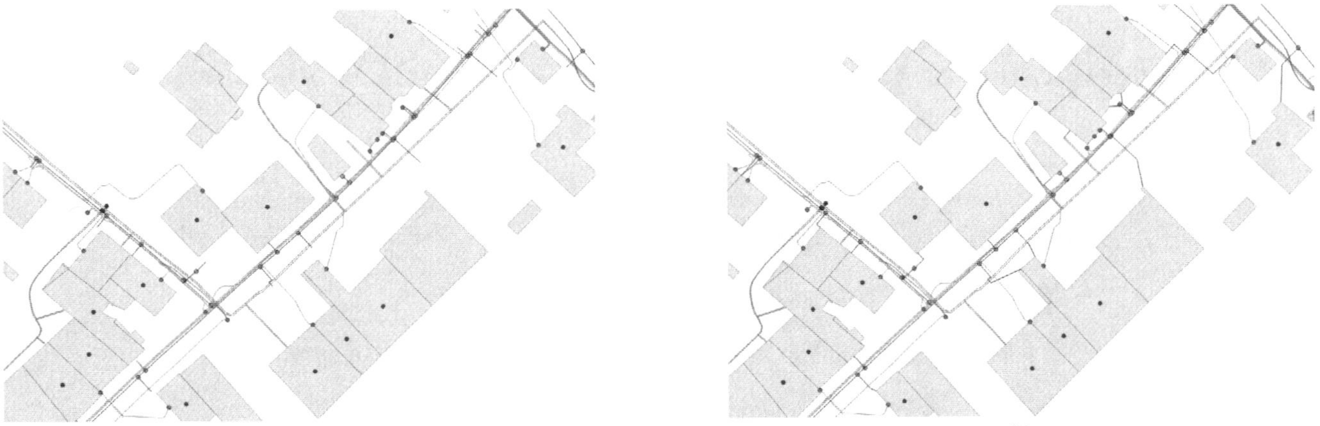

a) b)

Figure 8. Correcting dangling features: a) initial data containing not precisely connected features, b) Result of application of cleaning procedures

relations between features. These relations are especially important for application of spatial analysis algorithms. We considered one of the most frequent topological error in topologies of types line-line, line-point and line-polygon and suggested a method for their correction. This method allowed to restore lost connections in urban utility network data. Accurate topology is essential for a network analysis, that we work on using a graph-based model representing the geospatial topology of networks. However, as for each data cleaning algorithm, there is a danger of overcorrecting data and thus ending up with even worse data quality than before. In future, we are planning to improve our method taking into account further topology rules, domain knowledge and other characteristics of the dataset.

ACKNOWLEDGMENTS

The authors thank colleagues from SIG Geneva who provided case study data that greatly assisted our research. Moreover, the first author gratefully acknowledges the European Commission for providing financial support during conduct of research under FP7-PEOPLE-2013 Marie Curie Initial Training Network CI-NERGY project with Grant Agreement Number 606851.

REFERENCES

[1] S. C. Guptill and J. L. Morrison, Elements of spatial data quality. Elsevier, 2013.

[2] K.-t. Chang, Introduction to geographic information systems. McGraw-Hill Higher Education Boston, 2006.

[3] H. Hansen and L. Grondal, "Methods for cross-referencing, consistency check and generalisation of spatial data," SENSOR Project Deliverable Report 5.1.2, 2006.

[4] M. Siejka, M. Ślusarski, and M. Zygmunt, "Correction of topological errors in geospatial databases," International Journal of Physical Sciences, vol. 8, no. 12, 2013, pp. 498–507.

[5] S. S. Maraş, H. H. Maraş, B. Aktuğ, E. E. Maraş, and F. Yildiz, "Topological error correction of gis vector data," International Journal of the Physical Sciences, vol. 5, no. 5, 2010, pp. 476–583.

[6] H. Gu, T. R. Chase, D. C. Cheney, D. Johnson et al., "Identifying, correcting, and avoiding errors in computer-aided design models which affect interoperability," Journal of Computing and Information Science in Engineering, vol. 1, no. 2, 2001, pp. 156–166.

[7] A. A. Mezentsev and T. Woehler, "Methods and algorithms of automated cad repair for incremental surface meshing." in IMR, 1999, pp. 299–309.

[8] "GRASS: Geographic resources analysis support system," https://grass.osgeo.org/, accessed: 2016-02-20.

[9] "QGIS: A free and open source geographic information system," http://www.qgis.org, accessed: 2016-02-20.

[10] "ArcGIS: Commercial GIS software application," https://www.arcgis.com, accessed: 2016-02-20.

[11] ArcGIS, "ArcGIS geodatabase topology rules," http://help.arcgis.com/en/arcgisdesktop/10.0/help/001t/pdf/topology_rules_poster.pdf, accessed: 2016-02-20.

[12] M. Blakemore, "Part 4: Mathematical, algorithmic and data structure issues: Generalisation and error in spatial data bases," Cartographica: The International Journal for Geographic Information and Geovisualization, vol. 21, no. 2-3, 1984, pp. 131–139.

[13] G. Klajnšek and B. Žalik, "Merging polygons with uncertain boundaries," Computers & geosciences, vol. 31, no. 3, 2005, pp. 353–359.

[14] M. Schneider, "Fuzzy spatial data types for spatial uncertainty management in databases." Handbook of research on fuzzy information processing in databases, vol. 2, 2008, pp. 490–515.

[15] W. Shi and K. Liu, "A fuzzy topology for computing the interior, boundary, and exterior of spatial objects quantitatively in gis," Computers & Geosciences, vol. 33, no. 7, 2007, pp. 898–915.

[16] X. Tong, T. Sun, J. Fan, M. F. Goodchild, and W. Shi, "A statistical simulation model for positional error of line features in geographic information systems (gis)," International Journal of Applied Earth Observation and Geoinformation, vol. 21, 2013, pp. 136–148.

[17] M. Neteler and H. Mitasova, Open source GIS: a GRASS GIS approach. Springer Science & Business Media, 2013, vol. 689.

[18] A. S. Analyst, "Advanced gis spatial analysis using raster and vector data," An ESRI White Paper, ESRI (Environmental Systems Research Institute), Redlands, USA, 2001.

[19] "OSRM: routing engine for shortest paths in road networks," http://project-osrm.org/, accessed: 2016-02-20.

[20] "pgRouting: an extension for PostGIS and PostgreSQL providing geospatial routing functionality," http://pgrouting.org/, accessed: 2016-02-20.

[21] "Flowmap: a software package for analyzing and displaying spatial flow data," http://flowmap.geo.uu.nl/, accessed: 2016-02-20.

[22] "GeoTools: The open source Java GIS toolkit," http://geotools.org/, accessed: 2016-02-20.

Identification of Areas with Potential for Flooding in South America

Sergio Rosim, João Ricardo de Freitas Oliveira, Alexandre Copertino Jardim, Laércio Massaru Namikawa, Cláudia Maria de Almeida
Image Processing Division.
National Institute for Space Research, INPE
São José dos Campos, Brazil
e-mail: {sergio, joao, alexandre, laercio}@dpi.inpe.br, almeida@dsr.inpe.br

***Abstract*— This paper presents the drainage network extraction of South American region using Shuttle Radar Topography Mission (SRTM) data set with 90 meters of horizontal resolution. Using different thresholds, the user can generate different drainage networks. The Height Above the Nearest Drainage (HAND) procedure was applied for each drainage. HAND determines the variation of relief in relation to the nearest drainage segments of each point, considering a regular grid structure representing the relief of the study geographic region. HAND is used to determine the areas with greatest potential for flooding. Results for South American region were shown illustrating the potential use of HAND process. The main benefit of this work is to show the viability of using this tool to study and simulation areas that can be flooded by natural processes or anthropic actions.**

Keywords-flooding; drainage network.

I. INTRODUCTION

The American continent has 39.6% of the world's fresh water, and of this total, 61.3% is in South America [1]. The existence of this huge amount of water is directly related to various situations of floods, both in cities and in rural areas, fast and slows flooding. Those floods, which can cause major social and economic damage, must be studied and whenever possible prevented.

This work employed a procedure, called Height Above the Nearest Drainage (HAND) [2], for determining the areas with greatest potential for flooding. The whole South American geographic region was used, considering drainage networks with different densities. These networks were defined by the TerraHidro system [3], which is a distributed hydrological model system that is being developed at Image Processing Division of the National Institute for Space Research, located in the city of Sao Jose dos Campos, Brazil.

HAND determines the variation of relief in relation to the nearest drainage segments of each point, considering a great regular structure representing the relief from their study geographic region. A comparison between the HAND flood results and the real flooded areas extracted from Landsat 8 image was done to show the accuracy of the HAND prediction in a real situation.

The objective of this work is to show the usefulness of HAND to determine areas with the greatest potential for flooding, from a drainage network extracted by TerraHidro system. We chose to apply the HAND throughout the region of South America, to get a sense of this potential across the continent. The paper is organized as follows: Section II briefly presents works related to the proposed in this article, Section III describes TerraHidro system, Section IV shows HAND procedure, Section V presents the results and some discussions, and Section VI contains the conclusions.

II. RELATED WORKS

The flood has been the subject of study for many years. The appearance of Digital Elevation Models (DEM) [4] and of medium and high resolution satellite images made possible the creation of methodologies and systems for identification, simulation and analysis of floods. Works have related to DEM quality and resolution with occurrence of floods [5]. Systems have also been developed in order to produce and analyze floods in local [6] and global level [7].

The study of flood forecasting involves specific models and different types of data, such as slope, land use and land cover, characteristics of rivers, soil types, among others. HAND constitutes a tool to aid the expert in the identification of areas with greater potential for flooding, based only on altimetry and drainage network.

The aim of this paper is not to propose a new methodology to say whether there is flood at a given geographical location. Its goal is to show places with greater or lesser potential for flooding only using the existing relief information. The quality of the result depends on the quality and appropriate resolution of employee DEM.

III. TERRAHIDRO DESCRIPTION

This work was carried out using TerraHidro system and HAND process, also a TerraHidro process [8]. TerraHidro is a distributed hydrological system created to develop water resource applications. It uses regular grid (DEM) as the surface and elevation structure for drainage extraction. TerraHidro uses HAND procedure to identify these areas.

TerraHidro is a plugin of the geographic viewer TerraView that loads and stores data in a geographical library called TerraLib [9], an open source geographical library implemented in C++ language that has also been developed at the Image Processing Division. This approach has allowed TerraHidro project team of designers and programmers to keep focused on the development of TerraHidro system functionality. It calculates, for every DEM cell, the altimetry difference between this cell and the nearest cell belonging to the drainage network, following the local drain directions.

TerraLib is an open-source Geographical Information System (GIS) software library. TerraLib supports coding of geographical applications using spatial databases, and stores data in different database management system (DBMS) including MySQL, PostgreSQL and other databases.

TerraHidro functionality used here are: first, the definition of local flow, called of Local Drain Directions (LDD) extraction [10]. For each DEM grid cell, the LDD was defined considering the steepest downstream regarding the 8 neighbors grid cell. At the end of the task, a new grid was created with the same number of columns and rows of DEM and same resolution. Each grid cell received a code indicating the water flow from this cell. Figure 1 shows LDD functionality.

The grid DEM contains the altimetry of the study area. The slope is calculated for each cell in the grid, considering its eight neighboring cells. The result of this can be seen in the second grid (SLOPE). The third grid (CODIFICATION) shows the encoding rule represented by numbers $2^{(0,1,\ldots,7)}$ defining the flow direction. The last grid (LDD) shows the flow direction for this example.

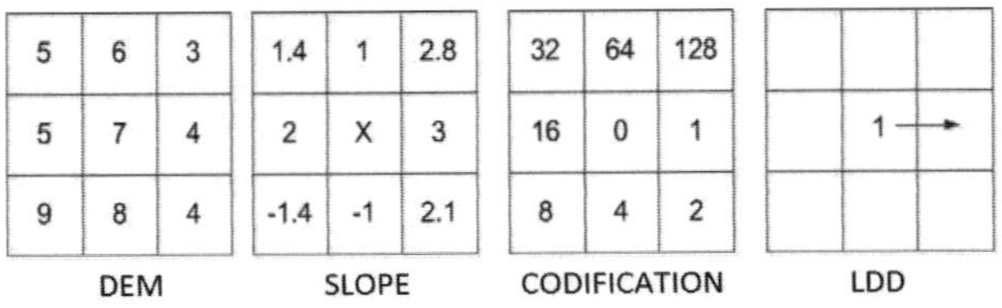

Figure 1. Local Drain Directions creation process.

Second, we consider the creation of the grid called contribution grid area. The user wants to work only with representative drainages regarding his application, not with drainages of all LDDs. Each cell of the contribution area grid receives a value that is the amount of the areas of all cells that participate in the path arriving at that cell. Figure 2 presents this concept.

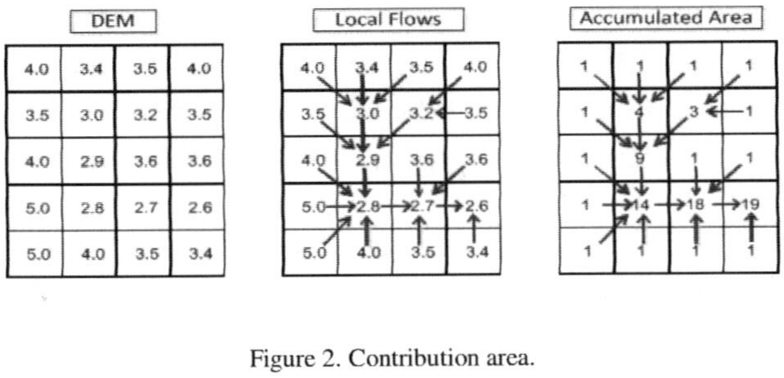

Figure 2. Contribution area.

Third, we consider the definition of a particular drainage network using a threshold value. The value of each cell of the contribution area grid is compared with the threshold value. If the value of contribution area grid is equal or greater than the threshold value the cell is selected as a drainage network cell. At the end of this process a new grid is created, defining the drainage network. Figure 3 presents a didactic example of drainage network and Figure 4 shows a South America drainage network extracted using threshold = 300000 and South America delimited by country and Brazil delimitates by States.

Threshold is an empirical value, defined by the user, in order to select a drainage containing the density necessary to their work. As a general rule, the threshold selects the most representative drainage because it contains the largest accumulated values.

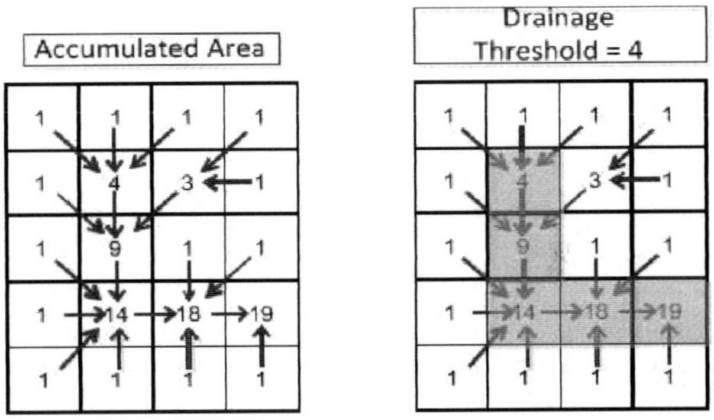

Figure 3. Drainage network (green), using threshold value = 4.

Another TerraHidro task is the watershed delimitation for drainage segments or for isolated points. A segment is a drainage path between water springs and junctions, between junctions, or between junctions and mouth of the drainage. A watershed point is a location defined by the user on a location containing drainage. Figure 5 presents an example of watersheds by segments.

Figure 4. America drainage network extracted using threshold = 300000.

Figure 5. Basin created for each drainage segment represented in blue color.

HAND will be described in the next session.

IV. HAND DESCRIPTION

TerraHidro uses the HAND procedure to identify flood potential areas. It calculates, for every DEM cell, the altimetry difference between this cell and the nearest cell belonging to the drainage network, following the local drain directions. As the HAND terrain descriptor is sensitive to drainage changes in the regions of sudden terrain variations, it was used as an attempt to determine critical drainage areas. Figure 6 shows a numeric example of the HAND process. The top figure shows the flow directions for each altimetry grid cell. This is shown in Figure 6. In this grid, the cell with value equal to 72 was highlighted. The calculation done by HAND was the subtraction of this value from the value of the cell found, according to the shortest flow direction path. The value to be considered is the 53, which was found according to the path highlighted in red, in the Local Drain Directions grid. Thus, in the resulting grid, HAND grid, the value of altimetry is equal to 19.

HAND process can only identify the areas with potential for flooding. This result allows the water resources manager to focus his efforts on the most susceptible areas to the occurrence of extreme events involving water. For a more refined study, hydrological models must be developed.

The materials used to develop this work were SRTM (Shuttle Radar Topography Mission) of 90 meters of horizontal resolution as surface elevation data set. TerraHidro and HAND used this data set to extract their information. A Landsat 8 image acquired on July/05/2014, showing a flooded region was compared with HAND result of a test area.

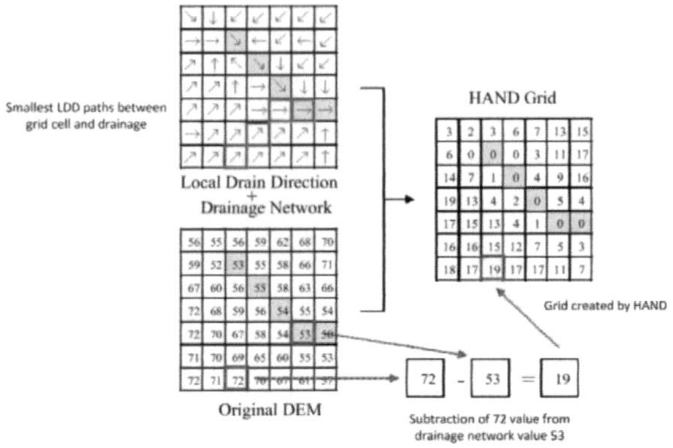

Figure 6. Hand process.

In the next session the results are shown.

V. RESULTS AND DISCUSSION

The HAND results extraction, based on TerraHidro drainage network definition, were shown as thematic maps containing altimetry tracks. Each altimetry track informs the greater or lesser proximity of its altimetry in relation to the nearest drainage altimetry. Different scenarios were created with the use of altimetry tracks with different relief intervals. Figure 7 shows HAND results for South America region from drainages extracted from threshold values of one and tree millions. The drainage networks were extracted from thresholds equal one and tree million. Only the differences between 0 and 15 meters divided into 5 slices were represented. Other areas have less potential for flooding.

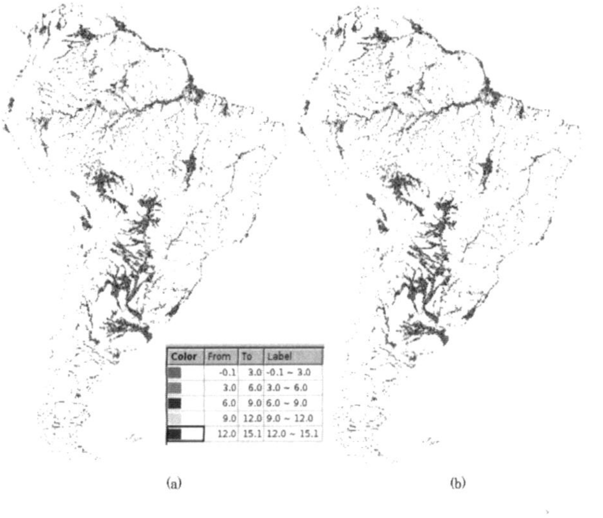

Figure 7. HAND results. (a) threshold = one million; (b) threshold = tree millions.

Figure 8 presents the details of the north and south regions of South American region. It is possible to see areas in the red color, with less than 3 meters of difference between the altimetry and the closer drainage. These areas have most potential flood than the other areas.

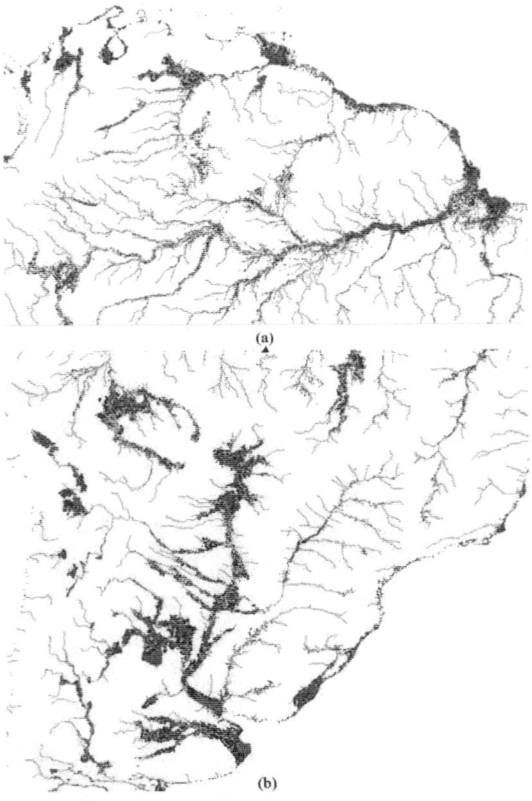

Figure 8. Zoom of (a) north and (b) south of South American region using threshold = one million.

This is the region Mariana, Minas Gerais state. The area reached by the rupture of a lake of mineral waste appears in red color. The result of the HAND appears in slices defined by lines of different colors, each color defines an altitude range. It may be noted that the blue line adequately represents the space occupied by waste. Figure 9 shows the area delimitated by HAND and the area extracted from image in a real flood situation.

The result discussion concerns the relationship between selected drainage and HAND results. Each drainage gives a particular potential flood region context. For quick floods, it is better to have dense drainages to grasp the water behavior in small regions. For slow floods the drainage can be less dense. HAND allows, in both situations, identifying the flood boundaries. Identifying how many meters the water will rise and how fast this will happen are parameters that are outside HAND scope.

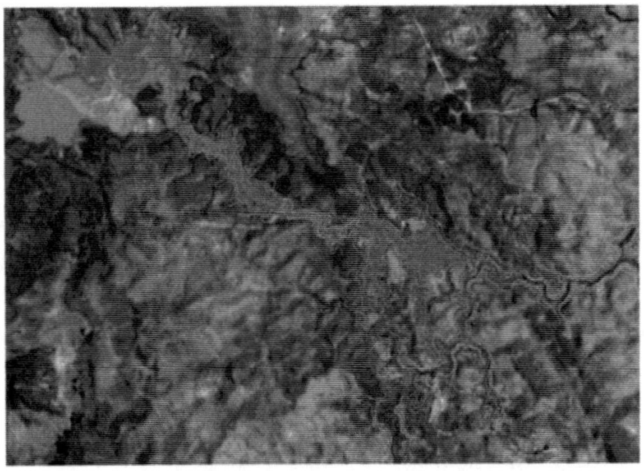

Figure 9. HAND real example. Brazil.

VI. Conclusion and Future Work

TerraHidro and HAND are used to define respective drainage network and potential flooded areas for the South America region. These areas serve as priority areas for experts to carry out studies on flooding.

This work shows a simple solution for identification of locations with more potential for flooding. It only employs relief and drainage extracted from this relief. This methodology aims to help the expert to focus on areas with the greatest potential for flooding. It doesn't say that there will be flooding. To be sure, experts should use appropriate hydrological models, which are beyond the scope of this proposal. We will carry out a job applying this methodology. This work will consist in the simulation of the areas that were flooded by the disruption of mineral waste lakes.

A comparison between HAND result and a real flooded area shown by a satellite image revealed coincidence between both flood areas. This confirms that the methodology shown in this paper is useful to determine critical areas for flooding.

Acknowledgment

This work has received financial support from 1022114003005 - MAS BNDES/CAMPO project, and from Permanent Protection Areas Project, EXAPP/FINEP – 01.13.0118.00.

References

[1] ANA, "A água no Brasil e no mundo", Agência Nacional de Águas, http://arquivos.ana.gov.br/institucional/sge/CEDOC/Catalogo/2014/AAguaNoBrasilENoMundo2014.pdf (site accessed 14/04/2016), (in portuguese).

[2] C. D. Rennó, et al. "HAND, a new terrain descriptor using SRTM-DEM: Mapping terra-firme rainforestenvironments in Amazonia," Remote Sensing of Environment v:(112) 3469–3481, (2008).

[3] S. Rosim, J. R. de F. Oliveira, A. C. Jardim, L. M. Namikawa, and C. D Rennó, " TerraHidro: a distributed hydrology modelling system with high quality drainage extraction". GEOProcessing 2013: The Fifth International Conference on Advanced Geographic Information Systems, Applications, and Services, February 24 - March 1, Nice, France. 2013.

[4] E. B. Brunson and R. W. Olsen, Data digital elevation model collection systems, Proc. Digital Terrain Models (DTM) Syrup. ASP/ACSM, St. Louis, Missouri, May 9-11, 1978, pp. 72-99.

[5] Y. Gorokhovich and A. Voustianiouk, "Accuracy Assessment of the Processed SRTM-Based Elevation Data by CGIAR using Field Data from USA and Thailand and its Relation to the Terrain Characteristics".Remote Sensing of Environment, vol. 104, no. 4, 2006, pp. 409–415.

[6] HR Wallingford, "Real_time forecasting and warning solutions". In: http://www.hrwallingford.com/expertise/flood-forecasting, (site accessed 15/01/2016).

[7] P. J. Smith, et al., "The Global Flood Awareness System", TIFAC-IDRiM Conference, 28th–30th, October 2015, New Delhi, India.

[8] S. Rosim, J. R. de F. Oliveira, J. de O. Ortiz, M. Z. Cuellar, and A. C. Jardim, "Drainage network extraction of Brazilian semiarid region with potential flood indication areas". Proc. SPIE 9239, Remote Sensing for Agriculture, Ecosystems, and Hydrology XV, September 22–25, Amsterdam, Netherlands, 2014.

[9] G Camara, R. C. M Souza, U .M Freitas, and J Garrido, "SPRING: Integrating remote sensingand GIS by object-oriented data modelling. Computers & Graphics, vol 20, n 3, May-Jun 1996, pp. 395-403.

[10] P. A. Burrough and R. A. McDonnell, "Principles of Geographical Information Systems". New York: Oxford University Press. 1998.

A Common Land-Use Change Model for Both the Walloon and Flanders Regions in Belgium

Benjamin Beaumont, Nathalie Stephenne, Eric Hallot
Remote Sensing & Geodata Cell
ISSeP
Liège, Belgium
email : (b.beaumont, n.stephenne, e.hallot)@issep.be

Lien Poelmans
Environmental Modelling Unit
VITO
Mol, Belgium
email : lien.poelmans@vito.be

Odile Close
Faculty of Bioengineering
UCL
Louvain-la-Neuve, Belgium
email : odile.close@gmail.com

Abstract—**Floods, urban heat islands, mobility issues and other environmental and health risks increase with urban growth. For a sustainable planning of their territory, authorities need operational decision support tools, which can assess short and long-terms impacts of current, intended or optional policies on land-use change. This paper considers the application of a constrained cellular automata land-use change model within both the Walloon and Flanders region in Belgium. Some methodological steps needed for this application are discussed. A national land-use change model is seen as a key asset for sustainable spatial planning.**

Keywords-land-use change; cellular automata; sustainable spatial planning; risk assessment.

I. INTRODUCTION

The Walloon region, south of Belgium, has an urbanization rate of 17 km² per year [1]. Driven by demographic projections of 200,000 more households between 2011-2026 [2], a further increase of the impervious surfaces is expected. In Wallonia, urban growth occurs typically under the form of rural ribbon development. This type of development is a major source of fragmentation of natural habitats and enhances rural-urban commuting. This, in turn, increases health and environmental risks by extending pollution sources distribution. Through spatial planning policies, the Walloon authorities try to fix a threshold at 9 km² per year of extra soil sealing towards 2040 [3]. Such policies require a holistic and dynamic vision of the fast changing urban environment. Current and historical land-use/land-cover (LULC) can be assessed with existing geodata and satellite images within Geographical Information Systems (GIS). Possible future impacts of policies can be simulated by means of model-based scenarios. Since Wallonia has an extensive catalog of geodata for current and historical trends [4], regional policy makers and city planners have expressed an interest in developing an operational framework for LULC change modelling in a project called SmartPop.

Over the last decades a broad range of LULC models have been developed to assist land management. LULC models can be static or dynamic, spatial or non-spatial, i.e., exploring patterns of change vs. rates of changes, inductive or deductive, i.e., with model parameters based on spatial correlation vs. explicit description of the process, agent-based or pattern-based, i.e., emulation of individual decision makers vs. inference of underlying behavior for the observation of patterns in the LULC [5]. A spatially explicit approach is needed to project and explore alternative scenarios [6]. Choosing one of these depends on the goals, inputs and validation data available and technical skills (developers/end-users). The model used in this study is a dynamic spatially explicit model based on an inductive pattern-based approach. Cellular automata (CA) are discrete, abstract computational systems defined by a regular grid of cells that are characterized by a finite number of states evolving through time according to rules namely related to neighboring cells [7]. CA have perhaps been the most popular way to model land-use change and spatially-explicit population density [8]-[11] because (i) they are intrinsically dynamic, (ii) they are able to deal with high resolutions and thus produce results with a useful amount of detail and (iii) they outperform other models in realistically modelling land-use change. In CA-based LU models, LU change is explained by the current state of a cell as well as by the changes within the neighboring cells.

At European Union level, the MOLAND LULC dynamics modelling framework initiated in the early 2000's the use of CA to forecast the sustainable development of urban and regional environments [12]. The constrained CA LU change model developed by [13] proposes a tool for assessing scenarios of LU policies in support of spatial planning in Belgium. It has been applied over Flanders at

100m resolution [14][15][16] and at the country scale at a coarser resolution (300m) [17][18]. In [18], an innovative travel time-based variable grid approach with transport network scenarios is used. Another approach is proposed by [19] where population projections [2] are used to model the suitability of constructible lands to host sustainable residential functions. The model compares space availability versus residential land demand. Residential land developments are prioritized using a multi-criteria GIS analysis.

Walloon administrations are interested in a high resolution (≤100m) predictive model of land-use change, for smart city monitoring of soil sealing expansion, risk assessment and sustainable planning. Such a model is currently not available in Wallonia.

This paper describes the steps in transposing the Flanders LU model to Wallonia. A similar approach will be applied in both regions since this is interesting for integrated planning. Homogenizing risks studies between regions is intended since natural hazards are not stopped by regional borders. However, model replication from one region to another is not straightforward. First of all different modelling goals, and geographical and social-economic contexts create a need for different parameter sets and scenarios. Secondly, availability, limited access, quality or semantic differences in existing data induce some model adaptations such as calibration, parameters and/or validation phases. Finally, knowledge of local and regional LU processes is required.

This paper will focus on the methodological choices and decisions taken through the application of the Flanders model to the Walloon region. Close collaboration between researchers from both regions as well as end-users commitment is needed during the process.

The paper is organized as follows: Section II briefly describes the constrained CA land-use change model initially developed for Flanders. The methodological steps for applying the model to Wallonia are detailed in Section III. A synthetic conclusion is proposed under Section IV.

II. LAND-USE CHANGE MODEL

The constrained CA land-use model is made up of three sub-models. These represent spatial dynamics that take place at three geographical levels: (1) 'global' level, i.e., the entire Walloon region, (2) 'intermediate' level, i.e., NUTS3 regions (Eurostat administrative units level 3, called arrondissements in Belgium), and (3) cellular level, i.e., a 100x100m grid [14][20]. At the global level, time series based on population growth and employment scenarios are needed. These global trends constrain the intermediate level, in which a spatial-interaction model is used to downscale the growth trends to the level of the intermediate level. At the local level, a CA-based model allocates to the grid cells the area needed for population and employment growth. This CA model simulates the evolving land-use until 2050 for each individual cell. The changing LU patterns result from spatial interactions that take place between the different land-uses within the immediate neighborhood around each cell. Besides this, the patterns are also constrained by institutional zoning status, physical suitability and transportation characteristics.

III. METHODOLOGICAL STEPS

The application of the constrained CA model to the Walloon region implies some particular contextual and methodological choices and decisions that are presented in this paper.

A. *Modelling goals and outcomes*

Identifying end-users and involving them closely in the model development process helps to precisely define the goals and outcomes. Surveys, meetings, workshops, etc. are various steps needed to create a decision makers group. Involvement of policy makers and international experts is done by integrating them into the project's steering committee or even directly as partners. This project also proposes an implementation group including scientists, data producers and decision makers from several administrations. Regional and local end-users have shown their willingness to participate in this group. They are involved in themes such as infrastructure (SPW-DGO1), mobility (SPW-DGO2), natural resources and environment (SPW-DGO3), spatial planning and geomatics (SPW-DGO4), health (SPW-DGO5), air and climate (AWAC), statistics (IWEPS) and cities (Liège) monitoring.

B. *Model inputs*

In Belgium, each region is responsible for its own geodata production and management. By consequence, data availability and data properties differ in Wallonia and Flanders. As an example, an important model input is a land-use map for the start year of the model simulation. The comparison and semantic adaptation between the existing LU maps for Flanders and Wallonia is necessary to define the land-use classes that are simulated by the model. A survey is currently being carried out to assess the users' satisfaction regarding the existing Walloon LU map, as well as the expectation towards the modelled products. This survey addresses the number of classes, their precision, the update time-step (each year), the coverage (2050) or the derived sub products (spatial indicator).

Facilitated by INSPIRE, geodata access still vary between thematic products in both regions. Input data differ in terms of content, extent, production date, spatial resolution and quality. Simulations further rely on social-economic data sets to define trends in land-use. Availability of historical and projected social-economic data also differs between regions. This information is not always produced at the same spatial and temporal resolution. For each of these inputs dataset, a discussion and a choice are needed.

Moreover, the regional significance is checked. Some geo-criteria may have high impacts on future land-use change in one region, such as harbors in Flanders or slopes in Wallonia, while they are less significant in the other one.

C. *Upcoming actions*

During model implementation, additional decisions must be taken together with end-users. These include calibrating the model and defining future scenario(s), e.g., using historical and projected LU or population data, as well as assessing model flexibility, i.e., what consequences if new data/studies/directive is published. Validation step will be discussed in detail, e.g., field work or use of authentic data sources such as buildings. Qualitative and quantitative assessment of model outputs will be made by comparing them to other relevant geodatabases, e.g., PICC, CadMap, BelMap, etc. Final choices will be made regarding authorities access right to the model and results publication.

IV. CONCLUSION

Generic and common land-use change models are key decision support tools for sustainable spatial planning in the whole of Belgium. Involving end-users in the model development and application guarantees future valorization and use of this model. Land-use change simulation will help drawing policies that limit risks caused by further urbanization.

ACKNOWLEDGMENT

Authors would like to thank the Walloon Region and IWEPS for providing all relevant data of interests as well as ULB, partner of the SmartPop project.

REFERENCES

[1] C. Cuvelier, "Land artificialization : inventory in Wallonia," DEMNA-SPW, Libramont Fair, July 2015.

[2] J. Charlier, I. Reginster, and M. Debuisson, "Demographic projections per municipalities up to 2026 and spatial planning : residential space consumption in Wallonia estimations exercise using three scenarios," IWEPS Working Paper 11, March 2013.

[3] SPW, Regional space development plan – A vision for Wallonia. SPW-Editions, 2013.

[4] SPW, WalOnMap, URL: http://geoportail.wallonie.be/en/home/ressources/geocatalogue.html [retrieved: March, 2016].

[5] J. F. Mas, M. Kolb, M. Paegelow, M. Camacho Olmedo, and T. Houet, "Modelling land use / cover changes: a comparison of conceptual approaches and softwares," Environmnetal Modelling and Software, Elsevier, 51, 2014, pp. 94-111.

[6] N. Stephenne and E. F. Lambin, "A dynamic simulation model of land-use changes in Sudano sahelian countries of Africa (SALU)," Agriculture, Ecosystems & Environment. Volume 85, Issues 1-3, June 2001, pp. 145-161.

[7] F. Berto and J. Tagliabue, "Cellular Automata", The Stanford Encyclopedia of Philosophy, Edward N. Zalta (ed.), URL: http://plato.stanford.edu/archives/sum2012/entries/cellular-automata/ [retrieved: March 2016].

[8] X. Li and A. Yeh, "Modelling sustainable urban development by the integration of constrained cellular automata and GIS," International Journal of Geographical Information Science, 14, 2, 2000, pp.131-152.

[9] G. Caruso, D. Peeters, J. Cavailhès, and M. Rounsevell, "Spatial configurations in a periurban city: a cellular automata-based microeconomic model," Regional Science and Urban Economics, 37, 2007, pp. 542-567.

[10] L. Poelmans and A. Van Rompaey, "Complexity and performance of urban expansion models," Computers, Environment, and Urban Systems, 34, 2010, pp. 17-27.

[11] E. Koomen and J. Borsboom – van Beurden, Land-Use Modelling in Planning Practice. Springer, Heidelberg, volume 101, 2011.

[12] C. Lavalle, J. Barredo, N. McCormick, G. Engelen, R. White, and I. Uljee, "The MOLAND model for urban and regional growth forecast. A tool for the definition of sustainable development paths," European Commission, JRC, 2004, pp. 1-22.

[13] R. White, G. Engelen, and I. Uljee, Modeling Cities and Regions as Complex Systems: From Theory to Planning Applications. MIT Press, 2015.

[14] G. Engelen, I. Uljee, J. L. de Kok, L. Van Esch, L. Poelmans, and H. van der Kwast, Integrated modelling of land use dynamics in support of Spatial Planning and Policy-making. VITO, 2011.

[15] J. L. de Kok, L. Poelmans, G. Engelen, I. Uljee, and L. van Esch, "Spatial dynamic visualization of long-term scenarios for demographic, social-economic and environmental change in Flanders," International Environmental Modelling and Software Society (iEMSs), July 2012, pp. 1984-1991.

[16] G. Engelen, C. Lavalle, J. I. Barredo, M. van der Meulen, and R. White, "The MOLAND modelling framework for urban and regional land use dynamics," Springer, chapter Modelling Land-Use change, Volume 90, 2007, pp. 297-320.

[17] R. White, I. Uljee, and G. Engelen, "Integrated modelling of population, employment and land-use change with a multiple activity-based variable grid cellular automaton," International Journal of Geographical Information Science, 26, 2012, pp. 1251-1280.

[18] T. Crols, R. White, I. Uljee, G. Engelen, L. Poelmans, and F. Canters, 2015. "A travel time-based variable grid approach for an activity-based cellular automata model," International Journal of Geographical Information Science, 29:10, 2015, pp. 1757-1781.

[19] Q. Jungers, A. Leclercq, P. Neri, J. Radoux, F. Waldner, and P. Defourny, Research note – Towards a sustainable zoning plan – General methodology of the model. CPDT, 58, April 2015.

[20] R. White and G. Engelen, "Cellular Automata as the Basis of Integrated Dynamic Regional Modelling," Environment and Planning B, 24, 1997, pp.235-246.

Brokered Approach to Federating Data using Semantic Web Techniques

Jeremy Siao Him Fa*, Geoff West†, David A. McMeekin‡ and Simon Moncrieff§
Cooperative Research Centre for Spatial Information
Department of Spatial Sciences, Curtin University, Bentley, Western Australia
Email: *jeremy.siao@live.com, †gawwwest@gmail.com ‡d.mcmeekin@curtin.edu.au §s.moncrieff@live.com

Abstract—**There are many situations when spatial datasets that have different data structures, schema, and format need to be combined (e.g., to form a national road network from those supplied by different regions). Current methods dealing with combining independent data sources are manually intensive, and time inefficient suggesting the need for a more automated, efficient, and user centric approach. The main problem is due to syntactic, schematic, and semantic heterogeneities as the same domain data can be represented in different ways. Although tools solving syntactic heterogeneities are numerous, there is still a requirement to solve the underlying semantic problems. One way to alleviate semantic issues is by sharing a global schema for data providers to adhere to, but this method is unlikely to be implemented in multi-governmental federated countries such as Australia, as the data sets are owned by the different States and Territories. An automated brokered approach is proposed in this paper as a solution. This approach allows domain knowledge sharing across existing and future data providers, making it plausible where data sets are diverse and owned by different parties.**

Keywords–integration; semantic; wfs; federation; broker; owl-s.

I. INTRODUCTION

In this era of big data, data integration and interoperability has become an important technical challenge to research. The current methods dealing with interoperability issues have been manually intensive and costly [1]. Furthermore, the same datasets can be combined multiple times by different parties, leading to duplicated effort and conflation issues. The need for better and faster access to more user centric spatial data is further enhanced as spatial information is becoming more crucial to everyday life [2]. Such desires can be accomplished by a more automated way to combine data from different sources. This problem is most relevant for countries such as Australia, where the unification of the datasets has to be done at various governmental levels; Local Government Agencies (LGAs), States and Territories, and Commonwealth levels.

Schematic, syntactic, and semantic heterogeneities occur in datasets due to independent representation of the same domain data [3][4], and a lack of communication regarding the internal workings of organisations. An example of syntactic heterogeneities would be organisations storing data in different formats. As for schematic heterogeneity, a straight road can be stored as two points, or as a starting point and, a length and direction. As for an example of semantic heterogeneity, an 'aircraft' in organisation A might be a small light aircraft, while an 'aircraft' in organisation B might be an intermediate aircraft.

In order to solve these issues, and facilitate access to relevant information, interoperability at technical, syntactic, schematic, and semantic levels is required [5]. From all these, semantic interoperability is the underlying goal. Semantic heterogeneities occur because the same entity can have more than one representation or meaning [5][6][7]. Dealing with semantic problems requires domain knowledge sharing and common vocabularies. Technical interoperability, on the other hand, is already widely used; Hypertext Transfer Protocol (HTTP) used on the Web allows multiple machines to communicate via the same protocol. As for syntactic interoperability, the use of same data formats (e.g., JavaScript Object Notation (JSON), Extensible Markup Language (XML), shape files [8], Resource Description Framework (RDF)) or seamlessly transforming one data format to another are possible solutions. Tools are already available for the latter task, such as Feature Manipulation Engine (FME) [9].

Schematic or structural heterogeneities happen when there is a difference in the data schema or structure [10][7]. This is due to the datasets being developed independently, and thus using different and varying structures for the same or similar concepts [3]. Although the main consensus to solve schematic heterogeneities is through standard schemas, it has been shown that they are limited in success unless there are strong incentives to use such standards [3]. Such an approach can be observed in the Infrastructure for Spatial Information in the European Community (INSPIRE) initiative in Europe [11]. This initiative requires the different data providers to change their existing business model to adhere to a global schema, as well as added legislations and policies to ensure data quality. Data providers are further required to translate their data (new and old) to the new format. The restrictions imposed could also potentially lead to loss of information.

As such, a unifying process not requiring much change from the data providers, and allowing them to keep their existing models would be preferable. Other possible approaches include (1) point to point, (2) centralised, (3) aggregated, and (4) brokered [5]. The point to point approach requires the user to do all the unification without any intermediary between them and the data providers. This approach is very time consuming, manually intensive, and leads to duplicated effort and conflation issues. The second approach happens where all the data is stored, handled, and provided by a single organisation, which is improbable to happen in federated countries. The aggregated approach requires a data warehouse to aggregate and store the datasets from the multiple data sources. The cost of such a method scales as more data sources are included, alongside the increased duplication of data. The fourth approach is explored in this paper and will be discussed in the next few paragraphs.

In Australia, where the datasets are owned by different private agencies, and States and Territories, data aggregation is the method currently used to share spatial data across the different jurisdictions. This approach though does not provide the most up to date information, being out of date anywhere from three months to six months [12]. As such, to provide the most up to date information, while not restricting data providers, a broker approach is explored in this research.

A brokered approach makes use of a centralised mediator to transform the data from multiple sources onto an agreed schema [5]. Its aim is to deal with semantic heterogeneities via translations of diverse conceptualisations [13]. Most of the effort is put on the mediator, while the providers do not have to change their existing business model as long as they publish their schema and when they are updated. As no change in business models is required, this pattern caters for both current and future data providers. An Example of such an approach can be found in EuroGEOSS [14] in Europe.

In this paper, we propose to use an automated brokered approach to seamlessly unify different datasets from different governmental levels. Using semantic web technologies, it discovers data suppliers and adapts to their schemas dynamically. Furthermore, it does not impose change in existing business models, and gives the most up to date unified information from multiple sources. The main outcome is to enable an easy way to access specific distributed spatial information, and alleviate users from manually translating datasets.

This paper first presents relevant background material in Section 2. Related work is presented in Section 3, followed by a description of the proposed automated brokering system in Section 4. Section 5 details the current state of the project, and is followed by conclusions and plans for future work in Section 6.

II. Background Information

A. Spatial Web Services

Web Feature Services (WFS) are popular to provide access to spatial features. WFS is an Open Geospatial Consortium (OGC) web standard interface to provide remote querying to a collection of geographic features [15]. Although WFS provide data interoperability by offering different data output format, it is not designed to support schematic interoperability [16].

A WFS uses a multi layered approach to allow for querying of its capabilities (GetCapabilities), the schema of a particular feature (DescribeFeatureType), and the actual data set of a feature (GetFeature).

B. Ontologies

Dealing with semantic heterogeneities with an automated brokered approach requires ontologies to represent knowledge so as to understand how to interpret semantic differences and enable transformations. An ontology is defined as a concept where domains of interest share their understanding with each other [17][18][19]. In more technological usage, ontologies are utilised for sharing, representing, and storing knowledge in the form of data [20]. Essentially, an ontology is a graph that can be traversed in order to find links between concepts of a given domain's knowledge. Ontologies can be serialised in multiple text formats such as XML, Turtle (ttl), and Notation 3 (n3). Any of these formats can be easily transformed to another, making them syntactically interoperable.

A common semantic web language for ontologies is that of the Web Ontology Language (OWL) [21]. OWL is a World Wide Web Consortium (W3C) standard whose main purpose is to provide formalisms and to allow for knowledge representation. In this research, we make use of OWL-S [22], which is based on OWL to describe web services.

OWL-S was primarily designed to describe any potential web services, and thus was adapted to cater for WFS specifically. The modification follows the works of Stock et al. [23] closely but as only a partial of their modified ontology was available, many changes had to be extrapolated from their work. The OWL-S ontology therefore, is an ontology that can be adapted to describe WFS.

The OWL-S ontology is made up of three main components [22]:

1) The service profile used to identify, advertise, and discover web services. This is where a program can find out if a particular web service is what is needed;
2) The process model explains how the service works, the input, output, and processes for its different functions. This is where a program finds out how to use a particular service, what to input, and what is received back after the process; and
3) The grounding provides details on how to interact with the web service via messages.

The OWL-S ontology gives a strong foundation to describing web services due to it being a standard, allowing for future planning, and for much shareability.

III. Related Work

Work related to this research include various adaptations of the OWL-S ontology, novel ways to approach the brokering method, proposals to facilitate sharing of data, and attempts at solving ambiguities regarding heterogeneous datasets.

An adaptation of OWL-S includes a cloud service broker [24]. The authors define a cloud broker as 'an entity that manages the use, performance, and delivery of cloud services, and negotiates relationships between cloud providers and cloud consumers'. Their challenge is the heterogeneous nature of multiple service providers. Cloud service specifications are not standardised and promote semantic heterogeneity. By using OWL-S, the cloud service broker is able to dynamically discover complicated services whose attributes are constrained. The constraints of the services are represented using the Semantic Web Rule Language (SWRL). The cloud service broker is able to solve varying tasks of different levels through multiple case studies but requires the different service providers to use a shared ontology. Ngan and Kanagasabai [24] state that ontology alignment and learning (part of this research) can address that issue.

Another adaptation related to this paper involves implementing the OWL-S ontology to describe WFS and Web Map Service (WMS) [23]. To cater for Open Geospatial Consortium (OGC) compliant web services, the OWL-S WSDL grounding was changed to a simplified OGC equivalent grounding. Their work shows detailed analysis of WFS and WMS, and an implementation of their OWL-S adaptation demonstrated its practicality in the marine domain. Although they state that OWL-S is a 'fairly cumbersome specification', their work showed that it can be adapted to OGC compliant web services.

Yue et al. [25] explores automated geospatial web services' composition using semantics. In their approach, the authors

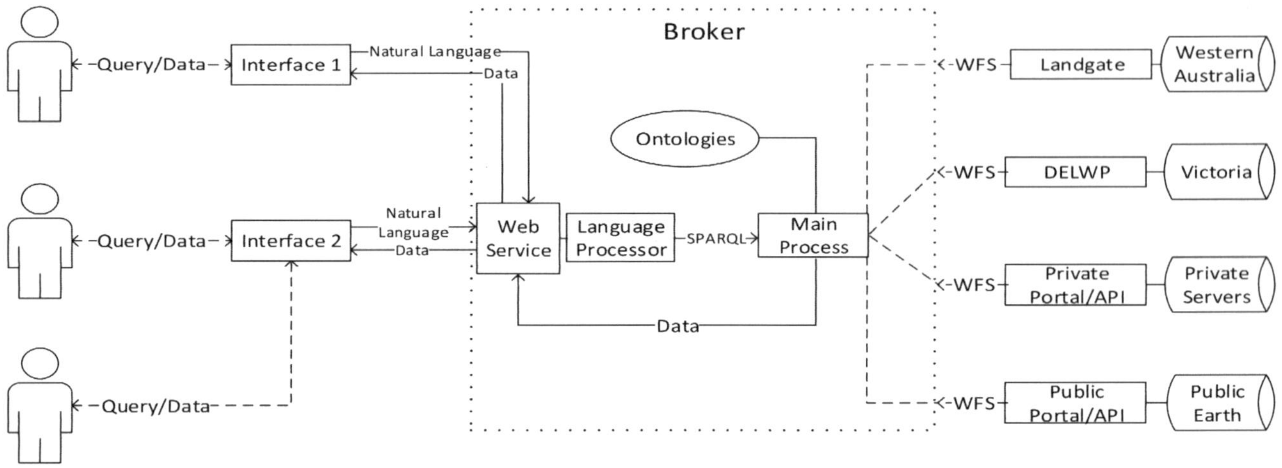

Figure 1. Automated Broker System.

designed and used 'DataType', 'ServiceType', and 'Association' ontologies as semantic schemas. Although some results have been obtained in their use case, the automation of the approach needs improvement; domain experts are required in some aspects and the ontologies' reasoning needs more work.

At a lower level, work has been done aiming at matching WSDL descriptions to that of OWL-S semantic annotations [26]. Facilitating the move from WSDL description of web services to OWL-S would promote web services discovery. Their method includes an ontology repository where a matching algorithm finds the best match related to specific concepts in a WSDL description. Using various designed and real examples, their algorithm has been validated to be practical and to reduce time and effort.

A different approach is using ontologies to provide a conceptual overview of the data sources. This particular approach is called Ontology-Based Data Access (OBDA). Its aim is to provide a query-able ontological view of the data sources to the users. Some of the current systems harnessing OBDA includes the Optique System [27] and MASTRO [28]. A recent evaluation the OBDA approach found that it can be 'orders of magnitude faster than standard triple stores' [29], using proper optimisation techniques. However, OBDA can also perform poorly if proper techniques are not used.

A real life application of OBDA using the MASTRO system was explored by Antonioli et al. [30]. They found that the mapping definition from the Italian Department of Treasury case study had to essentially be done manually. They had to manually analyze the structure of the data sources to understand the data semantics. This though, only had to be done once using the MASTRO system, and once for any new sources [30]. Some query rewriting optimisation has also been suggested, making previously impractical methods possible.

Regarding ambiguities and uncertainties, Yang et al. [31] studied the usage of Bayesian Networks. According to them, Bayesian theory provides a principled representation of uncertainty, alongside logic to unify observations with previous knowledge, and learning theory for refining ontologies. Although their approach had improved efficiency and accuracy in regards to geospatial web services, it targeted mainly discovery of such services. Furthermore, manual work in conjunction with venture capitalists were required to build the raw causal map needed in their work.

IV. Automated Broker for Unifying Uncontrollable Heterogeneous Data Sources

To facilitate access to multiple data sources, it is required to have (1) web services, (2) descriptions of the web services, and (3) a federated model.

For the web services, two WFS (described using OWL-S) are used - Landgate and Department of Environment, Land, Water and Planning (DELWP), Victoria.

As for a federated model, the Foundation Spatial Data Framework (FSDF) is used in this paper. The FSDF is a national level dataset developed by the Australian and New Zealand Land Information Council (ANZLIC) [5]. Its aim is to provide a number of foundational data themes (geocoded addressing, administrative boundaries, positioning, place names, land parcel and property, imagery, transport, water, elevation and depth, and land cover), each one consisting of a number of datasets [32]. A part of the administrative boundary dataset was used as a use case study.

Datasets in the FSDF have been modelled using the Unified Modelling Language. The FSDF was transformed to an OWL equivalent using the tool Protégé [33] to allow reasoning and querying using SPARQL Protocol and RDF Query Language (SPARQL) - an RDF query language.

As we are using semantic web concepts, ontologies will thus be the main technique used to describe the knowledge needed to link up the web services' descriptions, and federated model. This combination will be referred to as ontology Θ from here on.

A. Broker Architecture

The structure of the proposed brokered system has the federated model acting as a unified view for the user, the broker in the middle, and the web services at the end. Multiple users can use the broker system through a web service (e.g., WFS, Web Processing Service) and adapt it to their own needs. The exposed web service then translates the user query into a

SPARQL equivalent which is then used to query the ontologies and relevant data sources.

This structure is portrayed in Figure 1 where the left hand side are multiple users using the same or different interfaces plugged into an exposed web service, the middle is the broker with its various components, and the right hand side are the data providers and their web services. The broker system is the mediator between the users and the web services, and while the users can interact freely with the broker they do not need knowledge of the data sources.

In general, the broker has to deal with:

1) Querying relevant web services;
2) Processing the user's query; and
3) Combining the differing data sets from the relevant suppliers.

B. *Querying Web Services*

In this paper, web services used are WFS that adhere to standards (i.e., OGC), and have available URLs. The capabilities of the WFS can be analysed to identify if the serviced features are related to the query posed. For example, parsing the 'Title' and 'Description' tags, a matching algorithm can determine if a particular feature is relevant or not. If it matches, then the WFS capabilities are modelled and stored in the OWL-S ontology for future use. This can be observed in Figure 2 where the OWL-S ontology is connected to a GetCapabilities ontology.

After obtaining the capabilities of the service, and asserting that some typeNames (name given to a specific feature) are useful to the user, a DescribeFeatureType call is made to the service. This call returns a schema describing the meta-data stored by the web service. The schema is then parsed for meta-data related to the user's query (e.g., to see if a particular attribute 'name' is available in a particular feature). The DescribeFeatureType schema of that particular feature is then modelled and stored in the GetCapabilities node found in Figure 2.

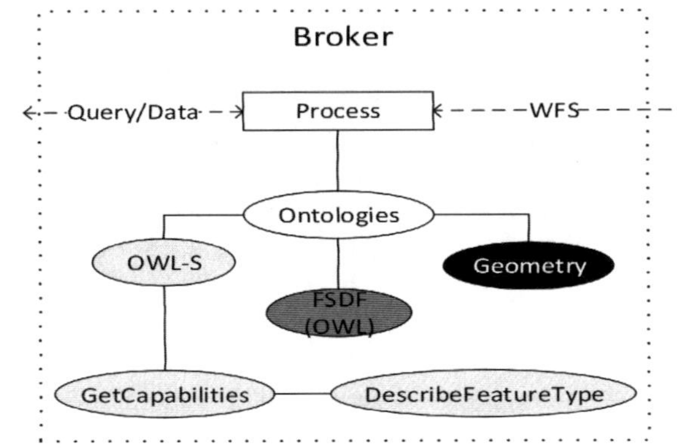

Figure 2. Ontologies Used.

To automate the process of linking the DescribeFeatureType schema to the broker's ontology, Extensible Stylesheet Language Transformations (XSLT) are used. These are programmable style sheets that allow transformation of XML to any other format. Doing so enable an automatic transformation of the DescribeFeatureType XML to RDF providing an on the fly linkage to the ontology used in the broker.

Figure 3 shows the sequence of actions needed to query a number of URLs.

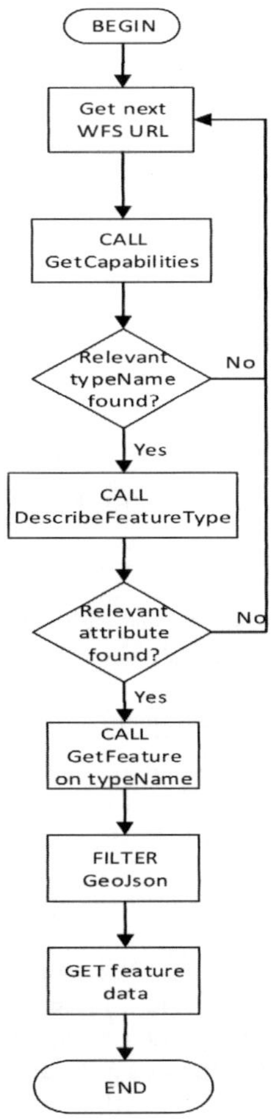

Figure 3. Overview Flowchart.

The resulting RDF triples are added to ontology Θ. Each WFS URL now has a feature, and that feature has its meta-data linked to it (Refer to 'slip:LGATE-069_Type' in Figure 6).

Once the required attribute has been obtained, the link is then added to the ontology as an equivalent term. For example, the term *StateElectoralDivision* from the FSDF, has an individual from Landgate's LGATE-069 feature type, and the attribute *fsdf_name* has an equivalent link to LGATE-069's name attribute in the ontology.

Figure 6 demonstrates the ontology developed. The light grey parts are from the OWL-S ontology, its extension can be seen from the white ellipses and rectangles, the black part is the start of an OWL implementation of the ISO 19107 Spatial Schema [34], and the dark grey part is the ontology adapted from the FSDF model.

To link up all three ontologies - pseudo-FSDF, OWL-S, and geometry, the following changes have been made:

1) Adding a WFSAtomicProcess class as a subclass of the OWL-S AtomicProcess;
2) Adding process name individuals for the WFSAtomicProcess (i.e., GetCapabilities, DescribeFeatureType, and GetFeature);
3) Linking the DescribeFeatureType and GetFeature WFSAtomicProcesses to the individual 'slip:LGATE-069' as input;
4) Adding WFSProfile as a subclass of OWL-S' ServiceProfile;
5) Adding DescribeFeatureType's meta-data as attributes for slip:LGATE-069_Type using the 'hasFeatureTypeComponent' object property; and
6) Using the geometry ontology for the 'bbox' meta-data for slip:LGATE-069_Type.

C. Processing the Query

A user's natural language query is transformed to a SPARQL equivalent query. SPARQL allows constraints to be placed upon the queries, enabling more detailed querying.

In order to process a SPARQL query, it is first required to extract its components. In this paper, the two main components will be the `SELECT` and `FILTER` clauses. The `SELECT` clause is assumed to mean a specific feature type, and the `FILTER` clause is assumed to mean a specific instance of the feature type found in the `SELECT` clause. By making this differentiation, there are thus two main types of queries: (1) a generic query (2) a detailed query:

1) The generic query is associated with the `SELECT` clause, and is assumed to refer to a specific feature type, not an instance. As such, a GetCapabilities call to the WFS is all that is required. From the retrieved XML, it is then possible to filter the title, description, and keywords to match the query's details; and
2) The detailed query is determined when a `FILTER` clause is found. The clause will determine which details the user is looking up in a particular feature type. The generic query needs to be processed first though. That is, only after getting the feature type associated with the `SELECT` clause, can a particular instance from that feature type be found.

From both generic queries and detailed queries, the base ontology Θ can then be expanded as the WFS is being explored.

1) Matching Generic Queries: A generic query is determined by the `SELECT` clause of a SPARQL query. That clause is assumed to be direcly linked to a feature type in a WFS. An example of such a query is depicted in Figure 4.

```
SELECT ?feature
   WHERE {
         ?feature rdf:type ex:StateElectoralDivision .
   }
```

Figure 4. SPARQL for Matching Generic Queries.

The query is assumed to be looking for a feature type that's related to 'StateElectoralDivision'. For such queries, the GetCapabilities of each WFS, has to be checked. The features' 'name', 'title', 'abstract', and 'keyword' go through a set of matching algorithms in order to find any similarity with the feature 'StateElectoralDivision' from this example. These particular fields have been chosen as the 'name' is a mandatory field in any feature, the 'title' is intended to 'briefly identify the feature type', the 'abstract' provides more descriptions about the feature type, while the 'keyword' is intended to aid catalog searching [15].

Figure 5 demonstrates the sequence of processes that take place, after retrieving the URL of the WFS from the ontology. After the URL is retrieved, a GetCapabilities call is made to the server. The returned XML is parsed for the fields described previously, and a matching algorithm is run to find any similarities. Given that a particular feature is found to be similar, the next step is to make a DescribeFeatureType call for the matching feature, giving back a schema (e.g., XML Schema Definition). The schema is parsed for any attributes, and these are created in the broker's ontology as 'hasFeatureTypeComponent' links to the feature (Refer to Figure 6).

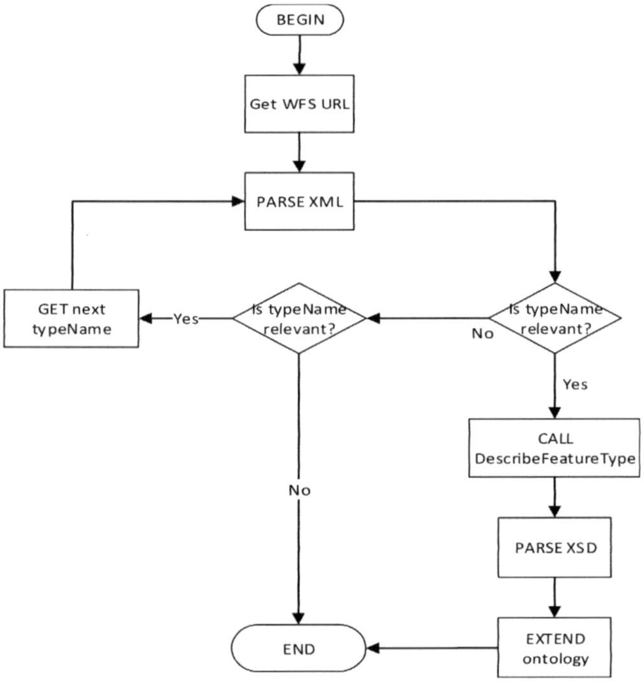

Figure 5. Generic Queries Matching Flowchart.

2) Matching Filter Queries: A filter query determined by the `FILTER` clause of a SPARQL query. That clause is assumed to be directly linked to a specific attribute in a WFS' feature. An example of such a query is depicted in Figure 7.

```
SELECT ?feature
   WHERE {
         ?feature rdf:type ex:StateElectoralDivision .
         ?feature ex:name ?name .
         FILTER( name = 'Albany' )
   }
```

Figure 7. SPARQL for Matching Filter Queries.

It is assumed the query is looking for a feature type whose attribute relating to 'name' is related to 'Albany'. After parsing the generic query of finding a typeName similar to

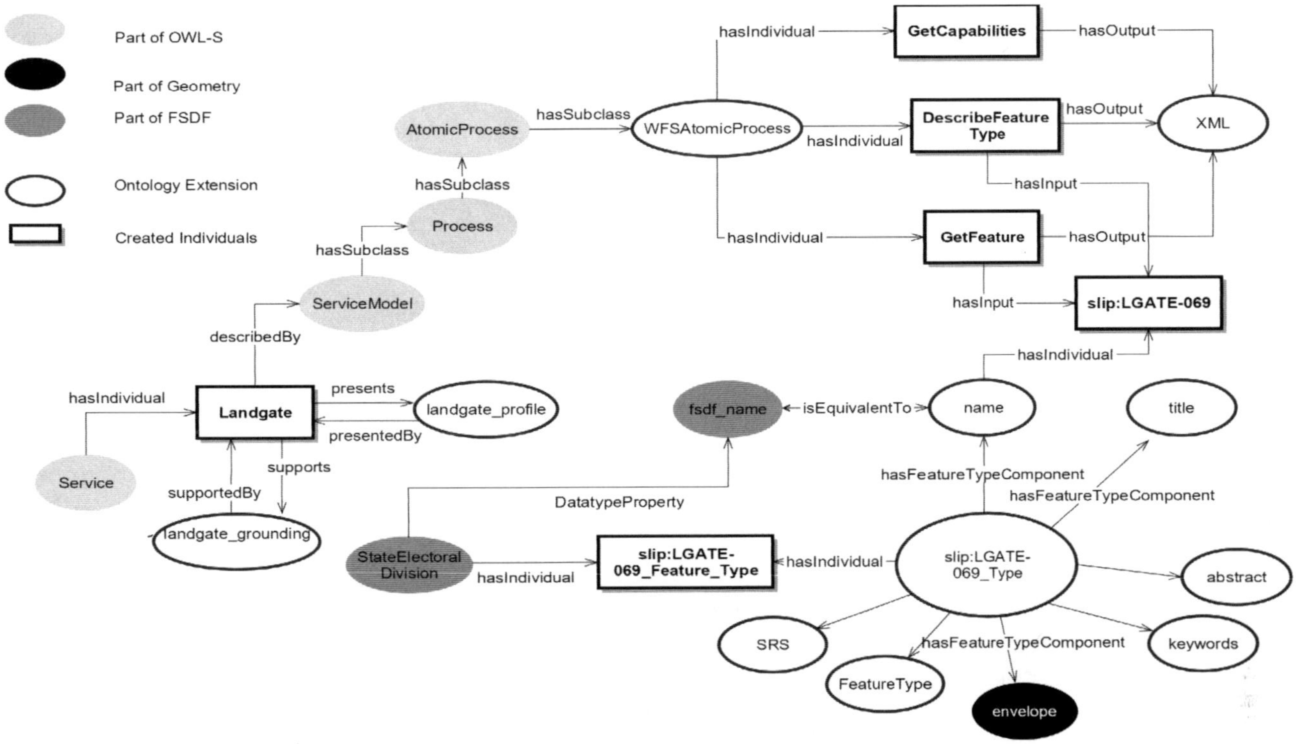

Figure 6. Adapted Ontology (Θ) Dependence.

TABLE I. COMPARISON OPERATORS.

WFS Comparison Operator	SPARQL Comparison Operator
PropertyIsEqualTo	=
PropertyIsNotEqualTo	!=
PropertyIsLessThan	<
PropertyIsGreaterThan	>
PropertyIsLessThanOrEqualTo	<=
PropertyIsGreaterThanOrEqualTo	>=
PropertyIsLike	N/A
PropertyIsBetween	< && >

'StateElectoralDivision', the next steps are demonstrated in Figure 8. All the meta-data are retrieved from the particular feature type using a DescribeFeatureType call. The schema obtained is filtered to find a similar attribute to that of 'name'. Given that a similar attribute is found, the next step is to call GetFeature from the web service, and filter the attributes for 'Albany'.

3) SPARQL Filters to WFS Filters: Instead of filtering a whole feature type within the broker, filter parameters can be passed into a GetFeature call to the WFS. It enables the provider's WFS to run a filter search server side, removing the filtering overheads from the broker. For simplicity, only comparison operators (e.g., =, !=, <, >, etc.) are discussed in this paper but others such as 'Union', and regular expressions are possible.

Table I shows the comparison operators that are common in WFS and SPARQL.

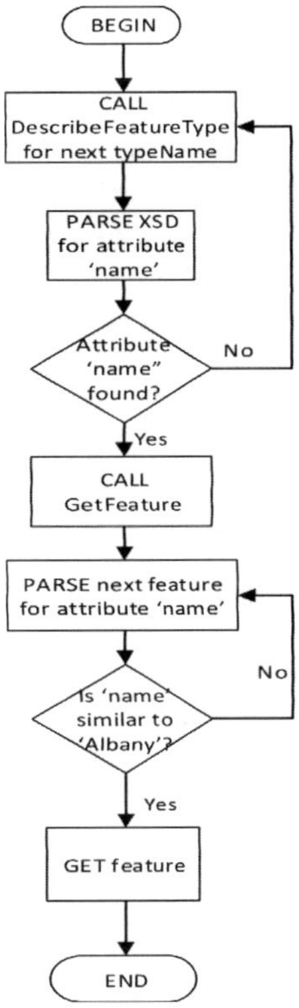

Figure 8. Filter Queries Matching Flowchart.

Considering Table I, the SPARQL query can thus be converted to a WFS call equivalent.

```
SELECT ?feature
   WHERE {
         ?feature rdf:type ex:StateElectoralDivision .
         ?feature ex:fsdf_name ?name .
         FILTER( name = 'Albany' )
   }
```

Figure 9. SPARQL Name matching.

Given that 'LGATE-069' is the equivalent to 'StateElectoralDivision', and that 'name' is the equivalent to 'fsdf_name', then the WFS call would be as depicted in Figure 10.

```
www2.landgate.wa.gov.au\ows\wfspublic_4283\wfs?SERVICE=WFS&
VERSION=1.0.0&REQUEST=getFeature&typeName=LGATE-069&Filter=

<Filter><PropertyIsEqualTo>
      <PropertyName>name</PropertyName>
      <Literal>Albany</Literal>
</PropertyIsEqualTo></Filter>
```

Figure 10. Name Filtering Call.

Using the 'PropertyIsEqualTo' though, is restrictive, as the precise name of the anything stored has to be known. This issue can be solved by using the 'PropertyIsLike' operator. The same query can then be rewritten as depicted in Figure 11.

```
www2.landgate.wa.gov.a\ows\wfspublic_4283\wfs?SERVICE=WFS&
VERSION=1.0.0&REQUEST=getFeature&typeName=LGATE-069&Filter=

<Filter><PropertyIsLike wildCard='*' singleChar='!'>
      <PropertyName>name</PropertyName>
      <Literal>*Albany*</Literal>
</PropertyIsLike></Filter>
```

Figure 11. PropertyIsLike Call.

That call would look for any name attribute that has 'Albany' in it.

For multiple word names such as 'Alfred Cove', a simple algorithm permutating all the ways it can be written can be used. For example, the identified ways 'Alfred Cove' can be written as are:

nolistsep

1) Upper case (ALFRED COVE);
2) Lower case (alfred cove);
3) Upper case with underscore (ALFRED_COVE);
4) Lower case with underscore (alfred_cove);
5) Camel case with space (Alfred Cove);
6) Camel case without space (AlfredCove); and
7) Camel case with underscore (Alfred_Cove).

All the seven ways of naming conventions can be programmed to fit in the WFS GetFeature filter URL as depicted in Figure 12.

```
www2.landgate.wa.gov.au\ows\wfspublic_4283\wfs?SERVICE=WFS&
VERSION=1.0.0&REQUEST=getFeature&typeName=LGATE-069&Filter=

<Filter><Or>
   <PropertyIsLike wildCard='*' singleChar='!'>
         <PropertyName>name</PropertyName>
         <Literal>*ALFRED COVE*</Literal>
   </PropertyIsLike>
   <PropertyIsLike wildCard='*' singleChar='!'>
         <PropertyName>name</PropertyName>
         <Literal>*alfred cove*</Literal>
   </PropertyIsLike>
   <PropertyIsLike wildCard='*' singleChar='!'>
         <PropertyName>name</PropertyName>
         <Literal>*ALFRED_COVE*</Literal>
   </PropertyIsLike>
   <PropertyIsLike wildCard='*' singleChar='!'>
         <PropertyName>name</PropertyName>
         <Literal>*alfred_cove*</Literal>
   </PropertyIsLike>
   <PropertyIsLike wildCard='*' singleChar='!'>
         <PropertyName>name</PropertyName>
         <Literal>*Alfred Cove*</Literal>
   </PropertyIsLike>
   <PropertyIsLike wildCard='*' singleChar='!'>
         <PropertyName>name</PropertyName>
         <Literal>*AlfredCove*</Literal>
   </PropertyIsLike>
   <PropertyIsLike wildCard='*' singleChar='!'>
         <PropertyName>name</PropertyName>
         <Literal>*Alfred_Cove*</Literal>
   </PropertyIsLike>
</Or></Filter>
```

Figure 12. ProperIsLike Name Filtering Call.

D. Converting XSD to RDF

Converting XML Schema (XSD) to RDF format automatically has been achieved in various disciplines [35][36][37].

In this research, XSLT was used as it enables the transformation of an XML Schema to RDF without needing any human assistance. The XSLT was developed and used successfully but does contain some limitations, such as data type duplicates. It was utilised to automatically link the DescribeFeatureType schema of a WFS to the broker's ontology as depicted in Figure 2 and Figure 5. The generic steps undertaken in the XSLT are:

1) For each XSD targetNamespace, convert them to RDF namespaces;
2) For each XSD complexType, convert them to OWL Class;
3) For each XSD extension within a xsd complexType, convert the class to a subclass of the extension;
4) For each sequence within a XSD complexType, convert them to an OWL subClassOf owl:Restriction;
5) For each XSD element within a XSD sequence, convert them to an owl:onProperty with the rdf:resource being the base URL plus the element's name;
6) For each minOccurs within the xsd:element, convert them to owl:minCardinality;
7) For each maxOccurs within the xsd:element, convert them to owl:maxCardinality; and
8) For each type within the xsd:element, convert them to an owl:Datatype with the rdfs:range being the xsd:type and the rdfs:domain being the owl:Class;

Following these mapping guidelines, an RDF file was automatically generated to extend ontology Θ.

E. Combining Differing Data Sets

The combination of varying data sets is done on a per-query basis. For each of the queries tasked by the user, the broker goes through the two steps above: (1) querying web services, and (2) processing the user query if relevant terms in the query are not already existent in the ontology. For example, if the general term in the query's `SELECT` is found to have an equivalence of 'LGATE-069' from Landgate in ontology Θ , then further processing is not required.

As these processes are carried out, the ontology would be growing to have further links and equivalent terms, rendering the processing less and less required as the results of previous queries are stored. That semantic aspect makes the broker highly scalable, as the system constructs more links as more queries are processed.

F. Error Handling

Given that an error occurs in calling a WFS service, an error message would be returned either in JSON or Hypertext Markup Language (HTML). A JSON error message can be parsed to find the specific cause of the problem (e.g., feature does not exist), while the error code of the HTML provides a direct indication of it (e.g., Error 500 for internal error). Depending on the cause of the problem, two main scenarios can happen (1) the ontology can be modified to cater for the changes, and (2) the error cannot be handled properly.

Case number one happens when the schema of the WFS is changed and thus the ontology is not up to date with the changes. This can be resolved by calling the WFS GetCapabilities, running a matching algorithm to the broker's ontology and compare for discrepancies. The mismatches can then be resolved by updating the ontology to reflect the new schema.

Case number two happens when there is either a problem on the provider's side (e.g., server is down), or a problem on the user's side (e.g., user input error). In both cases, nothing can be done as these are errors of the broker's scope, and the only solution is to notify the user about it.

G. Performance

Using the on-the-fly approach ensures that only a minimal amount of information is stored on the broker's side. Most of the main data (e.g., coordinates, shapes, etc.) remain at the data source. The only stored information are the ontologies that link data providers' gateway to the global ontology. With the example of WFS, only information up to the secondary level (DescribeFeatureType) is stored as part of the ontology, the third level, which contains most of the data, is left at the source. The ontologies can moreover be distributed on the cloud; various servers can be used, making storage problems minimal. Furthermore, caching will be used to speed searching, and parallel computing or torrenting technologies could be explored as well.

V. RESULTS

A case study of the broker system was implemented using the Python programming language [38]. Python has various libraries and frameworks already available for usage. For example, the library rdflib allows the usage of RDF and ontology graphs, to query, create, edit, and import. Reusability of such libraries is a forte of Python, especially for proofs of concept. Furthermore, the framework Django [39] has been used as it enables easy set up of a web interface to implement a server locally.

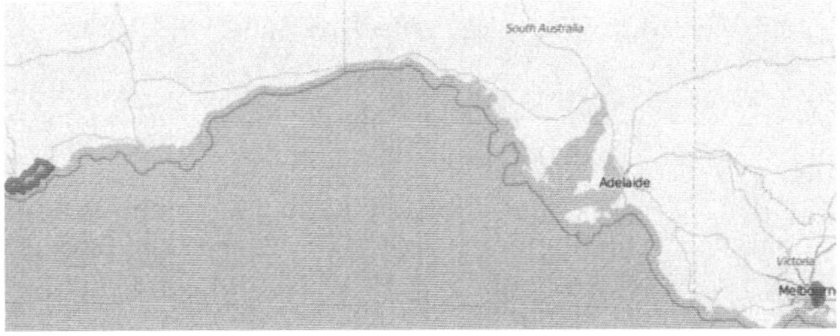

Figure 13. Brokered System Result.

A. Current State

The broker system implemented provides a visual display for the OWL implementation of the FSDF. A list of classes - with their respective attributes from the FSDF UML - is shown. Once a class and an attribute are selected, the user can specify a value to query. A start button starts the querying process, and a map is updated automatically when any result is obtained.

Figure 13 shows the returned result from two different States (Western Australia and Victoria). The red marks are the locations' boundaries, which can be seen clearer in Figure 14 and Figure 15 respectively.

VI. CONCLUSION AND FUTURE WORK

This paper discussed the development of a broker system to automatically and virtually consolidate various spatial data sources on a per-query basis for easier user consumption. The aim of the broker is to act as a mediator between users and data sources, to combine various heterogeneous data sources.

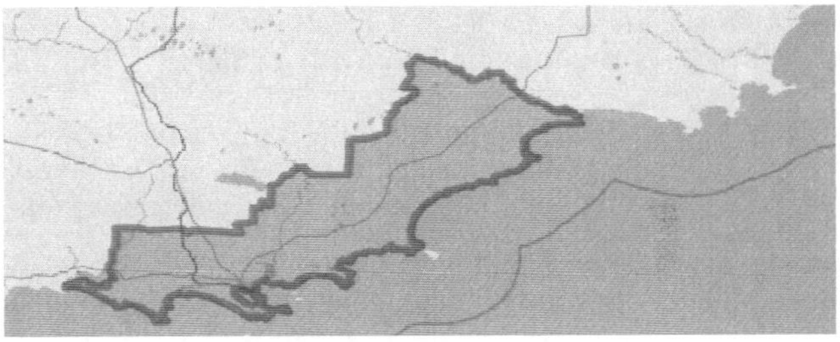

Figure 14. Albany (WA) Result.

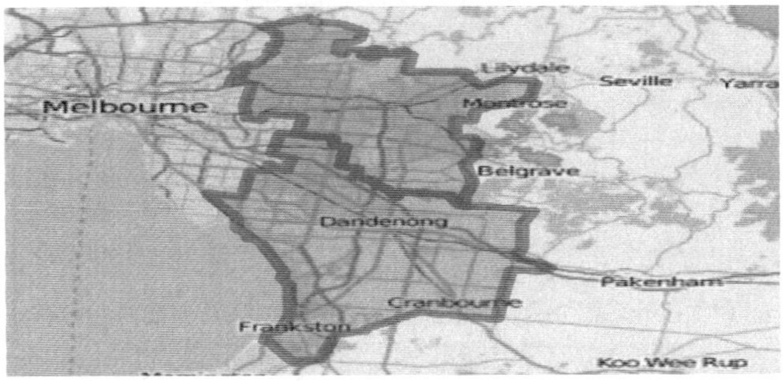

Figure 15. Eastern Metropolitan Area (VIC) Result.

This work is far from being completed and it is planned to extend the broker system to more agencies and operations. Reasoning to cope with ambiguities is to be developed, as well as, the automatic expansion of the ontologies. Other future work will include more filtering abilities such as intersection of regions, and queries regarding different data sets such as housing and forests. Furthermore, the web services to be implemented have to be diversified, as well as the themes from the FSDF.

ACKNOWLEDGEMENT

This work has been supported by the Cooperative Research Centre for Spatial Information, whose activities are funded by the Business Cooperative Research Centres Programme.

REFERENCES

[1] C. Terblanche and P. Wongthongtham, "Ontology-based employer demand management," *Software: Practice and Experience*, pp. 1–46, Apr 2015.

[2] ACIL Tasman, "The Value of Spatial Information," ACIL Tasman Pty Ltd, Tech. Rep. March, 2008.

[3] A. Halevy, "Why Your Data Won't Mix: Semantic Heterogeneity," *Queue*, vol. 3, no. 8, pp. 50–58, oct 2005.

[4] D. J. Abel, B. C. Ooi, K.-L. Tan, and S. H. Tan, "Towards integrated geographical information processing," *International Journal of Geographical Information Science*, vol. 12, no. 4, pp. 353–371, jun 1998.

[5] P. Box, B. Simons, S. Cox, and S. Maguire, "A Data Specification Framework for the Foundation Spatial Data Framework," CSIRO, Australia, Tech. Rep., 2015.

[6] J. X. He, "An Ontology-Based Methodology for Geospatial Data Integration," Ph.D. dissertation, uOttawa, 2010.

[7] I. F. Cruz and H. Xiao, "The role of ontologies in data integration," *Journal of Engineering Intelligent Systems*, vol. 13, pp. 245–252, 2005.

[8] Esri, "ESRI Shapefile Technical Description," *Computational Statistics*, vol. 16, no. July, pp. 370–371, 1998.

[9] S. Software, "Feature Manipulation Engine," 2016, URL: http://www.safe.org/ [accessed: 2016-04-06].

[10] A. Buccella, A. Cechich, and N. R. Brisaboa, "Ontology-Based Data Integration Methods : A Framework for Comparison," *Revista Colombiana de Computación*, vol. 6, no. 1, 2005.

[11] INSPIRE Thematic Working Group Utility and governmental services, "D2.8.III.6 INSPIRE Data Specification on Utility and governmental services - Draft Technical Guidelines," INSPIRE Thematic Working Group, Tech. Rep. March, 2004.

[12] Anzlic, *One ANZ Foundation Spatial Data Framework*. ANZLIC, 2012, no. November.

[13] H. Wache, T. Scholz, H. Stieghahn, and B. Konig-Ries, "An integration method for the specification of rule-oriented mediators," in *Proceedings 1999 International Symposium on Database Applications in Non-Traditional Environments (DANTE'99) (Cat. No.PR00496)*, no. 01. IEEE Comput. Soc, 1999, pp. 109–112.

[14] EuroGEOSS, "EuroGEOSS," 2016, URL: http://www.eurogeoss.eu/default.aspx [accessed: 2016-04-06].

[15] P. A. Vretanos, "Web Feature Service Implementation Specification," Open Geospatial Consorium Inc., Tech. Rep., 2005.

[16] P. Staub, "A Model-Driven Web Feature Service for Enhanced Semantic Interoperability," *OSGeo Journal*, vol. 3, no. December, pp. 38–43, 2007.

[17] J. Partyka, N. Alipanah, L. Khan, B. Thuraisingham, and S. Shekhar, "Content-based ontology matching for GIS datasets," in *Proceedings of the 16th ACM SIGSPATIAL international conference on Advances in geographic information systems - GIS '08*, no. c. New York, New York, USA: ACM Press, 2008, pp. 1–4.

[18] M. Uschold and M. Gruninger, "Ontologies: principles, methods and applications," *The Knowledge Engineering Review*, vol. 11, no. 02, pp. 93–162, jul 1996.

[19] T. R. Gruber, "A translation approach to portable ontology specifications," *Knowledge Acquisition*, vol. 5, no. 2, pp. 199–220, jun 1993.

[20] R. Megala and K. Nirmala, "Semantic Queries in Distributed Relational Database Using Global Ontology Construction." *ICTACT Journal on Soft Computing*, pp. 942–945, 2015.

[21] D. L. McGuinness and F. Van Harmelen, "OWL Web Ontology Language Overview," 2009, URL: https://www.w3.org/TR/owl-features/ [accessed: 2016-04-06].

[22] D. Martin, M. Burstein, J. Hobbs, O. Lassila, D. McDermott, S. Mcllraith, S. Narayanan, M. Paolucci, B. Parsia, T. Payne, E. Sirin, N. Srinivasan, and K. Sycara, "OWL-S: Semantic Markup for Web Services," URL: http://www.w3.org/Submission/OWL-S/ [accessed: 2016-04-06].

[23] K. Stock, A. Robertson, and M. Small, "Representing OGC Geospatial Web Services in OWL-S Web Service Ontologies," *International Journal of Spatial data Infrastructures Research*, vol. 6, 2011.

[24] L. D. Ngan and R. Kanagasabai, "OWL-S Based Semantic Cloud Service Broker," *2012 IEEE 19th International Conference on Web Services*, pp. 560–567, 2012.

[25] P. Yue, L. Di, W. Yang, G. Yu, and P. Zhao, "Semantics-based automatic composition of geospatial Web service chains," *Computers and Geosciences*, vol. 33, no. 5, pp. 649–665, 2007.

[26] T. A. Farrag, A. I. Saleh, and H. A. Ali, "Toward SWSs discovery: Mapping from WSDL to OWL-S based on ontology search and standardization engine," *IEEE Transactions on Knowledge and Data Engineering*, vol. 25, no. 5, pp. 1135–1147, 2013.

[27] P. Haase, I. Horrocks, and D. Hovland, "Optique System: Towards Ontology and Mapping Management in OBDA Solutions," *Second International Workshop on Debugging Ontologies and Ontology Mappings - WoDOOM13*, pp. 21–32, 2013.

[28] D. Calvanese, G. De Giacomo, D. Lembo, M. Lenzerini, A. Poggi, M. Rodriguez-Muro, R. Rosati, M. Ruzzi, and D. F. Savo, "The MASTRO system for ontology-based data access," *Semantic Web*, vol. 2, no. 1, pp. 43–53, 2011.

[29] D. Lanti, M. Rezk, G. Xiao, and D. Calvanese, "The NPD benchmark: Reality check for OBDA systems," *Proc. 18th International Conference on Extending Database Technology (EDBT)*, pp. 617–628, 2015.

[30] N. Antonioli, F. Castanò, C. Civili, S. Coletta, S. Grossi, D. Lembo, M. Lenzerini, A. Poggi, D. F. Savo, and E. Virardi, "Ontology-Based Data Access: The Experience at the Italian Department of Treasury," *CAiSE Industrial Track*, vol. 1017, pp. 9–16, 2013.

[31] X. Yang, W. Cui, Z. Liu, and F. Ouyang, "Study on uncertainty of geospatial semantic Web services composition based on broker approach and Bayesian networks," in *Geoinformatics 2008 and Joint Conference on GIS and Built Environment: Geo-Simulation and Virtual GIS*

Environments, L. Liu, X. Li, K. Liu, X. Zhang, and A. Chen, Eds., vol. 7143, oct 2008, pp. 714 305–714 305–8.

[32] Anzlic, “ANZLIC - the Spatial Information Council is the peak intergovernmental organisation providing leadership in the collection, management and use of spatial information in Australia and New Zealand.” 2016, URL: http://www.anzlic.gov.au/ [accessed: 2016-04-06].

[33] “Protégé,” 2016, URL: http://protege.stanford.edu/ [accessed: 2016-04-06].

[34] S. J. D. Cox, “OWL representation of ISO 19107 (Geographic Information - Spatial Schema),” 2015, URL: http://def.seegrid.csiro.au/isotc211/iso19107/2003/geometry [accessed: 2016-04-06].

[35] Rhizomik, “ReDeFer,” 2016, URL: http://rhizomik.net/html/redefer/#XSD2OWL [accessed: 2016-04-06].

[36] Brishniz, “XML2OWL Demonstration Platform,” 2016, URL: http://xml2owl.sourceforge.net/index.php [accessed: 2016-04-06].

[37] Incunabulum, “XsdImport - Convert XSD schemas to OWL,” 2016, URL: http://www.incunabulum.de/projects/it/xsdimport/ [accessed: 2016-04-06].

[38] “Python,” 2016, URL: https://www.python.org/ [accessed: 2016-04-06].

[39] “Django: The web framework for perfectionists with deadlines.” 2016, URL: https://www.djangoproject.com/ [accessed: 2016-04-06].

Underground Monitoring Systems using 3D GIS for Public Safety

Kwangsoo Kim, Dong-Hwan Park, Jaeheum Lee, and Inwhan Lee
UGS Convergence Research Division
Electronics and Telecommunications Research Institute
Daejeon, Republic of Korea 34129
Email: {enoch, dhpark, ljh, ihlee}@etri.re.kr

***Abstract*—This paper describes an underground monitoring system using 3D geographic information system (GIS), which models and shows real world objects. This system consists of wireless sensor networks, middleware, and a 3D visualizer. The wireless sensor networks provide sensing values on the state of underground facilities, the middleware provides an abstraction layer for various sensing devices and communication protocols, and the 3D visualizer shows the shapes, the locations, the states, and the risk indexes associated with facilities. The 3D GIS used in the visualizer provides a powerful tool that enhances the ability to monitor the underground environment for public safety and helps improve the efficiency of the underground safety management.**

Keywords–Underground facility; underground safety; 3D GIS.

I. Introduction

In the industrial society, human beings moved from rural to urban areas during the industrial revolution. Therefore, the number of people living in cities increased dramatically. According to the Economist, about 86 percent of the developed world and 64 percent of the developing world will be urbanized by 2050 [1]. More than 50 percent of the world's human population now live in cities. This portion will increase continually. As the population of large cities has increased, people have developed an infrastructure to support their comfortable life. The infrastructure was very important for the urban dwellers. In the cities, water pipes were constructed underground to transport water from water sources to individual homes, factories and buildings. Sewer systems were also constructed underground to transport sewage from places to the outside of the cities. Roads were constructed to better facilitate the movement of important things such as food, soldiers, vehicles, and other goods. In modern cities, underground railway systems have been constructed to solve the heavy traffic congestion.

Presently, the outdated and aging infrastructure is becoming a problem. As many cities depend on extensive infrastructure systems that support them, the aging infrastructures are in need of repair or replacement to provide urban dwellers with a comfortable life. For example, New York City's water mains were installed underground more than 100 years ago. Breaks occur frequently. Since 1998, more than 400 main water breaks have happened every year in New York City [2]. Also, road subsidence in downtown areas in Korean cities has emerged as a serious social problem. For example, the road subsidence in Seoul City occurred at 3,328 locations from 2010 to 2014 [3]. In addition, subsidence occurred in Busan, Incheon, Gwangju, Suwon, Suncheon, Andong, and elsewhere. Unfortunately, it is very hard to recognize the place and time of the leaks or breaks of water pipes and road subsidence. Furthermore, failures of underground facilities are invisible. Sometimes we do not know where they are under the roads.

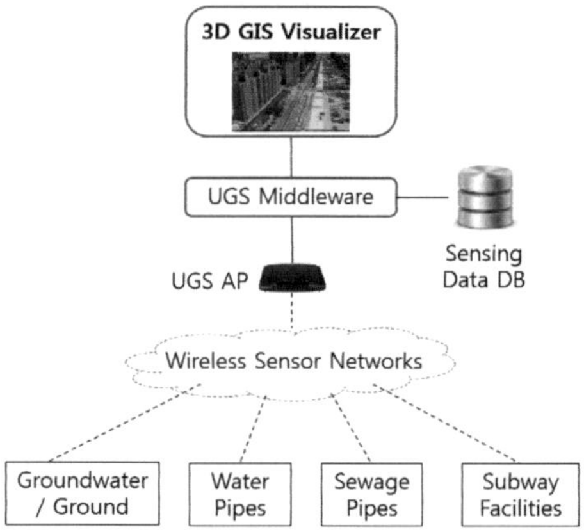

Figure 1. Architecture of Underground Monitoring System.

We have developed an underground monitoring system to detect the state changes occurring in the underground space. To achieve this goal, the underground monitoring system monitors the state changes of sewer pipes, water pipes, subway lines, subway stations, and groundwater levels. Figure 1 shows the architecture of the underground monitoring system. We install various types of sensing devices underground and acquire the sensing data related to the state of the utilities. In Figure 1, the underground safety middleware (UGS-M) is responsible for collecting data from those sensing devices. The locations of all utilities and sensing devices are shown in the 3D visualization module using GIS technologies. Thus, the purpose of this paper is to introduce the underground monitoring system for public safety.

The remainder of this paper is organized as follows. Section II describes wireless sensor networks that transmit sensing data. Section III describes the UGS middleware that collects and manages sensing data. Section IV describes the visualizer that displays geographic data and sensing data. Finally, Section V concludes this paper.

II. Wireless Sensor Networks

We use a wireless sensor network (WSN) to sense the states of underground utilities and transmit them to a centralized data collector. A WSN consists of two types of devices: an access point (AP) and a sensor node. They form a star topology. An AP interconnects a WSN and an IP-based network. An AP provides diverse communication schemes such as wireless fidelity (WiFi), Ethernet, and long term evolution (LTE). A sensor node includes one or more transducers. One transducer

Figure 2. Underground facilities.

Figure 3. Node overlay and risk index.

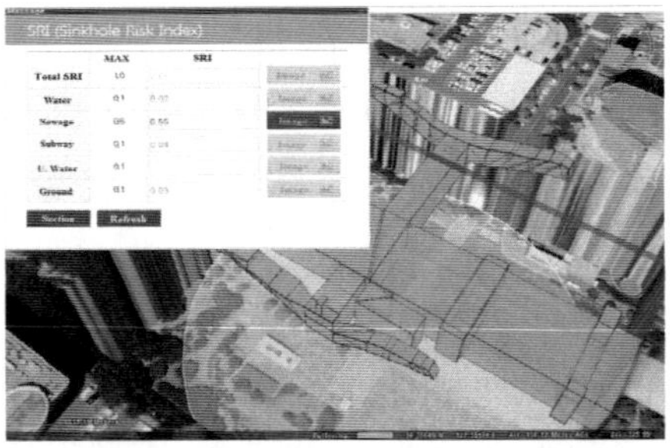

Figure 4. Visualization of underground space.

is either a sensor or an actuator. Sensor nodes are installed in tube wells and subway stations and attached to sewer pipes, water supply pipes, and subway lines.

Underground utilities generate different sensing types according to their properties. Sewer pipes generate videos and still images that show their states. Water pipes generate pitch (lateral axis), roll (longitudinal), and leak noise. Subway lines and stations generate videos, still images, the amount of influent water, stress, and acceleration. Tube wells generate water level, water temperature, water conductivity, water turbidity, soil temperature, soil conductivity, and soil moisture. Most of the sensing values are transmitted to the UGS-M through the WSN; however, videos and still images are uploaded into the database through the Internet.

III. UGS MIDDLEWARE

UGS-M is located between several applications including the 3D GIS Visualizer and the sensor networks consisting of a larger number of sensor nodes. UGS-M is responsible for collecting data from the sensor networks, transmitting it to the applications, and providing an abstraction on diverse specifications of sensing devices and communication protocols. To achieve this goal, UGS-M consists of a communication manager, data translator, resource manager, sensing data manager, monitoring manager, and a service interface [3].

IV. 3D GIS VISUALIZER

The 3D GIS visualizer creates a detailed model that describes the objects both from above and below the city ground. The model includes maps, imagery, and subsurface features such as water pipes, water supply manholes, sewage pipes, sewage manholes, subway lines, and subway stations. The GIS data, which represents the underground features, is provided by Daejeon Metropolitan City. Daejeon generates the underground features as the two-dimensional objects. Water pipes, sewage pipes, and subway lines are represented as the line object. Manholes are represented as the point object. Subway stations are represented as the polygon object. Daejeon City uses the GIS data to recognize which underground facilities are located within construction areas. Therefore, we convert two dimensional objects into three dimensional objects by adding its depth and pipe size to display the state information such as cracks on the facilities, slopes of sewage pipes, etc. The underground facilities represented as three dimensional objects are shown in Figure 2. Figure 3 shows the sensor nodes overlaid on the map. Since the road subsidence has occurred mainly along the subway lines, the sensor nodes were installed near the subway lines (marked in green) initially. Figure 3 also shows the risk indexes represented as circles in different colors. Green, yellow, and red indicate low, medium, and high risk, respectively. The sinkhole risk index (SRI) in Figure 4 shows the degree of risk associated with each underground facility. Its value is between 0 and MAX. Figure 4 shows the underground space with the dangerous sewage pipes that are marked in red.

V. CONCLUSION

This paper presents the brief features of an underground monitoring system to detect the risks related to underground utilities. We implemented and evaluated the prototype of this system. It senses the states of utilities, collects the sensing values, and visualizes those values and objects. We are going to implement several analysis schemes that extract both state changes and risks from the sensing values associated with underground utilities. Then, we will apply this system to a testbed to be implemented in Daejeon. In the testbed, the correctness and efficiency of those schemes will be evaluated.

ACKNOWLEDGMENT

This work was supported by the National Research Council of Science & Technology (NST) grant by the Korea government (MSIP) (No. CRC-14-02-ETRI).

REFERENCES

[1] Economist, "Urban life: Open-air computers," Special reports: Technology and geography, 2012.

[2] A. Forman, "Caution Ahead," Report published by the Center for an Urban Future, Mar. 2014.

[3] K. Kim, D. H. Park, J. Lee, and S. I. Jin, "UGS Middleware for Monitoring State of Underground Utilities," in Proceedings of the International Conference on Information and Communications Technology Convergence (ICTC) Oct. 28–30, 2015, Jeju, Rep. of Korea. The Korean Institute of Communcations and Information Sciences, Oct. 2015, pp. 1186–1188.

Image Based-Localization on Mobile Devices Using Geometric Features of Buildings

Hasinarivo Ramanana
Dept. of Mathematics and Computer Science
University of Antananarivo
Madagascar
e-mail: hasram006@gmail.com

Andriamasinoro Rahajaniaina
Department of Mathematics, Computing and Applications,
University of Toamasina,
Madagascar
e-mail: hajatoam@gmail.com

Jean-Pierre Jessel
IRIT, VORTEX
Paul Sabatier University
Toulouse
France
e-mail: jessel@irit.fr

***Abstract*—Outdoor localization is problem that many people are facing in everyday life. One way to determine the location of a user is to use an image-based localization method. In this paper, we propose an approach based on geometric features of buildings to address image-based localization in an outdoor environment. Our proposed method can be described as follows: first, we compute descriptors of buildings façades at the scene by using cross-ratios, then, we match them to images in a database, we get the length of the façade retrieved and we estimate the location of the user's camera. We use cross-ratios to compute the descriptors of buildings because it is a projective invariant. Our method is tested with buildings within a campus and a Geographic Information System (GIS) created from OpenStreetMap.**

Keywords-Image-based localization; cross-ratios; GIS; building recognition.

I. INTRODUCTION

Image-based localization consists in determining the position of a user's camera, i.e., the user 's position, by computing the distance from known objects in an image. We use it in many domains such as robot localization and landmark recognition. In urban environment, tall buildings may interfere with the Global Positioning System (GPS) signal which results in an inaccurate localization. In such a situation, an image-based localization may replace the GPS-based solution [16]. Generally, image-based localization is composed of three main steps: performing image matching of the scene from all geo-referenced images stored in database and retaining the best candidate to find the approximated area of the user, computing the length of a known object in the scene and estimating the position of the user.

For the first step, special features of the image are extracted to differentiate it from the other images in the database. In [1], [7] and [8] authors used Scale Invariant Feature Transform (SIFT) descriptor [2] for image matching. In [3], Roberto Cipolla utilized Harris-Stephens detector [4]. In [5], C. Card and W. Hoff used Oriented FAST and Rotated BRIEF (ORB) [6] descriptor for image matching.

This article proposes an image-based localization approach in an urban environment which extracts geometric features of building's façade as keypoints. We use cross-ratios to describe buildings images. The remainder of this paper is organized as follows. Section 2 lists the related works. Section 3 describes our proposed method in detail. Section 4 explains our experiences and our results. Section 5 is the conclusion.

II. RELATED WORKS

There are many researchers who proposed techniques for outdoor localization in a city using mobile devices. In this section, we will list some works relating to image-based localization and building recognition.

Johansson and Cipolla [13] proposed a technique that uses the parallel planes in buildings to find the homography of images in a city scene and hence predict the location of the camera. This approach reduces the amount of memory allocated for images in the database but it is lacking in precision.

N. Haala and J. Bohm [14] presented a system for locating a building in a city using a database of 3D models of buildings. They convert a 3D model into a single 2D view per orientation for each image and apply a 2D to 2D matching. To recognize the building, the system extracts the edges and corners by the Generalized Hough Transform. The telepointing device used for their approach is composed of a camera, a GPS receiver, an electronic compass, a tilt sensor and a laptop. All of these materials are hard to bear.

In 2004, D. Robertson and R. Cipolla [15] worked on a localization technique in urban environment using a smartphone. The user takes a picture of his surroundings and sends this image as a query to a server which searches in a database of façades. The system computes the vanishing points of query image in horizontal and vertical directions by extracting lines in theses principal directions. These vanishing points will be used for camera pose estimation. After this process, the system computes descriptors in query image based on Harris corner detection. The descriptor is defined by a vector of 8x8 matrix of Red Green Blue (RGB) pixel values centered in each interest point. The detection of interest points is repeated with different image scales using a pyramid of scaled images. After that, the matching is executed at each level of scale in the pyramid to achieve robust matching. This approach requires additional memory space and high computation time, which decreases the speed of the process.

In 2009, N. Yazawa and H. Uchiyama [17] developed a system for estimating user position by matching a captured image from a camera equipped with a compass and GPS into a database of 104 panoramas. The method Speeded Up Robust Features [20] (SURF) is used for image comparison and triangulation for estimating the camera pose. The matching of SURF features with panoramas in the database and the captured image took 400 seconds. In our system, we want to improve this time computation.

In 2013, M. Donoser and D. Schmalstieg [18] introduced a discriminative classification problem for matching interest points detected in the query image and the 3D point in the known world. They compared their method with the standard Nearest Neighbor, Random Fern and Random Forest. The result proves that their proposed method gives the highest value of mean classification accuracies and standard deviations.

In 2015, B. Zeisl and T. Sattler [19] presented a voting-based pose estimation strategy for matching images in the database and query image. They wanted to compare spatial verification and appearance-based filtering.

III. Proposed Method

A. System overview

Our proposed method is divided in two parts: server side and client side. The client side is composed of a smartphone

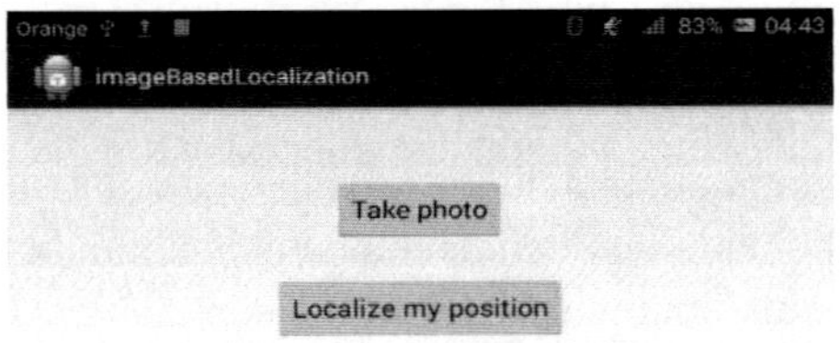

Figure 1. Home interface of the application

with which the user takes pictures of his surroundings with the smartphone's camera. The photo of buildings will serve to estimate his position. The user interface of our application is shown in Figure 1.

The information related to a specific building is shown on the smartphone's screen after identification. After that, the system estimates the user position. The server stores all images of buildings that people have already taken, computes descriptors of images according to the general structure of buildings and saves them in a YAML file (acronym of "YAML Ain't Markup Language") which will be uploaded to the smartphone. Figure 2 shows this process.

This YAML file is coupled with 2D GIS which contains the spatial disposition of buildings within the campus that we are interested in our experience. We use OpenStreetMap and Quantum Geographic Information System [21] (QGIS) software to create these data. A screenshot of the spatial data of the campus, exploited within QGIS, is shown in Figure 3.

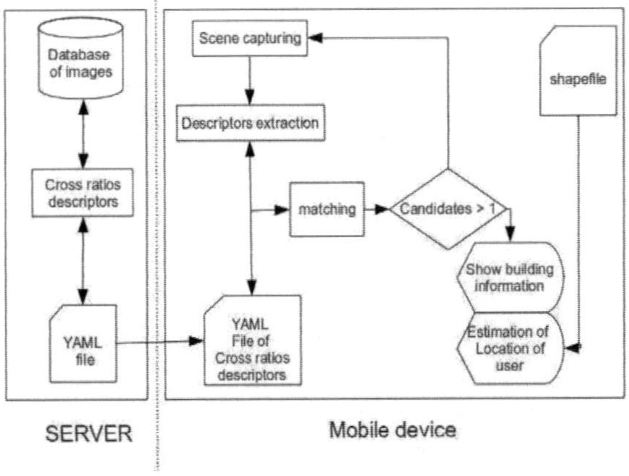

Figure 2. Flow chart diagram describing the system

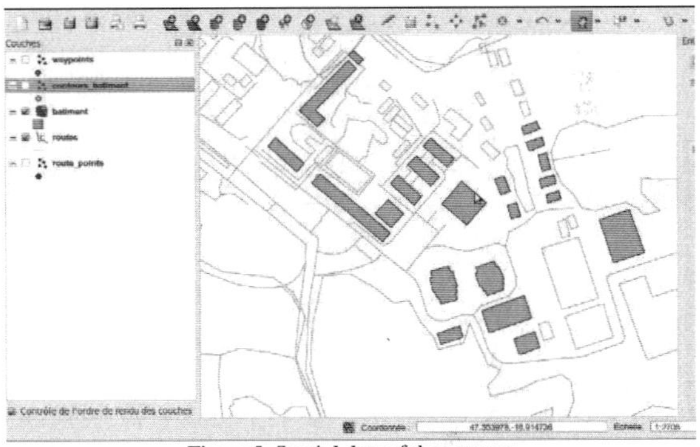

Figure 3. Spatial data of the campus

B. Descriptors extraction and images matching

Most of the time, buildings present a lot of linear structures resulted from windows, doors and facades' arrangement (Figure 4-a). We will use these linear structures to detect keypoints of buildings. To extract these keypoints, we follow this algorithm:

Step 1: Convert the image into greyscale and apply the Canny Edge detector [23] on that image (Figure 4-b).

Step 2: Apply Standard Hough line Transform [24] and keep only the lines in the vanishing directions (horizontal and vertical). Figure 4-c shows vertical (green) and horizontal (red) lines after Hough lines Transform.

Step 3: Extract the intersections of lines (blue dots in Figure 4-d) in different vanishing directions and keep them as relevant keypoints for that image.

From these keypoints, we compute the descriptor of each keypoint as the cross-ratio of four collinear points and store them in a matrix. We choose cross-ratio because it is a projective invariant. The cross ratios CR of four collinear points P1, P2, P3 and P4 is defined as follows:

$$CR = \frac{(x3 - x1)(x4 - x2)}{(x3 - x2)(x4 - x1)} \quad (1)$$

where x1, x2, x3 and x4 are respectively the values of x-coordinates of points P 1, P2, P3 and P4.

To match cross ratios descriptors of two images, we use Fast Library for Approximate Nearest Neighbour (FLANN) [9] based matching to find the best matches of descriptors in the scene. FLANN is enhanced by Random Sample Concensus [22] (RANSAC) algorithm to remove outliers. We build indexes using Locality Sensitive Hashing (LSH) [10] [11] which is robust in high dimension of data. With LSH, we can achieve faster matching.

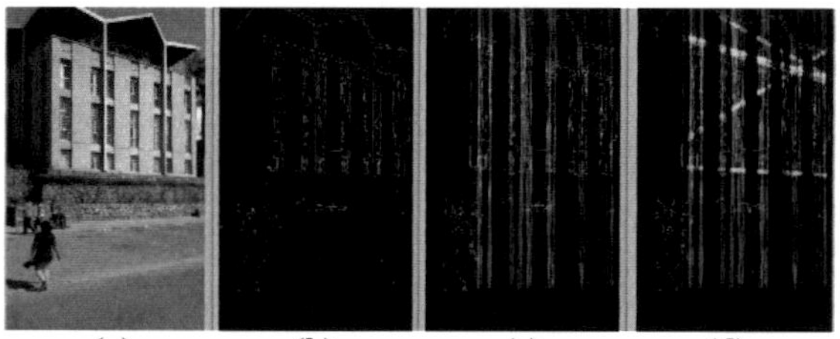

(a) **(b)** **(c)** **(d)**

Figure 4. Extraction of intersections of lines.

C. *First pose estimation from GIS query*

The list of intersections from the descriptor extraction step will serve as input to track the edges of the building. The coordinates of intersections are stored in a matrix called intersections matrix. This matrix is shown in Figure 5. We keep the external coordinates of intersections to mark the building location in the image. We query the name of the building into the GIS to find its position on the map. At this step, the approximated user position is estimated as near the facade of this building.

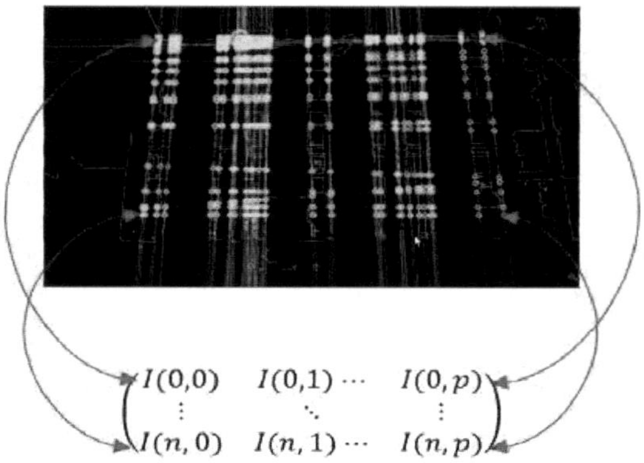

Figure 5. Matrix of coordinates of intersection points. Blue dots represent the extracted intersections of line in step 3.

D. *Camera pose estimation from camera parameters and GIS*

After retrieving the approximated user position, we would like to know his exact position by camera pose estimation. For that, we compute the homography of the facade extracted in the camera's screen and the facade in the GIS. We can compute the camera pose estimation by computing the camera intrinsic by this matrix as in [12]:

$$\begin{pmatrix} u \\ v \\ 1 \end{pmatrix} = A[RT] \begin{pmatrix} x_{wc} \\ y_{wc} \\ z_{wc} \\ 1 \end{pmatrix} \quad (2)$$

where A is the camera intrinsic matrix, *R* and *T* are the extrinsic parameters (Rotation and Translation matrix) of the camera, x_{wc}, y_{wc} and z_{wc} are world coordinates of one point and $[u\ v]^T$ are the coordinate of one point in pixel coordinates.

$$A = \begin{pmatrix} \alpha_x & \gamma & u_0 \\ 0 & \alpha_y & v_0 \\ 0 & 0 & 1 \end{pmatrix} \quad (3)$$

where γ is the skew (often we take 0 as its value), $[u_0, v_0]^T$ is called the principal point, usually the coordinates of the image center, α_x and α_y are the scale factor in the x and y coordinate directions, and are proportional to the focal length f of the camera:

$$\begin{cases} \alpha_x = k_x f \\ \alpha_y = k_y f \end{cases} \quad (4)$$

k_x and k_y are the number of pixels per unit distance in x and y directions.

The coordinates of camera in the world coordinates are given by:

$$C = -R^{-1}T \quad (5)$$

IV. Experience and Results

We perform our experiment with an Alcatel One Touch Pixi 3, a low cost smartphone, which has the following specifications: processor: MediaTek MT6572M - 1 GHz Dual Core, OS: Android 4.4.2 KitKat, RAM 512 Mbyte. Our database is composed of 100 images of buildings taken around a campus.

Here, we show some matching results of images in the scene and from the database: the images on the left are query images taken from smartphone, and the images on the right are those stored in the database. Here we prove that our cross-ratio descriptors are perspective invariant. Thus, the result image from the database can be the image of the same building but in different view point as we see in the first line of result images (Figure 6-b).

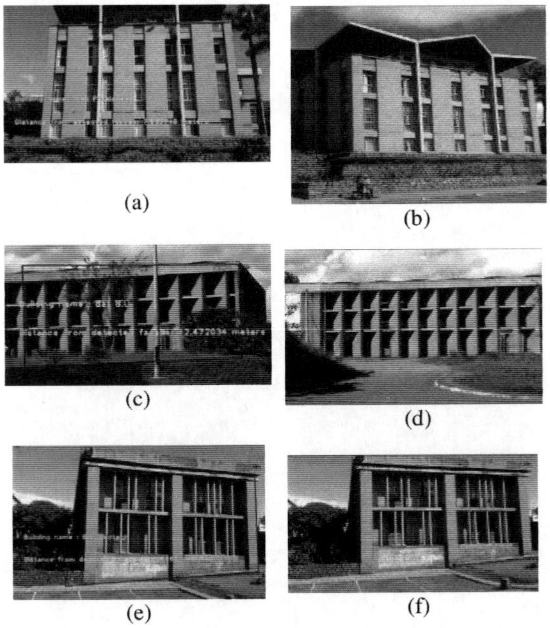

Figure 6. Some results of building localization

In addition, the building's façade is marked within a blue rectangle and the information about its position is printed on the screen. The time computation for matching 100 images in the database took about 120 milliseconds with this low cost smartphone. This demonstrates the accuracy of our approach for real-time application.

V. Conclusion and Future Work

In this paper, we present an urban image-based localization method in mobile devices using cross-ratios of feature points detected on the building's facades. These feature points are intersections of horizontal and vertical lines in the vanishing directions after applying Hough transform. All coordinates of intersections are kept in a matrix which will be used to detect the edges of building's facade. We compute the pose estimation of user in the world coordinates in keeping with length of facade in the GIS and length of the same facade in the smartphone's screen.

Our approach can be used for other images with linear structure such as trains or buses, in order to classify them. We can improve the method presented in this paper by segmenting the building in the image. In addition, we can use parallel computing and middleware to perform our technique in order to reach a better performance.

References

[1] W. Zhang and J. Kosecka, "Image Based Localization in Urban Environments," Proceedings of the Third International Symposium on 3D Data Processing, Visualization, and Transmission, June 2006, pp. 33-40.

[2] D. G. Lowe, "Object recognition from local scale-invariant features". Proc. 7th International Conference on Computer Vision (ICCV'99), Corfu, Greece, 1999, pp. 1150-1157.

[3] D. Robertson and R.Cipolla, "An image-based System for urban Navigation", In Proc. British Machine Vision Conference, Kingston, UK, 2004, pp. 819-828.

[4] C. G. Harris and M. Stephens. "A combined corner and edge detector". In Proc 4th Alvey Vision Conf, Manchester, 1988, pp. 147–151.

[5] C. Card and W. Hoff, "Qualitative Image-Based Localization in a Large Building", IPCV, 2015, pp. 338-344.

[6] E. Rublee, V. Rabaud, K. Konolige, and Gary Bradski, "ORB: an efficient alternative to SIFT or SURF", Computer Vision (ICCV), 2011 IEEE International Conference on. IEEE, 2011, pp. 2564-2571, doi: 10.1109/ICCV.2011.6126544.

[7] L. Carozza, F. Bosché, and M. Abdel-Wahab, "Image-based localization for an indoor vr/ar construction training system". CONVR, 2013.

[8] Y. Huang et al.,"Image-based Localization for Indoor Environment Using Mobile Phone",The International Archives of the Photogrammetry, Remote Sensing and Spatial Information Sciences, 2015, pp. 211-215.

[9] M. Muja and D. G. Lowe, "Fast Approximate Nearest Neighbors with Automatic Algorithm Configuration", In Proc. International Conference on Computer Vision Theory and Applications (VISAPP'09) (Lisbon, Portugal), 2009, pp. 331—340.

[10] P. Indyk and R. Motwani, "Approximate Nearest Neighbors: Towards Removing the Curse of Dimensionality", In Proceedings of 30th Symposium on Theory of Computing, 1998

[11] A. Andoni and P. Indyk, "Near-optimal hashing algorithm for approximate nearest neighbour in high dimensions", Communications of the ACM, Vol. 51, 2008

[12] P. R. S. Mendonca and R. Cipolla, "A Simple Technique for Self-Calibration", Computer Vision and Pattern Recognition (CVPR). IEEE Computer Society Conference on. (Volume:1), June 1999., pp. 500-505

[13] B. Johansson and R. Cipolla, "A system for automatic pose-estimation from a single image in a city scene", In IASTED Int. Conf. Signal Processing Pattern Recognition and Applications, Greece., 2002

[14] Haala, N., Böhm, J."A multi-sensor system for positioning in urban environments". ISPRS J PHOTOGRAMM, 58 (1-2), 2003 pp. 31-42. doi:10.1016/S0924-2716(03)00015-7.

[15] D. Robertson and R.Cipolla, "An image-based System for urban Navigation", In Proc. British Machine Vision Conference, Kingston, UK., pp. 819-828, 2004

[16] N. Bioret, G. Moreau and M. Servières , "Urban Localization based on Correspondences between Street Photographs and 2D Building GIS Layer",CORESA, Toulouse, France, 2009.

[17] N. Yazawa and H. Uchiyama, "Image based view localization system retrieving from a panorama database by surf", in Proc. of the IAPR Conference on Machine Vision. Applications, 2009, pp. 118-121.

[18] M. Donoser and D. Schmalstieg, "Discriminative Feature-to-Point Matching in Image-Based Localization", Conference CVPR , IEEE, 2014, pp. 516-523.

[19] B. Zeisl, T/ Sattler, and Marc Pollefeys, "Camera Pose Voting for Large-Scale Image-Based Localization",Conference ICCV, IEEE, 2015, pp. 2704-2712.

[20] H. Bay, T. Tuytelaars, and L. V. Gool, "SURF: Speeded up robust features", In Proc. ECCV, 2006, pp. 404–417.

[21] Quantum GIS, http://qgis.org/.

[22] M. A. Fischler and R. C. Bolles, "Random sample consensus: a paradigm for model fitting with applications to image analysis and automated cartography", Communications of the ACM, 1981, pp.381–395.

[23] J. Canny, "A Computational Approach to Edge Detection", IEEE Trans. Pattern Analysis and Machine Intelligence, 1986, pp.679–698, 1986.

[24] D. H. Ballard, "Generalizing the Hough Transform to Detect Arbitrary Shapes". Pattern Recognition, 1981, pp. 111–122, doi:10.1016/0031-3203(81)90009-1.

A Method for Calculating Shape Similarity among Trajectory of Moving Object Based on Statistical Correlation of Angular Deflection Vectors

Alexandre Altair de Melo
Glaucio Scheibel
and Fabiano Baldo

Computer Science Department
Santa Catarina State University (UDESC)
Joinville, Santa Catarina, Brazil
Email: {dcc6am, dcc6gs}@joinville.udesc.br
Email: fabiano.baldo@udesc.br

Fernando José Braz

Informatic Department
Federal Institute Catarinense (IFC)
Araquari, Santa Catarina, Brazil
Email: fernando.braz@ifc-araquari.edu.br

***Abstract*—In recent years, mobile electronic devices, specially smartphones, are gaining more attention in people's daily life. These devices provide many features that often process information gathered by one of their several built-in sensors. Among them, one of the most popular is the Global Positioning System (GPS) receiver. This sensor allows the systematic and chronological collection of location information that represents the trajectory of the moving object that carries it. Trajectories are considered valuable sources of information for analysing and understanding of moving objects' behavior. There is a considerable number of researches that analyse objects' trajectories. Among them, the identification of trajectories' similarity is one where researches are currently being developed. The similarity of trajectories may indicate common behaviors inside groups of individuals that can be useful in various application areas. However, how to measure the trajectories' similarity even though they are in different directions and far from each other? Trying to solve this problem, this paper proposes a method to calculate similarity among trajectories applying statistical correlation over their vectors of angular deflections. Preliminary results indicate that the method can identify similarity in shape among trajectories. However, when their shapes are too complex, it can not reach suitable results.**

Keywords*—*Moving Objects; Trajectory; Similarity; Correlation Statistics.

I. Introduction

In the last years, mobile electronic devices have gained an increasing importance in peoples daily life. Among all the types of existing mobile devices, smartphones are the most widespread. Their use is pushed by the amount of functions that they provide. These features are performed by applications that often process information collected by one of its several built-in sensors. Concerning the sensors that they have, one of the most popular is the Global Positioning System (GPS) [1]. This sensor allows individuals that carry mobile devices to record their movement in time. The chronological collection of location information can represent the trajectory of an object [2]. Trajectories are considered valuable sources of information for analysing and understanding the behavior of moving objects. An increasing volume of trajectories makes it possible to find patterns inside the movement of an individual or group of individuals. Research areas such as bio-monitoring, logistics, navigation systems, car and pedestrians traffic planning are examples of trajectory analysis applications.

Commonly, a trajectory (T) is represented by one id (T_{id}) and a set of points composed of x, y and t values, where x and y are geographic coordinates, and t is a timestamp [3], [4]. Additionally, a trajectory and its points can be enriched with information like: speed, direction, acceleration, etc. [5], [6]. In order to gather such information, it is necessary to use other sensors instead of only GPS. But in the case of smart phones, it is not a big deal considering its number of embedded sensors.

The moving objects' trajectories data is a vast source for data analysis and its analysis can bring direct contribution to people's life. There is a considerable number of researches that analyse objects' trajectories in order to find answers for phenomena observed into the cities' environment. Among them, the identification of trajectories' similarity is one where researches are currently being developed. Works like those carried out by [7]-[14] present approaches for identifying similarity considering some trajectory aspects.

The similarity of trajectories' shape may indicate common behaviors inside groups of individuals that can be useful in various application areas like identifying drivers behaviour, cargo stolen, dangerous places and so forth.

Each proposed approach presents a specific set of metrics to identify similarity among trajectories. However, according to Pelekis et al. [12], trajectories can be considered similar if some of the following aspects are true: (i) fully or partially overlapping in space; (ii) similar shapes in different places; (iii) same start and/or end position; (iv) partially or fully synchronous movement behavior; (v) or totally separated in time, but with similar dynamic behavior (speed, acceleration, etc.). This work focuses on the aspect (ii) considered by [12], where the similarity of shape is used as criterion for defining trajectories similarity. However, shape is just one of the aspects that must be tackled when dealing with trajectory similarity as a whole. It must be combined with time and length characteristics, as well as with extra semantic aspects, when possible.

In the literature, there is a considerable number of works that propose different metrics to calculate the shape similarity among trajectories. Examples include works conducted by [9], [13] who present solutions to identify the degree of similarity for sets of nearby trajectories. Another example is the work of [15], which presents a compendium of distance metrics used to evaluate similarity. Another research [16] proposes a qualitative trajectory calculus (QTC^c), to find distances and directions of trajectories.

Despite their important contribution, none of the related works address the problem of identifying shape similarity between trajectories that are far from each other or that have different moving direction. For example, trajectories that are in different cities or even in different countries and from objects moving from West to East and objects moving from South to North. There are several situations where this proposal can be useful. The dissemination of diseases could be represented by a trajectory. In this case, the behaviour of the similarity trajectories considering the shape of trajectories, regardless of orientation, could be a very important knowledge in order to take actions to interrupt the dissemination of the diseases. Another possible context of the use of similarity, regardless of orientation, is the observation of behaviour of a set of individuals of an animal species. The similarity of trajectories independent of orientation could be an indicator of an occurrence of a particular event in that group. In this case, the same event could be identified in different regions, regardless of distance and orientation. A damaged ship also is an example; the movement of a ship in this situation has some characteristics that could be identified in other trajectories. In this case the shape of the trajectory is more important than its orientation.

Therefore, this paper tackles the following problem: How to measure the trajectories' shape similarity even though they are in different directions and far from each other? Trying to solve this problem, this paper proposes a method to calculate the shape similarity among trajectories applying statistical correlation over their angular deflections.

The measurement of geometry angles can be encompassed into the topography area, where azimuth (*AZ*) and deflection (*DF*) angles are considered the basic measures to calculate segment orientation. Considering that a trajectory is a geometry formed by several points, after calculating the deflection of each one of its segments, this will result in a sequence of $n - 1$ deflection angles, where n is the number of points in the trajectory. Having the sequence of angular deflections of both compared trajectories, it is possible to reduce the problem of calculating shape similarity of two trajectories by calculating the statistical correlation coefficient of both deflection sequences. Statistical correlation analysis is a discipline that aims to measure the coefficient of relationship or association between two variables [17].

Therefore, this work addresses the hypothesis that it is possible to identify the shapes similarity between two trajectories calculating the statistical correlation coefficient over the sequences of segment deflection (*DFV*) from both trajectories.

The paper is structured as follow: Section II reviews the related work. Section III lists definitions concepts found in the literature. Section IV lists the steps of the proposed method. Section V reports the experiments. Section VI presents the findings and results of the experiments. Finally, Section VII provides a view of future works.

II. Related Works

Shape similarity between trajectories is a field with many challenges and several works have proposed solutions to this problem, as mentioned in section I. In this section, we detail some approaches found in the literature.

Vlachos et al. [7] propose the longest common subsequence (LCSS). Its main idea is to match two sequences by allowing them to stretch, without rearranging the sequence of the elements, and allowing some elements to be unmatched. This method has reached great effectiveness in the presence of noise. However, it does not penalize unmatched sub-sequences, given no information of how to separate the unmatched subsequences. In addition, its original concept does not consider the direction, and may fail to separate two trajectories near in space with very different directional behavior.

Chen et al. [8] propose the Edit Distance on Real Sequences (EDR) function. This function is based on the Edit Distance Function which has been used to quantify the similarity between two strings. Given two strings, the Edit Distance function calculates the minimum number of insertions, deletions and replacements needed in order for both to become identical. Like LCSS, this function also assumes that the trajectories have the same length and sampling rate.

Van de Weghe et al. [16] propose the Qualitative Trajectory Calculus (QTC_c), that is a qualitative approach to represent two vectors by means of a 4-tuple representing the orientation of both vectors with respect to each other. The relative movement of two objects are represented by a four-component label, where the first two components describe the tendency of distance changing of an object to the current position of another object, and the other two components describe the relative orientation of the object movements with respect to the reference line that connects them. One problem of this approach is that it does not present a quantitative measure of similarity. Another point that needs to be considered is the time consumed caused by calculating the Shape Matrix for every trajectory to be compared.

Frentzos et al. [9] propose a Dissimilarity Metric (DISSIM). The DISSIM between two trajectories Q and R is calculated by the integration of their Euclidean Distance over a definite time interval when both Q and R are valid. So it takes into account the time dimension in both trajectories. Moreover, DISSIM can be used for trajectories with different sampling rates, because non-recorded points are approximated by linear interpolation. Its main drawback is the high computation consumption.

Dodge et al. [10] developed a conceptual and methodological framework focused on the analysis of similarities in dynamic behavior of moving objects. They also proposed to pre-process the data, resampling the data to a regular time interval, using linear interpolation of fixed time intervals. It means that the authors concentrate on tracks, rather than sample points, and the method is limited to a fixed sampling period.

Liu and Schneider [11] proposed an approach to calculate the similarity of trajectories that not only considers the geographical issues, but also the semantic aspects of the trajectories' movement. For the geographical part, the authors consider the following aspects: bearing, distance between trajectories, center of mass, smaller distance between the initial and final point of the trajectory and angle cosine to find subtrajectories. The problem is that the method for calculating the semantic similarity depends on the similarities obtained in the geographical part.

Pelekis et al. [12] define a method that groups trajectories using various distance functions such as GenLIP, GenSTLIP and others. Based on motion properties such as spatial location, speed, acceleration and direction, the similarity is calculated. This work use a clustering approach to group trajectories.

Sankararaman et al.[13] present a framework to rate the trajectories according to their similarities based on distance. To calculate this similarity, they use the following algorithms: DTW (Dynamic Time Warping), euclidean distance and direction of the segment. Their main contribution is to find equal and not equal parts of the trajectories. The authors include the time as an extra dimension, allowing their model to be extended to spatio-temporal data.

Xie [14] proposes a metric to calculate the distance called Edit Distance on Segment (EDS). This metric is used to check the similarity in sub-trajectories using their segments. To calculate this similarity metric the authors define the cost of a segment-wise transformation, i.e., the cost of changing a segment to another one. The idea is that given two segments it is possible to transform them by displacing, stretching and rotating properly, in order to identify the similarity of sub-trajectories.

Concerning the reviewed works, it is possible to see that most of them also assume that the trajectories should have the same number of points to perform the shape comparison. Also, some of them replace points as well as estimate approximated points in order to perform the comparison. Besides that, other works produce qualitative results, instead of quantitative ones. Finally, even using bearing and azimuth, none of the reviewed works identifies shape similarity among trajectories in different directions and far from one another.

III. Basic Concepts and Definitions

This section details concepts and definitions about trajectories, angular measurements and statistical correlation, based on the following works [17], [18], [19], [20], [21].

A. Trajectories of moving objects

Below, we present the definitions concerning the trajectory of moving object.

Definition 1: A coordinate (c) is a tuple (x, y), where x is a latitude and y is a longitude. A coordinate defines a georeferenced position on the earth surface.

Definition 2: A point (p) is a tuple (c, t), where c is a coordinate and t is a time-stamp that represents the time when the coordinate c has been taken.

Definition 3: A trajectory (T) is composed of a sequence of points and can be defined as $T = [p_i, p_{i+1}, ..., p_n]$, where p_i is the start point, p_n is the end point, $p_i < p_{i+1}$ and n is the number of points.

Definition 4: A segment (S) is a sequence $S = [p_i, p_{i+1}, p_{i+2}, ..., p_f]$, where p_i is the initial point of the segment, p_f is the final point of the segment, $0 \leq p_i < p_f$, $p_i < p_f \leq p_n$ and $p_i < p_{i+1}$, n is the number of trajectory points. It means that $S \subset T$.

B. Angular measurements

Below, we present the definitions of angular measurements used in this work.

Definition 5: Bearing is the angle formed between the North-South meridian and a line to West or East. It varies from $0°$ to $90°$. In order to represent the bearing direction, it is necessary to define in which quadrant it is placed: North-West (NW), North-East (NE), South-East (SE), South-West (SW) [20].

Definition 6: Azimuth is the angle that begins on North and turns clockwise until it reaches the desired line. It varies from $0°$ and $360°$. Instead of bearing, azimuth does not indicate the direction of the line because this is implicit [20].

The azimuth can be obtained by (1).

$$AZ_i = \arctan 2 \left(\frac{\sin \Delta\lambda . \cos \phi_{i+1}}{\cos \phi_i . \sin \phi_{i+1} - \sin \phi_i . \cos \phi_{i+1} . \cos \Delta\lambda} \right) \tag{1}$$

Where $\Delta\lambda = (\lambda_i - \lambda_{i+1})$, and λ and ϕ indicate the longitude and latitude of a point p_i, respectively. An azimuth vector is a sequence of azimuths such as $AZV = [AZ_i, AZ_{i+1}, AZ_{i+2}, ..., AZ_j]$, where $0 < i < j$ and $i < j \leq n - 1$, n is the number of points in the trajectory.

Definition 7: Deflection angle is calculated by the difference between the azimuths of two consecutive lines. It varies from $0°$ to $\pm 180°$. It is positive if the azimuth of the first line was greater than the second one and vice-versa [20].

The deflection can be obtained by (2).

$$DF_i = (AZ_i - AZ_{i+1}) \tag{2}$$

Where AZ_i represents the azimuth of a point inside the trajectory and AZ_{i+1} is the azimuth of the subsequent point in the same trajectory. A deflection vector is a sequence of deflections such as $DFV = [DF_i, DF_{i+1}, DF_{i+2}, ..., DF_k]$, where $0 < i < k$ and $i < k \leq j - 1$, j is the number of azimuths in the trajectory.

C. Statistical Correlation Coefficients

Correlation is a bi-variate analysis that measures the statistical relationships between two variables. The value of the correlation coefficient may vary between $+1$ and -1, where 0 indicates no correlation, and $+1$ and -1 indicate positive and negative correlation, respectively [17]. Nevertheless, when the coefficient varies between ± 0.10 and ± 0.29, it indicates weak correlation; between ± 0.30 and ± 0.49, medium correlation; and between ± 0.50 and ± 1, strong correlation [21]. Among the coefficients found in the literature, Pearson (*r*) [22], Spearman (*rho*) [23] and Kendall (τ) [24] are the most remarkable ones.

In statistics, p-value (p) represents the probability of a statistical test be considered valid, instead of random. It defines the probability to reject the null hypothesis (H_0) [21]. The level of significance is represented by α and it has the following general rule: if $p > \alpha$ then H_0 is accepted, but if $p \leq \alpha$ then H_0 is rejected.

IV. Proposed Method

As mentioned before, this work assumes that the shape similarity between two trajectories can be identified through the application of statistical correlation coefficient in their deflection vectors. This assumption is based on the idea that if the trajectories have similar shapes, then they would have similar deflection angles (DF) between their segments.

The method for identifying trajectories with similar shape is composed of the following steps:

1) **Selecting the reference trajectory:** The first step is to select which trajectory (or trajectory segment) to use as reference for comparing its shape with the others;
2) **Selecting the set of compared trajectories:** After that, it is necessary to select the set of trajectories (or trajectories segments) to be used in shape comparison with the reference one.
3) **Compacting the compared trajectories:** As each selected trajectory (or segment) must have the same number of points as the reference one, it is necessary to compact them using a compression algorithm that maintains a predefined and fixed memory size of points in the compressed trajectory. Examples of algorithms that use this approach are Spatiotemporal Trace (STTrace) [25] and Spatio Quality Simplification Heuristic (SQUISH) [26].
4) **Segmenting the trajectories:** For computing the azimuths and deflections, every trajectory should be segmented into segments composed of two points. So, for each trajectory, an array $S = [s_1, s_2, ..., s_s]$ of segments will be generated, where $0 < s \leq n - 1$ and n is the number of trajectory points.
5) **Computing azimuths for trajectories segments:** For every trajectories segment we calculate its azimuth using (1). The azimuth is used as requirement to compute the deflection that is the main value used to identify trajectories shape similarity. This information is stored in an array of j positions, where $j = s$ and s is the number of trajectory segments.
6) **Computing deflection between two consecutive trajectory segments:** After calculating the segments azimuth it is possible to compute the trajectory segments deflection using (2). In order to decrease their variance, the values of deflection are modularized. This information is stored in an array of k positions, where $k = j - 1$ and j is the number of trajectory azimuths.
7) **Applying statistical correlation coefficient:** Having the deflection vectors it is possible to calculate the correlation coefficient between the reference trajectory and each one presented in the compare set (two by two). In this work, we applied the following correlation methods: Pearson (r) [22] and Spearman (rho) [23].

V. Experiments

To test the proposed method, we created four scenarios, one with synthetic data manually made (Figure 1) and three with real data collected in the city of Joinville - Brazil (Figure 2, Figure 4 and Figure 6). The collected data were produced by Costa and Baldo [27] in a work aimed at generating digital road maps. Three scenarios are used to assess whether the method can identify trajectories similarity in shape and one is used to assess whether the method does not identify similarity when the trajectories are considerably different in shape. This last scenario was proposed to ensure that the method rejects the H_0.

Based on the literature review, the ideal scenario would reach -1 or +1 correlation, with a p-value (p) ≤ 0.05, and this value of p, is a possible value to discard the null hypothesis (H_0), since the chance of the α error is small. However in these experiments it has been assumed that two trajectories are considered similar when the *correlation coefficient* is ≥ 0.30 and the *p-value* is ≤ 0.10. The values for these parameters are more flexible, and this is why we consider several scenarios of different complexity for analysis. Looking at similarity based on the correlation, the medium values are between ± 0.30 and ± 0.49 and strong values are between ± 0.50 and ± 1 [21].

All compared trajectories have the same size and a minimum of 10 points. This size has been chosen in order to produce a valid statistical analysis, because this is a requirement for statistical correlation, where the analyzed variables must be the same size. At last, in order to reach better linear distribution among the deflection values, they were normalized as follows: $T = [|DF_1|, |DF_2|..., |DF_k|]$, where k is the number of trajectory deflections.

The first experiment, presented in Figure 1, uses data manually created and tries to depict the ideal case where two trajectories, even though in different orientation (one North-South and another West-East) and relatively far from one another, with similar segment deflections should have high correlation coefficient and thus high similarity using a statistical correlation approach. As seen in Table I, this experiment reaches 0.99 of Pearson correlation with 0.01 of p-value which means that they can be considered equal in shape.

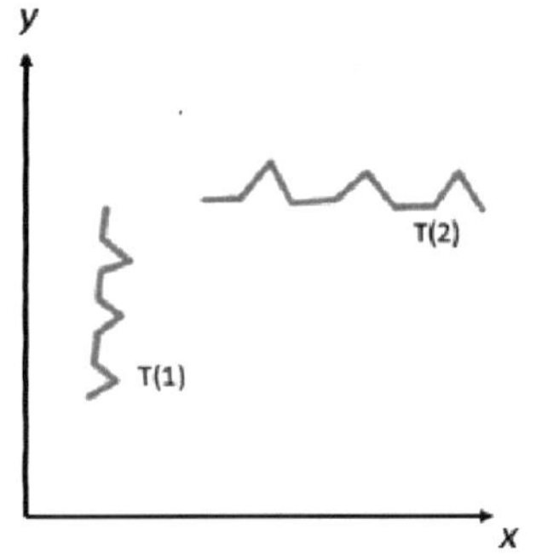

Figure 1. Experiment 1, same shapes and different directions, T_1 is the reference trajectory.

TABLE I. EXPERIMENT RESULTS.

Experiment	Points	Pearson	Spearman	p-value
1	10	0.99	0.83	0.01
2	11	0.38	0.63	0.07
3	32	0.08	0.24	0.19
4	33	-0.03	0.02	0.88

The second experiment presented in Figure 2 has been created based on trajectories extracted from the sample presented in Figure 3. This experiment tries to analyse if the method can be applicable and reaches high correlation value (similarity) comparing real trajectories collected by GPS receivers. As can be seen in Table I, this experiment reaches 0.63 of Spearman correlation with a p-value of 0.07. It means that the trajectories have relatively high correlation, so it would be said that they have similar shapes too, as expected.

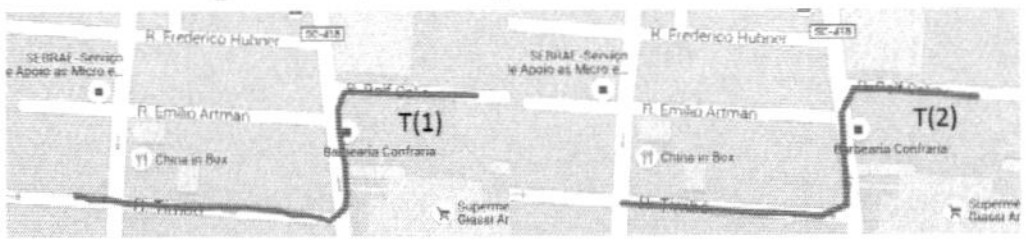

Figure 2. Experiment 2, same shapes and directions, T_1 is the reference trajectory.

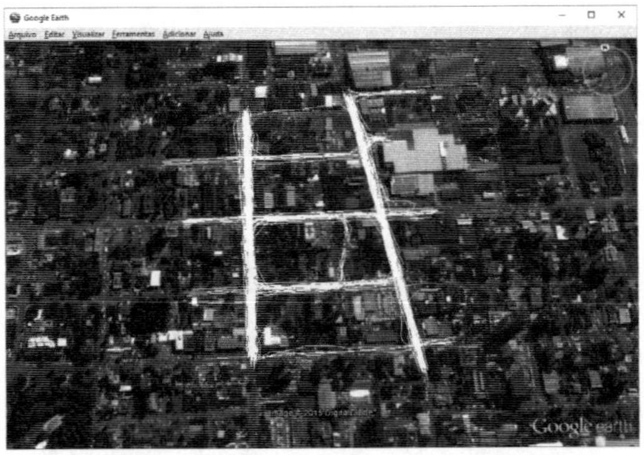

Figure 3. Trajectories collected in downtown.

The third experiment presented in Figure 4 has been created based on trajectories extracted from the sample presented in Figure 5. This experiment tries to analyse if the method can reach high correlation value (similarity) comparing trajectories with complex shapes as, for instance, a turn of 360°. As can be seen in Table I, this experiment reaches only 0.08 of Pearson and 0.24 of Spearman, correlation with a p-value of 0.19. It means that not only they do not have correlation, but also that the H_0 can not be rejected, so this result can be considered aleatory. Analysing this results it is possible to see that the proposed method can not be applied to identify similarity to every kind of trajectory's shape. Depending on the shape complexity it would not reach the expected result, even with trajectories that have visually similar shapes.

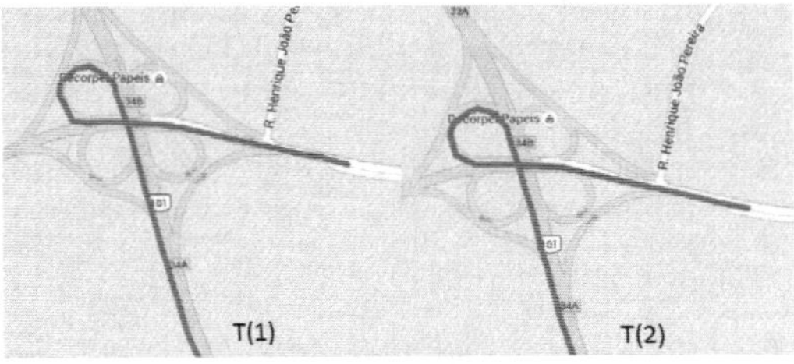

Figure 4. Experiment 3, same complex shapes and directions, T_1 is the reference trajectory.

The fourth experiment presented in Figure 6 has been created based on trajectories extracted from the sample presented in Figure 3. This experiment tries to analyse if the method can recognize when two trajectories do not have similarity in shape. This situation is represented by low correlation (value near to 0) and high p-value (value near to 1). As can be seen in Table I, this experiment results in a -0.03 Pearson value and a 0.02 Spearman value with a p-value of 0.88. It means that the trajectories do not have correlation among their deflection

Figure 5. Trajectories collected in the Joinville Industrial District.

vectors which ensure that they are not similar in shape. Besides that, as the p-value is high, it is not possible to reject the H_0 (null hypothesis), which contribute to ensure their no similarity in shape.

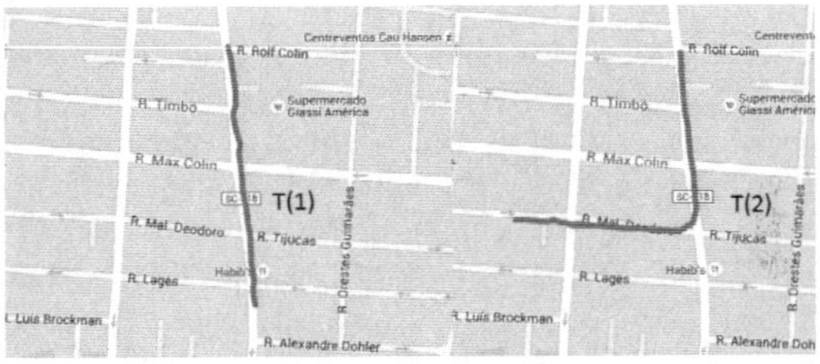

Figure 6. Experiment 4, different shapes and directions, trajectories T_1 is the reference trajectory.

VI. Result Analysis

Analysing the results showed in Table I, it is possible to see that the proposed method has potential to identify the similarity between trajectories. This can be answered by the experiments 1 and 2. However, it can not be applied in every trajectory shape with suitable results, as presented in experiment 3. This occurs because the array of trajectory deflections does not follow a linear distribution and this is a requirement for applying statistical correlation methods.

By nature, statistical correlation methods work with linear distributed variables. It means that if one variable is increasing the other one should have the same behavior in order to present high correlation coefficient. Concerning the two applied correlation methods, Pearson suffers more impact in its results than Spearman due to the fact that variables do not have linearity. This can be seen in Table I, experiments 2 and 3, where Spearman reaches high correlation values. This is explained because Spearman takes into account not only the variables linear distribution, but also the similarity of values in the same position in both variables. So, for Spearman even if the variables are not linearly distributed, if they have similar values in the same position they can be considered similar.

Although positive results have been obtained using Spearman, as this is an ongoing work, it was not decided which one to elect as the statistical method to be adopted. The next steps will include a bench of massive tests where it will be decided which method to adopt as well as which range of correlation values will be considered enough to ensure similarity in shape among trajectories.

VII. Conclusion

Trajectories are considered valuable sources of information for analysing and understanding of moving objects' behavior. Despite the considerable progress concerning the measurement of trajectories' shape similarity, the literature does not present a method able to measure similarity among trajectories in different directions and/or far from each other.

The main goal of this research is to develop a mechanism to calculate similarity among trajectories applying statistical correlation over their vectors of angular deflections. Another objective is to use correlation methods in order to calculate the level of similarity. Considering the preliminary results it is possible to affirm that the proposal is able to find similarity trajectories considering the angular deflexions. However, the developed work is not conclusive in order to identify the more adequated correlation method to analyze those sets of angular deflexions. By using the proposal to analyze complex shapes the method does not reach suitable results. This can be explained by the non-linearity of the deflections array.

As future work, we plan to start a bench of massive tests in order to decide which statistical method to adopt (Pearson or Spearman), as well as which range of correlation values to consider suitable to ensure similarity in shape among trajectories. A very important point of research is the improvement of data quality. Actually, current research does not check the relationship between data dispersion level and accuracy of the results of similarity. This investigation, considering several different compression algorithms in order to decrease the data dispersion, than those mentioned in this article, is an additional point to execute in the future works.

Furthermore, the presented approach does not consider other characteristics in order to calculate similarity. However, this proposal is just a piece of a method to find groups of similar trajectories. The method considers two additional characteristics: spent time and length of trajectories. Therefore, the proposal of the future method is to find groups of similar trajectories considering several gradients of shape, duration and length of the trajectories.

References

[1] F. Giannotti et al., "Unveiling the complexity of human mobility by querying and mining massive trajectory data," The VLDB JournalThe International Journal on Very Large Data Bases, vol. 20, no. 5, 2011, pp. 695–719.

[2] S. Mehta, R. Machiraju, and S. Parthasarathy, "Towards object based trajectory representation and analysis," OSUCISRC-03/06-TR30, Tech. Rep., 2006.

[3] G. Andrienko et al, "Space, time and visual analytics," International Journal of Geographical Information Science, vol. 24, no. 10, 2010, pp. 1577–1600.

[4] E. Frentzos, Y. Theodoridis, and A. N. Papadopoulos, "Spatio-temporal trajectories," in Encyclopedia of Database Systems. Springer, 2009, pp. 2742–2746.

[5] C. Parent et al., "Semantic trajectories modeling and analysis," ACM Computing Surveys (CSUR), vol. 45, no. 4, 2013, p. 42.

[6] V. Bogorny, C. Renso, A. R. Aquino, F. Lucca Siqueira, and L. O. Alvares, "Constant–a conceptual data model for semantic trajectories of moving objects," Transactions in GIS, vol. 18, no. 1, 2014, pp. 66–88.

[7] M. Vlachos, G. Kollios, and D. Gunopulos, "Discovering similar multidimensional trajectories," in Data Engineering, 2002. Proceedings. 18th International Conference on. IEEE, 2002, pp. 673–684.

[8] L. Chen, M. T. Özsu, and V. Oria, "Robust and fast similarity search for moving object trajectories," in Proceedings of the 2005 ACM SIGMOD international conference on Management of data. ACM, 2005, pp. 491–502.

[9] E. Frentzos, K. Gratsias, and Y. Theodoridis, "Index-based most similar trajectory search," in Data Engineering, 2007. ICDE 2007. IEEE 23rd International Conference on. IEEE, 2007, pp. 816–825.

[10] S. Dodge, R. Weibel, and E. Forootan, "Revealing the physics of movement: Comparing the similarity of movement characteristics of different types of moving objects," Computers, Environment and Urban Systems, vol. 33, no. 6, 2009, pp. 419–434.

[11] H. Liu and M. Schneider, "Similarity measurement of moving object trajectories," in Proceedings of the Third ACM SIGSPATIAL International Workshop on GeoStreaming. ACM, 2012, pp. 19–22.

[12] N. Pelekis, G. Andrienko, N. Andrienko, I. Kopanakis, G. Marketos, and Y. Theodoridis, "Visually exploring movement data via similarity-based analysis," Journal of Intelligent Information Systems, vol. 38, no. 2, 2012, pp. 343–391.

[13] S. Sankararaman, P. K. Agarwal, T. Mølhave, and A. P. Boedihardjo, "Computing similarity between a pair of trajectories," arXiv preprint arXiv:1303.1585, 2013.

[14] M. Xie, "Eds: a segment-based distance measure for sub-trajectory similarity search," in Proceedings of the 2014 ACM SIGMOD international conference on Management of data. ACM, 2014, pp. 1609–1610.

[15] K. Toohey and M. Duckham, "Trajectory similarity measures," SIGSPATIAL Special, vol. 7, no. 1, 2015, pp. 43–50.

[16] N. Van de Weghe, G. De Tré, B. Kuijpers, and P. De Maeyer, "The double-cross and the generalization concept as a basis for representing and comparing shapes of polylines," in On the Move to Meaningful Internet Systems 2005: OTM 2005 Workshops. Springer, 2005, pp. 1087–1096.

[17] P. Y. Chen and P. M. Popovich, Correlation: Parametric and nonparametric measures. Sage, 2002, no. 137-139.

[18] F. J. Braz and S. Orlando, "Trajectory data warehouses: Proposal of design and application to exploit data." in GeoInfo. Citeseer, 2007, pp. 61–72.

[19] Y. Zheng, L. Zhang, X. Xie, and W.-Y. Ma, "Mining interesting locations and travel sequences from gps trajectories," in Proceedings of the 18th international conference on World wide web. ACM, 2009, pp. 791–800.

[20] B. F. Kavanagh and S. G. Bird, Surveying: Principles and applications. Prentice Hall Upper Saddle River (NJ), 2000.

[21] J. Cohen, Statistical power analysis for the behavioral sciences. Academic press, 2013.

[22] C. Spearman, "Demonstration of formulae for true measurement of correlation," The American Journal of Psychology, 1907, pp. 161–169.

[23] ——, "The proof and measurement of association between two things," The American journal of psychology, vol. 15, no. 1, 1904, pp. 72–101.

[24] M. G. Kendall, "A new measure of rank correlation," Biometrika, 1938, pp. 81–93.

[25] M. Potamias, K. Patroumpas, and T. Sellis, "Sampling trajectory streams with spatiotemporal criteria," in Scientific and Statistical Database Management, 2006. 18th International Conference on, 2006, pp. 275–284.

[26] J. Muckell, J.-H. Hwang, V. Patil, C. T. Lawson, F. Ping, and S. S. Ravi, "Squish: An online approach for gps trajectory compression," in Proceedings of the 2Nd International Conference on Computing for Geospatial Research & Applications, ser. COM.Geo '11. New York, NY, USA: ACM, 2011, pp. 13:1–13:8. [Online]. Available: http://doi.acm.org/10.1145/1999320.1999333

[27] G. H. Costa and F. Baldo, "Generation of road maps from trajectories collected with smartphone–a method based on genetic algorithm," Applied Soft Computing, vol. 37, 2015, pp. 799–808.

Advanced Association Processing and Computation Facilities for Geoscientific and Archaeological Knowledge Resources Components

Claus-Peter Rückemann
Westfälische Wilhelms-Universität Münster (WWU),
Leibniz Universität Hannover,
North-German Supercomputing Alliance (HLRN), Germany
Email: ruckema@uni-muenster.de

***Abstract*—Creating sustainable multi-disciplinary knowledge resources and enabling advanced features for processing of associations is one of the major goals of long-term knowledge development and discovery. This paper presents the main results from the development of resources allowing the use of advanced association processing and computation facilities. With the support from such resources, the paper also presents respective association processing results on exploiting the geoscientific and archaeological knowledge resources components. The practical application scenario is based on content from a natural sciences and archaeology research and studies campaign at the ancient city of Kameiros, Greece. The created resources are providing content, structures, and features for exploiting computation facilities, especially a multitude of reference types. The focus is to support knowledge resources with a set of features, which allow to extend and exploit long-term content discovery and gain new insights.**

Keywords–Knowledge Discovery; Association Processing; Scientific Knowledge Resources; Universal Decimal Classification; Advanced Computing.

I. INTRODUCTION

This paper presents the research conducted for creating knowledge resources and developing application components for supporting and providing advanced integrated systems for geoscientific, multi-disciplinary, and multi-lingual application scenarios. Existing data collections, unstructured and structured, combine a number of insufficient features and drawbacks, missing long-term aspects, support for multi-disciplinary conceptual knowledge, for classification, and for advanced and fuzzy methods like associations.

The purpose of the developed resources and components is to provide advanced knowledge object features, especially association processing features and computation in context with long-term multi-disciplinary and multi-lingual knowledge documentation and discovery. The new resources and application developments presented here are based on selected frameworks and resources, which have been created over the last two decades. The knowledge resources and Collaboration house framework [1] allowed for the implementation of multi-disciplinary, long-term knowledge resources and application components, for dynamical use as well as for complex and high end computation. The resulting components are used for universal and consistent documentation of knowledge and scientific research, and for consequent long-term purposes. These components are created using a universal classification [2], a flexible and portable all-purpose programming environment [3], and appropriate international standards [4].

In this case, for advanced association processing, new workflows had to be created and dynamically integrated into the framework components. Such implementation is possible if on the one hand the components' workflows allow a flexible integration of workflows, e.g., via scripting and compiled sources and on the other hand that structured knowledge resources can be extended for allowing a multitude of references types. The combination allows to create associations by making use of the available structures, processing, and computation facilities. For these purposes the object and media knowledge resources and the framework components were basically extended to support a data-centric approach.

This paper is organised as follows. Section II presents the state of the resources and frameworks. Sections III and IV introduce the new integration of workflows and reference types of the knowledge resources. Sections V and VI discuss the creation and processing of associations and how computational facilities can be exploited. Sections VII presents a geosciences and archaeology case study and implementation. Sections VIII and IX give an evaluation, present the main results and summarise the lessons learned, conclusions and future work.

II. STATE OF RESOURCES AND FRAMEWORKS

The resources and implementations are based on three major components: An architecture framework, long-term, multi-disciplinary knowledge resources, and a mostly widely used universal classification framework. The architecture implemented for an economical long-term strategy is data-centric and based on development blocks. Figure 1 shows the three main columns: Application resources, knowledge resources, and originary resources. The central block in the "Collaboration house" framework architecture [5], is represented by the knowledge resources, scientific resources, object collections, containers, databases, and documentation (e.g., LX [6], collections, containers). These resources provide multi-disciplinary content, context, and references, including structured and unstructured data, factual and conceptual knowledge.

The resources also refer to originary resources and sources (e.g., textual data, media data, photos, scientific data, literature). The knowledge resources are used as a universal component for compute and storage workflows. This feature can also be applied for supporting dynamical and ontology-based multi-agent, e.g., for production management as with the implementation supported by the European Framework Programme 7 (FP7) [7]. Application resources and components (Active Source, Active Map, local applications) are implementations for analysing, utilising, and processing data and making the information and knowledge accessible. The related information, all data, and algorithm objects presented are copyright the author of this paper, LX Foundation Scientific Resources [6], all rights reserved. The structure and the classification references based on the LX resources and UDC, especially mentioning the well structured editions [2] and the multi-lingual features [8], are essential means for the processing workflows and evaluation of the knowledge objects and containers. Both provide strong multi-disciplinary and multi-lingual support.

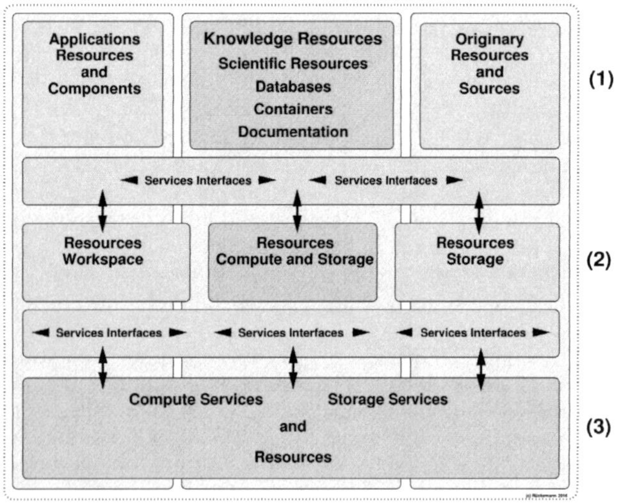

Figure 1. Architecture: The knowledge resources are the central component within the long-term architecture. Three major layers labelled (1), (2), (3).

The three blocks are supported by services' interfaces. The interfaces interact with the physical resources: In the local workspace, in the compute and storage resources where the knowledge resources are situated, and in the storage resources for the originary resources. The layers or 'levels' are labelled (1), (2), and (3) within the architecture. (1) is associated with the disciplines creating and using knowledge resources, application resources, and originary resources, 'realia'. (2) is associated with the tasks and contributions of services providers. (3) is associated with the computer and storage resources provided by resources providers. The framework allows to create any collaboration required for the development and operation of knowledge resources, required services, and High End Computing resources like compute and storage.

III. Integration of Workflows

The integration of association processing workflows with the workflows for creating arbitrary result matrices is most flexible and efficient and was based on the organisation and object features (Figure 2) in the knowledge resources. Object details and definitions have been discussed with computational views [9]. The illustration shows that object information is gathered from the objects and references in collections and containers. Configurable algorithms like filters and mapping are then used to compute a result matrix. The result matrix is considered "intermediate" because any of such workflows can be used in combination with other workflows, workflow chains or further processing.

(a) Geoscientific Association Processing Workflow Request: A request for geoscientific knowledge resources is initiated from within a discovery workflow. Such request is created in level (2) within the architecture.
(b) Geoscientific Knowledge Resources: The respective resources are initialised for the workflow. The knowledge resources are located in level (1).
(c) Collections and containers: The collections and containers within the resources are provided.
(d) Association Processing Algorithms and Definitions: The algorithms and definitions for the association processing are called. The processing involves (1), (2), and (3), especially the last two.
(e) Association Processing Intermediate Result Matrix: An intermediate result matrix is created by the algorithms and definitions. The matrix creation involves (1), (2), and (3), especially (2).
(f) Geoscientific Association Processing Workflow Reply: Such reply is created in level (2) within the architecture.

Figure 3 illustrates selected knowledge resources' objects, focussing on references in collections and containers.

IV. Implementation of Reference Types

Objects can carry any type of references. Objects can be grouped, e.g., in collections or containers. When larger groups are created then also these groups can carry their references. These references may occur in any combination but in practice these references will be a subset or a complementary set to the objects' references. Objects can be created by manual, automated, and hybrid means. Therefore, any type of references of that kind may exist.

Tables I and II show excerpts of the references, which were added to be used within the knowledge resources for two types of object groups, namely collections and containers.

TABLE I. GEOSCIENTIFIC KNOWLEDGE RESOURCES' COLLECTION AND CONTAINER REFERENCES TYPES USED FOR PROCESSING (EXCERPT).

References Types	*Group and Implementation Example*	
Classification	O & C	UDC
Concordance	O & C	UCC
In-object documentation	O & C	Text
Factual data	O & C	Text, data
Georeference	O & C	Geocoordinates
Keyword	O & C	Text
See	O & C	Text
Reference link	O & C	URL
Reference media	O & C	Link
Citation	O & C	Cite, bib
Content Factor	O & C	CONTFACT
Realia	O & C	Text
Language	O & C	EN, DE
Content-linked formatting	O & C	Markup, LaTeX

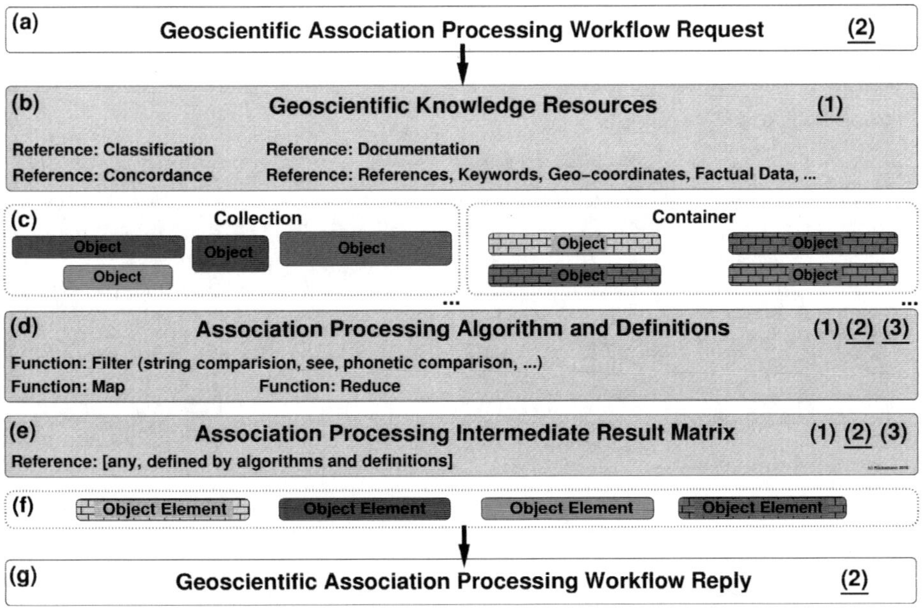

Figure 2. Geoscientific association processing workflow: Creation of intermediate result matrices from geoscientific resources and references (collections and containers) in reply to workflow requests (layers labelled with numbers, primary layers underlined, workflow steps labelled with lowercase letters).

TABLE II. GEOSCIENTIFIC KNOWLEDGE RESOURCES' IMPLEMENTED EXTERNAL REFERENCES TYPES USED FOR PROCESSING (EXCERPT).

References Types		*Group and Implementation Example*
Tag	E	Text, tags
Index Entry	E	Text, idx
Glossary Entry	E	Text, glo
List Entry	E	Text, lis

The reference types are organised in three major groups: Object collections (O), object containers (C), and external or externally created references (E).

This case study primarily addresses geoscientific and archaeological resources. The resources were extended for using a multitude of references types of creating associations (Tables II and I). Therefore, the resources especially contain georeferences, UDC classifications for any object, including complex conceptual knowledge, geoclassification, concordances like Universal Classified Classification (UCC) [10], and Content Factors in order to describe the content. Many reference types are part of the objects. Nevertheless, in practice, the organisation of references is more uniform within containers.

The reference types shown provide a lot of information regarding content and context, which could otherwise not be deducted from the object data itself. In addition, all reference types may exist in multiple views, multiple languages, and multiple context – any of which can be added in instances created by manual, automatic, and hybrid means.

V. CREATION OF ASSOCIATIONS AND PROCESSING MEANS

As far as the algorithms implemented in components carry essential information for processing and computation, e.g., for creating new results and output, they should be documented with the knowledge resources themselves. As associations can be created by arbitrary workflows, it is most important to know, which components can carry which facilities and how to exploit, e.g., in a multi-disciplinary context like geosciences and archaeology. Geocoordinates' data can be part of any knowledge objects, containers, container objects, and references (e.g., knowledge resources' references or Google Maps references). Conceptual knowledge data can be part of knowledge objects, containers and container objects, but it can also be contained in unstructured data, mostly used with automated processes with lower quality results. Associations can especially result from any constellation of content and context in object collections and containers, as well as from in-text references (e.g., comparisons, see), and external sources. Supporting methodologies and technologies, which were exploited for the creation and processing of associations are, e.g., string comparisons, transliterations, phonetics, statistics, metadata, Content Factor, object elements rhythm, and dynamical data. Associations were used for developing knowledge resources, optimising result matrices, e.g., within knowledge discovery workflows, creating concordances, creating references, improving knowledge objects and resources, gaining new knowledge. The combination allows various degrees of precision and fuzzyness as required for spanning multi-disciplinary and multi-lingual data. An optimisation can improve the quality of data, especially the quality of associations introduced for automated classification of unstructured data.

VI. EXPLOITATION OF COMPUTATION FACILITIES

Within the layers, there are three kinds of facilities, which are targets to be exploited by computation.

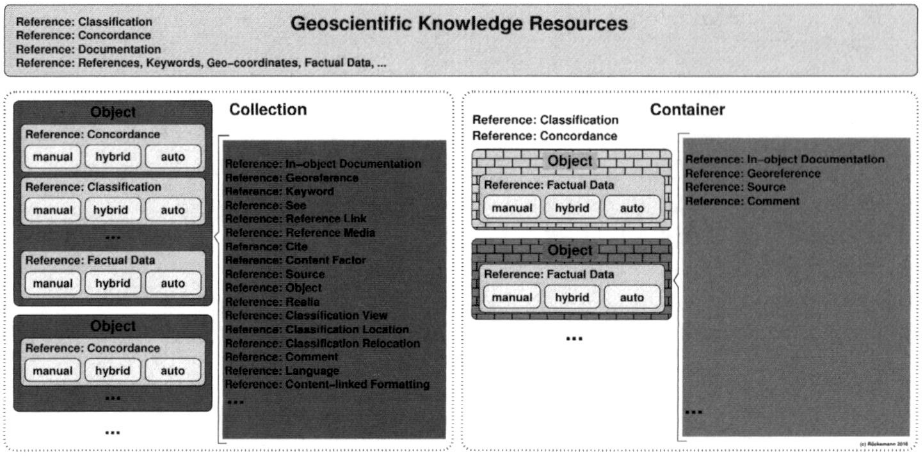

Figure 3. Geoscientific knowledge resources and objects: Selected knowledge resources' objects containing references for concordances and classifications in collections and containers. The excerpt illustrates a distinct handling of manually, hybrid, and automatically created data.

(1) **First block:** Knowledge resources.
- Purpose: Data.
- Implementation: Editing components, versioning tools, and high end text editors are used together with automation tools and scripting. The knowledge resources themselves are based on fully portable structures and markup.

(2) **Second block:** Services and interfaces.
- Purpose: Workflows.
- Implementation: Perl, Tcl, and C are used for an implementation.

(3) **Third block:** Processing.
- Purpose: Individual and parallelised processes and tasks as well as dynamical and interactive processes.
- Implementation: Here, Portable Batch System (PBS), Torque, and Moab are used, formerly also IBM LoadLeveler and Condor. As far as required for a certain scenario also dynamical or interactive jobs can be executed.

The exploitation of computation facilities is mostly based on these three featured component groups and the described implementation. This realises the purposes of extracting data and information, utilising workflow scripts, and submitting dynamical and batch jobs.

VII. Geosciences and Archaeology Case Study

The implementation has been done according to the described architecture and enabling the required association processing workflows based on the available components. Therefore, the major implementation tasks concentrated on the content related facilities, especially the geoscientific and archaeological knowledge resources, and the application components. The respective features were created in the knowledge resources' objects, which were under continuous development over the last decades in the LX knowledge resources. The application components have been extended and configured to work with the required application scenario. This includes the dynamical components from the Geo Exploration and Information (GEXI) project, e.g., the actmap components, based on Perl and Tcl scripting, C and Fortran.

The implementation for the case study was integrated with the the base for this case study, the long-term knowledge resources (LX Foundation Scientific Resources), which were developed and used over several decades, including geoscientific and archaeological objects and containers. A case study example based on the created resources is presented with the following workflow.

A. *Volcano and Rhodos association discovery workflow*

The workflow starts with the target to find possible associations and links between "Vesuvius" and "Rhodos".

1) Entry nodes: Vesuvius – Rhodos (Rhodos/Rhodes etc.).
2) Criteria and definition set.
3) Filter association processing criteria.
4) Filter association processing.
5) Selection and generation of compute instructions.
6) Sorting.
7) Formatting.
8) Selection.
9) First level association - both nodes.
10) Second level association.
11) Common object 1 and 2 (level 1).
12) Common object 11 and 22 (level 2).

Steps 2) to 5) of the workflow analyse and implement the criteria and definitions for the request and prepare the appropriate compute instructions. Steps 6) to 8) handle the sorting, formatting, and the selection of the intermediate result matrix. Steps 9) and 10) generate a first level association and after that a second level association. The concluding steps 11) and 12) generate the common objects for levels 1 and 2.

B. Resources and content

As an example, an object excerpt for one of the entry nodes is shown in Figure 4, which shows a referenced Vesuvius collection object containing factual and conceptual knowledge.

```
Vesuvius [Volcanology, Geology, Archaeology]:
         (lat.) Mons Vesuvius.
         (ital.) Vesuvio.
         Volcano, Gulf of Naples, Italy.
         Complex volcano (compound volcano).
         Stratovolcano, large cone (Gran Cono).
         Volcano Type: Somma volcano,
         VNUM: 0101-02=,
         Summit Elevation: 1281\UD{m}. ...
         Syn.: Vesaevus, Vesevus, Vesbius, Vesvius
         s. volcano, super volcano, compound volcano
         s. also Pompeji, Herculaneum, seismology
         compare La Soufrière, Mt. Scenery, Soufriere
         %%IML: UDC:[911.2+55]:[57+930.85]:[902]"63"(4+37+23+24)=12=14
         %%IML: GoogleMapsLocation: http://maps.google.de/maps?hl=de&gl=de&vpsrc
          =0&ie=UTF8&ll=40.821961,14.428868&spn=0.018804,0.028238&t=h&z=15
```

Figure 4. Workflow entry node: Knowledge resources collection object "Vesuvius" (LX resources, geoscientific collection, excerpt).

The object carries names and synonyms in different languages, dynamically usable geocoordinates, UDC classification and so on, including geoclassification (UDC:(37), Italia. Ancient Rome and Italy). The listing in Figure 5 shows an instance of a container entry excerpt from a volcanological features container.

```
CONTAINER_OBJECT_EN_ITEM: Vesuvius
CONTAINER_OBJECT_EN_PRINT: Vesuvius
CONTAINER_OBJECT_EN_COUNTRY: Italy
CONTAINER_OBJECT_EN_CONTINENT: Europe
CONTAINER_OBJECT_XX_LATITUDE: 40.821N
CONTAINER_OBJECT_XX_LONGITUDE: 14.426E
CONTAINER_OBJECT_XX_HEIGHT_M: 1281
CONTAINER_OBJECT_EN_TYPE: Complexvolcano
CONTAINER_OBJECT_XX_VNUM: 0101-02= ...
```

Figure 5. Processed instance of a simple knowledge resources container entry (LX resources, geoscientific container, excerpt).

The container component contains a large number of volcanic features and volcanoes, like Vesuvius, Thera, and Santorini. The excerpts have been processed with the appropriate `lx_object_volcanology` and `lx_container_volcanology` interfaces, selecting a number of items and for the container also items in English and German including a common formatting. The resources' access and processing can be done in any programming language, assuming that the interfaces are implemented. For example, combining scripting, filtering, and parallel programming can provide flexible approaches. The criteria and definitions are given by variables (Figure 6).

```
MATRIXLX
MATRIXRESLEV1
MATRIXRESLEV11
MATRIXRESLEV2
MATRIXRESLEV22
```

Figure 6. Criteria and definitions: Variables (LX Resources, excerpt).

The resource levels instruct the routines to execute two levels, one primary plain discovery each and a secondary in-depth discovery considering the primary results. The filter, selection, and processing instructions are handled by generators. The internal sequence is shown in Figure 7.

```
gen_matrix_pipe_level0_level1 "Vesuvius"
gen_pipe_reslevel1 | \
gen_grep_formatrix | \
gen_grep_forstrip
...
gen_matrix_pipe_level0_level1 "Rhod.s"
gen_pipe_reslevel11 | \
gen_grep_formatrix | \
gen_grep_forstrip
...
gen_matrix_pipe_level0_level2 "Kameiros"
gen_matrix_pipe_level2_level22 "Pozzolan"
gen_pipe_reslevel22 | \
gen_grep_formatrix | \
gen_grep_forstrip
```

Figure 7. Sequence of association routines for discovery workflow, dual-level (LX Resources, excerpt).

The sort, formatting, and selection are done with the function calls (`forstrip`). The "Vesuvius – Rhodos" association delivers "Kameiros", "Thera", "Santorini", and further intermediate result matrix elements from the secondary in-depth discovery.

C. The Kameiros' material results

The case study integrates the geoscientific and archaeological collection and container context and English entries. Figure 8 shows an excerpt of a referenced Kameiros object entry with UDC classification, media, and citation references, including geoclassification (UDC:(38), Ancient Greece).

```
Kameiros [Archaeology, Geophysics, Remote Sensing, Seafaring]:
         Greek city, Rhodos Island, Dodekanese, Greece.
         Modern location name Kámiros, Greece.
         ...
         Object:             Ancient architecture, stone, cement.
         Object-Keywords:    water tank, cement, lower area
         Object-Type:        Realia object.
         Object-Location:    Kameiros, Rhodos, Greece.
         Object-FindDate:    2011-10-27
         Object-Photo:       Claus-Peter Rückemann, ...
         %%SRC: 2013 CPR
         %%IML: media: YES 20130922 {LXC:DETAIL----} {UDC:(0.034)(38)770}
          LXDATASTORAGE://.../img_1342.jpg
         %%IML: UDC-Object:[902+903.2]+691.54+720.32+(38)+(4)
         ...
         %%IML: cite: YES 19980000 {LXK: concrete; pozzolan; Kameiros; Rhodos;
          Rhodes; Greece; Archaeology; Geosciences} {UDC:...} LXCITE://
          Koui:1998:Kamirian
         %%IML: cite: keyword: object: water storage tank
         %%IML: cite: keyword: material: concrete; Santorine earth mixed;
          natural cement; volcanic earth; lime
         %%IML: cite: keyword: location: Kameiros; Kamiros; Rhodos; Rhodes;
          Thera; Santorine; island of Yali; island of Nisyros
         ...
         %%IML: cite: YES 20120000 {LXK: cement; pozzolan; Kameiros; Rhodos;
          Rhodes; Greece; Archaeology; Geosciences} {UDC:...} LXCITE://
          Snellings:2012:Cementitious
         ...
         %%IML: cite: YES 20110000 {LXK: concrete; pozzolan; Kameiros; Rhodos;
          Rhodes; Greece; Archaeology; Geosciences} {UDC:...} LXCITE://
          Courland:2011:Concrete
         ...
         %%IML: cite: YES 20110000 {LXK: Archaeology; Geosciences; Vesuvius;
          Pompeji} {UDC:...} LXCITE://Hartge:2009:Vesuvius
         vgl. Rhodos, Iálissos, Lindos, Akandia
```

Figure 8. Association result matrix element, object "Kameiros" (LX resources, archaeological collection, excerpt).

The association processing "Vesuvius – Rhodos" revealed the reference to Vesuvius / (via Kameiros-associated citations) Pozzuoli / pozzolan. The excerpt also delivers a number of associated references on ancient concrete technology [11], cementitious materials [12], history of concrete [13], and evolution of concrete [14]. Looking for secondary documentation on eruptions being associated with Pozzuoli, e.g., the 1631 eruption of Vesuvius, delivers bibliographic sources like [15], which provides a lot of unique context information from an original source. This means there are several associations linking Vesuvius with Rhodos and one link is a technology, based on material from geoscientific context, documented in an archaeological site. The above sequence of association routines was used for the creation of a result matrix (routines implemented in `lxgrep_in_depth`). The listing in Figure 9 shows an excerpt of the result matrix for this case example.

```
MATRIXentry{Vesuvius}
MATRIXcitekeywords{location: Vesuvius, Italy}
MATRIXindex{pozzolan}
MATRIXindex{Campi Flegrei}
MATRIXindex{Pozzolana}
MATRIXindex{Pozzuoli}
MATRIXindex{Puteoli}
MATRIXkeywordcontext{keyword-Context: KYW :: 1634-1676 Polyhistor ...}
MATRIXkeywordcontext{keyword-Context: KYW REF S. 62 :: Vesuvius; pyroklastischer
  Strom; Aschewolke; Pozzuolo; Dreißigjähriger Krieg}
MATRIXkeywordcontext{keyword-Context: TXT :: 1631/1632 16xx, terra motus,
 fogelius}
MATRIXkeywordcontext{keyword-Context: TXT :: Fogelius, Historici Pragmatici
 universal, Terrae motus, Physica}
MATRIXkeywordcontext{keyword-Context: TXT REF S. 175 :: Pozzuolo ...}
MATRIXseealso{phlegra, Solfatara}
MATRIXsynonym{Vesaevus, Vesevus, Vesbius, Vesvius}
...
MATRIXentry{pozzolan}
MATRIXindex{diatomaceous earth}
MATRIXindex{Kameiros}
MATRIXindex{Kamiros}
MATRIXindex{Phlegraean Fields}
MATRIXindex{Pozzolana}
MATRIXindex{pozzolanic material}
MATRIXindex{Pozzuoli}
MATRIXindex{Puteoli}
MATRIXindex{Rhodes}
MATRIXindex{Vesuvius}
...
MATRIXentry{Kameiros}
MATRIXcitekeywords{material: concrete; Santorine earth mixed; natural cement;
 volcanic earth; lime}
MATRIXcitekeywords{material: pozzolan}
MATRIXcitekeywords{material: stone called Santorini}
MATRIXcitekeywords{object: water storage tank}
MATRIXcompare{Rhodos, Iálissos, Lindos, Akandia}
MATRIXindex{Pozzolana}
MATRIXindex{Pozzuoli}
MATRIXindex{Puteoli}
MATRIXindex{Vesuvius}
MATRIXobjectkeywords{Object-Keywords: water cistern, top area}
MATRIXobjectkeywords{Object-Keywords: water pipeline, clay, upper area}
MATRIXobjectkeywords{Object-Keywords: water tank, cement, lower area}
MATRIXtextintext{Kámiros, Greece}
```

Figure 9. Intermediate result matrix output, groups (excerpt).

If we extend the discovery and integrate chronological and associated objects and locations from the resources then the result matrix also includes years with volcanic, geological, geophysical, and technological context. The listing in Figure 10 shows a representation of additional result matrix entries associated for this case when these attributes were integrated.

```
MATRIXtextintext{date: -300000 Vesuvius, volcanic activity, oldest deposits}
MATRIXtextintext{date: -001800 Vesuvius, volcanic activity, Avellino eruption}
MATRIXtextintext{date: -001680 Santorin, Aegean, volcanic eruption, Thera}
MATRIXtextintext{date: -000700 Vesuvius, volcanic activity}
MATRIXtextintext{date: -000227 Rhodos, seismic activity}
MATRIXtextintext{date:  000062 Vesuvius, seismic activity, earthquake, Pompeji
 destruction}
MATRIXtextintext{date:  000079 Vesuvius, volcanic activity, explosive eruption,
 ash cloud, tuff, Pompeji destruction, Herculaneum, Stabiae}
MATRIXtextintext{date:  000142 Rhodos, seismic activity}
MATRIXtextintext{date:  000202 Vesuvius, volcanic activity}
MATRIXtextintext{date:  000345 Rhodos, seismic activity}
MATRIXtextintext{date:  000472 Vesuvius, volcanic activity}
MATRIXtextintext{date:  000512 Vesuvius, volcanic activity}
MATRIXtextintext{date:  000515 Rhodos, seismic activity}
...
MATRIXtextintext{location: Kameiros, island Rhodes, Greece; Kamiros, Greece;
 Rhodos; Rhodes}
MATRIXtextintext{location: Thera; Santorine; island Yali; island Nisyros}
MATRIXtextintext{location: Vesuvius}
MATRIXtextintext{location: Solfatara, Vesuvius}
MATRIXtextintext{location: Pantheon, Rome}
MATRIXtextintext{location: Caesarea Maritima}
MATRIXtextintext{location: Hagia Sophia, Konstantinopel}
...
MATRIXtextintext{material: pozzolan}
MATRIXtextintext{material: volcanic tuff}
MATRIXtextintext{material: Opus caementitium}
MATRIXtextintext{material: pozzolanic activity}
MATRIXtextintext{material: pozzolanic earths}
MATRIXtextintext{material: pozzolanic material}
MATRIXtextintext{material: volcanic ashes}
MATRIXtextintext{material: diatomaceous earth}
```

Figure 10. Additional result matrix entries for intermediate result matrix associated with integrated resources (excerpt).

The result is a very rich matrix. With its elements, the matrix links different content and context from hundreds of objects and sources. The listing depicts the content of the result matrix in a readable formatting and excerpts some elements. The matrix also contains references to the source data within the knowledge resources and also refers to many other data, e.g., terms, names, locations, georeferences, bibliographic data, citations, classification, and media data.

D. The Kameiros' media references results

The following photo data (Figure 11) from the media references for "Kameiros" was delivered by the result matrix.

Figure 11. Media photo objects associated with the knowledge object "Kameiros", referring to pozzolane and Vesuvius (LX resources, excerpt).

The photos have been taken in 2013 by the Knowledge in Motion (KiM) natural sciences and archaeology sections in the ancient city of Kameiros on Rhodos, Greece, during the GEXI Eastern Mediterranean research and studies campaign. Today, the location on the western coast of the island of Rhodes is named Kamíros.

The data shows the 'pozzolan' cement material, the water tank, and the water pipelines – the objects providing the missing link. These references from ancient Kameiros are also associated with Vesuvius volcano and refer to the later Roman adoption of comparable cement 'pozzolan' technology. Continuation of the case study [16] has conceived the documentation available and planning the additional research and development and the data to be collected and added to the knowledge resources.

VIII. Evaluation

The structure and the aggregation of references increases the flexibility of possible workflows. Increasing the quality of data in the described type of long-term knowledge resources – by including references – can increase the quality of result matrices from discovery processes.

The examined case showed that a technology and material, which have not been explicitely documented in context of a knowledge resources object, can be associated with the context of different objects. Here, the Greek origin of the "pozzolan" technology was associated, which was named after the later use in Roman times.

Association processing can support discovery processes even when references are not explicitely available in text and documentation, and would therefore be unexpected or unknown. Association processing can use multi-level discovery in order to gain additional information, which is not visible from an otherwise isolated documentation.

The developed structures and methods can be widely beneficial for knowledge development and discovery as well as for creating implementations for advanced discovery components. The methodology allows to extend and exploit long-term multi-disciplinary content documentation and discovery and gain new insights from otherwise not associated data.

IX. Conclusion

This paper presented the research on advanced features for processing of associations and some major results from the resources side and from the case study on geosciences and archaeology. First, the research showed that structured knowledge resources can be successfully extended for allowing a multitude of integrated references types, e.g., geoclassification and media. Second, the implementation showed that new workflows, e.g., association workflows can be integrated very flexibly and efficiently.

The elements from associations contained in the result matrix are not procurable when using only plain methods like simple string search or plain discovery. Furthermore, the integration of methods, e.g., association, classification, and phonetic algorithms, allows any degree of precision and fuzziness. From the structural and knowledge point of view, the extended features are least invasive to the described type of knowledge resources and procedures.

From the geoscientific and archaeological side the factual results are most notable because the methodology integrates multi-disciplinary and multi-lingual knowledge beyond conventional means and shows a large number of associations, which cross multiple disciplines and languages. The flexibility of the knowledge processing benefits from the advanced organisation of the data, which enables various scalable computational means for implementing directed graphs to fuzzy links, for which High End Computing resources can be deployed. Future work will be focussed on further developing the multi-disciplinary knowledge resources and creating advanced methods for describing the content and context of objects. The new method should carry facilities for supporting long-term knowledge development and analysis as well as for enabling automation and high end computing.

Acknowledgements

We are grateful to all national and international partners in the GEXI cooperations for their support and contributions. Special thanks go to the "Knowledge in Motion" (KiM) long-term project, DIMF, and to its scientific members, for prolific discussion, inspiration, and practical case studies, especially to Dr. Friedrich Hülsmann, Gottfried Wilhelm Leibniz Bibliothek (GWLB) Hannover, to Dipl.-Biol. Birgit Gersbeck-Schierholz, Leibniz Universität Hannover, and to Dipl.-Ing. Martin Hofmeister, Hannover, for practical multi-disciplinary case studies and the analysis of advanced concepts. We are grateful to the KiM long-term project and its members from the Unabhängiges Deutsches Institut für Multi-disziplinäre Forschung (DIMF) for the contributions to the research on the development and application of classifications and for funding the respective research campaigns.

References

[1] C.-P. Rückemann, "Integrated Computational and Conceptual Solutions for Complex Environmental Information Management," in The Fifth Symposium on Advanced Computation and Information in Natural and Applied Sciences, Proceedings of The 13th International Conference of Numerical Analysis and Applied Mathematics (ICNAAM), September 23–29, 2015, Rhodes, Greece, Proceedings of the American Institute of Physics (AIP). AIP Press, 2016, ISSN: 0094-243X, (in press).

[2] "Multilingual Universal Decimal Classification Summary," 2012, UDC Consortium, 2012, Web resource, v. 1.1. The Hague: UDC Consortium (UDCC Publication No. 088), URL: http://www.udcc.org/udcsummary/php/index.php [accessed: 2016-01-10].

[3] "Tcl Developer Site," 2016, URL: http://dev.scriptics.com/ [accessed: 2016-01-10].

[4] "ISO 14000 - Environmental management," 2016, URL: http://www.iso.org/iso/iso14000 [accessed: 2016-01-10].

[5] C.-P. Rückemann, "Enabling Dynamical Use of Integrated Systems and Scientific Supercomputing Resources for Archaeological Information Systems," in Proceedings INFOCOMP 2012, Oct. 21–26, 2012, Venice, Italy, 2012, pp. 36–41, ISBN: 978-1-61208-226-4.

[6] "LX-Project," 2016, URL: http://www.user.uni-hannover.de/cpr/x/rprojs/en/#LX (Information) [accessed: 2016-01-10].

[7] D. T. Meridou, U. Inden, C.-P. Rückemann, C. Z. Patrikakis, D.-T. I. Kaklamani, and I. S. Venieris, "Ontology-based, Multi-agent Support of Production Management," in The Fifth Symposium on Advanced Computation and Information in Natural and Applied Sciences, Proceedings of The 13th International Conference of Numerical Analysis and Applied Mathematics (ICNAAM), September 23–29, 2015, Rhodes, Greece, Proceedings of the American Institute of Physics (AIP). AIP Press, 2016, ISSN: 0094-243X, (in press).

[8] "UDC Online," 2016, URL: http://www.udc-hub.com/ [accessed: 2016-01-10].

[9] C.-P. Rückemann, "From Multi-disciplinary Knowledge Objects to Universal Knowledge Dimensions: Creating Computational Views," International Journal On Advances in Intelligent Systems, vol. 7, no. 3&4, 2014, pp. 385–401, ISSN: 1942-2679.

[10] C.-P. Rückemann, "Creation of Objects and Concordances for Knowledge Processing and Advanced Computing," in Proceedings of The Fifth International Conference on Advanced Communications and Computation (INFOCOMP 2015), June 21–26, 2015, Brussels, Belgium. XPS Press, 2015, pp. 91–98, ISSN: 2308-3484, ISBN-13: 978-1-61208-416-9.

[11] M. Koui and C. Ftikos, "The ancient Kamirian water storage tank: A proof of concrete technology and durability for three millenniums," Materials and Structures, Nov. 1998, Vol. 31, Issue 9, 623–627, DOI: 10.1007/BF02480613, ISSN: 1359-5997 (print), ISSN: 1871-6873 (online), Kluwer Academic Publishers.

[12] R. Snellings, G. Mertens, and J. Elsen, "Supplementary Cementitious Materials," Reviews in Mineralogy & Geochemistry, vol. 74, 2012, pp. 211–278, ISSN: 1529-6466, DOI: 10.2138/rmg.2012.74.6.

[13] R. Courland, Concrete Planet: The Strange and Fascinating Story of the World's Most Common Man-Made Material. Prometheus Books, 2011, ISBN: 978-1-61614-482-1, 396 p.

[14] A. A. Camões and R. M. Ferreira, "Technological evolution of concrete: from ancient times to ultra high-performance concrete," Structures and Architecture, 2010, pp. 1571–1578, ISBN: 978-0-415-49249-2.

[15] R. Hartge, Vesuvius, . . ., An Eye Witness Report of the Vesuvius Eruption 1631/1632 (in German: Vesuvius, Kulturgeschichtliche Betrachtungen zu den vulkanischen Kräften und Der aufgeschobene Weltuntergang? Ein Augenzeugenbericht vom Vesuv-Ausbruch 1631/1632). Verlag Die Blaue Eule, Essen, 2009, ISBN: 978-3-89924-233-1, 292 p.

[16] C.-P. Rückemann and B. Gersbeck-Schierholz, "Ancient Water Systems and Ageless Knowledge and Technologies," KiM Sky Summit, Knowledge in Motion, September 19, 2013, Sky Summit Meeting, "Unabhängiges Deutsches Institut für Multi-disziplinäre Forschung (DIMF)", Rhodos, Greece, 2013.

Differential Morphological Profile for Threat Detection on Pipeline Right-of-Way

Heavy Equipment Detection

Katia Stankov
R&D department
Synodon Inc.
Edmonton AB Canada
e-mail: katia.stankov@synodon.com

Boyd Tolton
R&D department
Synodon Inc.
Edmonton AB Canada
e-mail: boyd.tolton@ synodon.com

Abstract - **Unsupervised construction on the pipeline may provoke pipe rupture and consequently gas leaks. Monitoring the pipeline right-of-way for heavy equipment is important for environmental and human safety. Remotely sensed images are an alternative to expensive and time consuming foot patrol. Existing image processing methods make use of previous images and/or external data. Both are not always awailable. We propose a new method for image processing to detect heavy equipment without the need of auxiliairy data.**

Keywords-remote sensing image processing; right-of-way threats detection; histograms of oriented gradients; differential morphological profile.

I. INTRODUCTION

The periodic surveillance of gas and oil pipeline's Right-of Way (ROW) is vital to protect the human safety and to prevent ecological damage. Though most of the incidents are related to gas leaks, excavation activities are considered responsible for as much as two third of the incidents for the period 2002-2008 [1]. Recognizing dangerous activity, such as digging or construction equipment (threats), near the pipeline and undertaking prompt measures to insure safety is an obligation of oil and gas companies. Pipeline networks span thousands of kilometers and may be located in remote areas. Walking the ROW to survey for heavy equipment and dangerous activity is highly costly and time consuming. The alternative is the use of remotely sensed imagery.

So far the most reliable interpretation of these images remains the one made by human. The automation of the process faces difficulties from different origin: great variety of vehicles; uneven flight altitude; different view and orientation of the images; variable illumination conditions; occlusion by neighboring objects, and others [2]. In addition, construction vehicles are sometimes very similar to transportation vehicles. All these make the development of pattern recognition algorithms for ROW threat detection a challenging remote sensing image processing task.

Existing methods extract characteristic features to decrease the differences between construction vehicles (decrease the inter-class heterogeneity), while increasing the intra-class heterogeneity, i.e. make heavy equipment more distinguishable from other objects. In [3], scale-invariant feature transform was applied on previously defined scale invariant regions to receive object descriptors and detect vehicles. Presuming that local distribution of oriented gradients (edge orientations) is a good indicator for the presence of an object, Dalal [4] proposed the accumulative Histogram of the Oriented Gradients (HOG). In [5], the authors mapped HOG to Fourier domain to achieve rotation invariance and used kernel Support Vector Machine (SVM) to classify the data and identify construction vehicles. Using local textural descriptors and adaptive perception based segmentation, the authors in [2] sequentially eliminate background objects from the image, such as buildings, vegetation, roads, etc. The remaining potential threat locations are divided into several parts to extract and evaluate descriptive features and match them against template data. Extraction of local phase information allowed the separation between structure details and local energy (contrast) [6]. Afterward based on a single image template, the authors created a voting matrix to detect construction vehicles. To avoid the need of image template, potential threats locations are assessed with the aid of change detection in [7], next auxiliary data is used to decide upon the presence of a threat. Synthetic aperture radar images provide all weather coverage and together with optical images are used to produce a time sequenced image analysis for change detection and threat localization [8]. Existing methods need image templates or previous images and auxiliary data. Such external data is not always available.

We present a new methodology that avoids both, the need of image template and the need of auxiliary data. In addition to increased flexibility, it also makes the performance of the method nondependent on the quality of the external data. The rest of the paper is organized as follows. In section II we describe the method. Section III presents some results and validation, and conclusion is given in section IV.

II. DESCRIPTION OF THE METHOD

To build our method we take advantage of the fact that construction vehicles have non-flatten surfaces, which creates inequality in the intensity of surface pixels and together with their outer edges make that they appear as areas of high frequency in the image. Therefore, potential threat locations may be found by identifying areas of high frequencies that are in the range of heavy equipment size. Further, different descriptors may be used to discriminate between heavy duty vehicles and other objects. This is the general workflow of our method. To account for different illumination conditions we compute the color invariant of the blue band. To highlight frequencies (edges) we compute the gradient of the color invariant and retain only high frequencies applying a threshold on the gradient. The gradient is computed using the Otsu's method. To identify only areas in a certain size range we apply the Differential Morphological Profile (DMP). DMP is an iterative algorithm that performs opening/closing by reconstruction with a structuring element (SE). The size of the SE is increased in the consecutive iteration and the result is extracted from the result of the previous iteration. When the SE size exceeds the object size, the background intensity values are assigned to the object. Thus, by extracting two consecutive results, only objects that correspond to the SE size are retained. We derived the size of the SEs from the

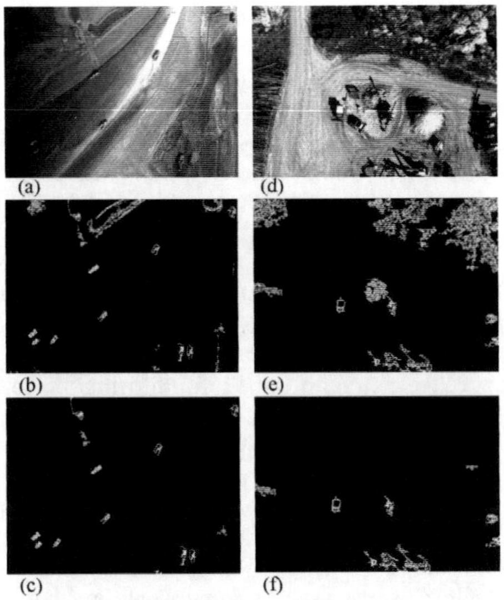

Figure 1. ROW threats detection. (a) and (d) - Original images; (b) and (e) - Results of the application of DMP: SE = 8x8 pixels; (c) and (d) - Detection.

size of the target object and the pixel size of the image. For each received object we compute the following descriptors: HOGs, curvature, the ratio between the major and minor axis of the object. Using all the descriptors simultaneously rather than thresholding each one separately, allows better assessment of the similarity between objects. To find the most similar objects we retain objects that maximize the ratio between the first and the last eigenvalue of the vectors. Finally, we apply spectral and shape constraint to discriminate between construction equipment and cars.

III. RESULTS AND VALIDATION

We present some results in Fig. 1. The first column (a) represents the original image, the second column (b) - the result of DMP, and the third column (c) - the detection.

To validate the accuracy of the method we compared the results to the results of manually detected threats. We refer to the latter as ground truth data. A test from 3 flight days (more than 1000 images of 1200x800 pixels with average pixel resolution of 9 cm) reveals a detection rate of 82.6% - heavy equipment machines that are present in the ground truth data and were detected by the algorithm. At this stage of the development of the algorithm we are less concerned with the rate of false recognition, as the results are reviewed by an operator. We consider including additional descriptors to reduce the number of false positives events while increasing the detection rate.

An advantage of using DMP is that by changing the shape of the SE different shape may be detected, changing the size of the SE allows the detection of construction machines with different size and also to account for the changing height of the flight. As the height of the flight in our case does not change a lot we derived the size of the SE empirically. If this height changes a lot, an automate way to choose the size of the SEs in accordance with the flight parameters should be adopted. In our opinion the method may have limited performance when applied on images with much lower spatial resolution, more than 1 meter for example, as it relies explicitly on information taken from an increasing neighborhood.

IV. CONCLUSION

We presented a new methodology for the detection of threats on the pipeline ROW that does not involve the use of external data. The initial results are promising and we believe that the method has the potential to replace the manual processing of the images.

REFERENCES

[1] S. Chastain, "Pipeline Right-of-Way Encroacment: Exploring Emerging Technologies that Address the Problem", Right-of-Way, May/June 2009, pp. 22-27.

[2] V. Asari, Vijayan, P. Sidike, C. Cui, and V. Santhaseelan. "New wide-area surveillance techniques for protection of pipeline infrastructure", SPIE Newsroom, 30 January 2015, DOI: 10.1117/2.1201501.005760

[3] G. Dorko and C. Schmid, "Selection of scale-invariant parts for object class recognition". IProceedings of the 9th International Conference on Computer Vision, Nice, France, pp 634–640, 2003.

[4] N. Dalal and B. Triggs, "Histograms of oriented gradients for human detection", IEEE Conference on Computer Vision and Pattern Recognition, pp. 886-893, 2005

[5] A. Mathew and V. K. Asari, "Rotation-invariant Histogram Features for Threat Object Detection on Pipeline Right-of-Way", in Video Surveillance and Transportation Imaging Applications 2014, edited by Robert P. Loce, Eli Saber, Proc. of SPIE-IS&T Electronic Imaging, SPIE Vol. 9026, pp. 902604-1-902604-1, 2014 SPIE-IS&T doi: 10.1117/12.2039663

[6] B. Nair, V. Santhaseelan, C. Cui, and V. Asari, "Intrusion detection on oil pipeline right of way using monogenic signal representation", Proc. SPIE 8745, 2013, p. 87451U, doi:10.1117/12.2015640

[7] M. Zarea, G. Pognonec, C. Schmidt, *et al.*, "First steps in developing an automated aerial surveillance approach", Journal of Risk Research, vol.13(3–4): pp. 407–420, 2013 doi:10.1080/13669877.2012.729520

[8] Roper, W. E. and Dutta, S. "Oil Spill and Pipeline Condition Assessment Using Remote Sensing and Data Visualization Management Systems." George Mason University, 4400 University Drive, 2006

SOLAP_Frame: A Framework for SOLAP using Heterogeneous Data Sources

Tiago Eduardo da Silva
Federal Institute of Education, Science and Technology of Pernambuco, Palmares, Brazil
e-mail: tiagoes@gmail.com

Daniel Farias Batista Leite, Cláudio de Souza Baptista
Federal University of Campina Grande, Campina Grande, Paraíba, Brazil
e-mail: danielfarias@copin.ufcg.edu.br, baptista@computacao.ufcg.edu.br

Abstract—**Business Intelligence (BI) and Geographic Information System (GIS) technologies have been used by several organizations aiming to improve the decision making process. In this context, the term Spatial OLAP (SOLAP) emerged for spatial multidimensional data sources. Nevertheless, it is very complex to integrate heterogeneous data sources into a SOLAP solution. This paper proposes a framework that enables the analysis of multidimensional spatial data from heterogeneous data sources. This goal is accomplished through an implementation of a SOLAP framework that contains interfaces and abstract classes that can be implemented and extended to support new data sources. To validate the proposed ideas, a case study was conducted.**

Keywords- Business Intellignce; GIS; SOLAP.

I. Introduction

With the increasing volume of data coming from a large variety of sources, there has been a considerable increase in investments on technologies capable of extracting information from these data and, consequently, help managers in the decision making process. *Business Intelligence* (BI) tools provide a historical, updated and predictive view of business operations of a company, enabling the identification of patterns, the availability of new functionalities and products, and improving the relationship with costumers. *On-line Analytical Processing* (OLAP) is one of the most used BI tools. An OLAP tool enables rapid exploration and analysis of data stored in multiple aggregation levels, according to the multidimensional approach. In this context, most companies are adopting BI tools in order to become more competitive in the marketplace [1].

In addition to this, most companies heavily deal with the spatial dimension in their datasets. Hence, it is important to investigate how to explore that dimension in order to improve the decision making process. However, traditional BI technologies do not take advantage of spatial data. On the other hand, Geographical Information Systems (GIS) were designed to work on georeferenced data using the Online Transaction Processing (OLTP) approach, and thereby prevents an efficient and deep data analysis.

More recently, corporations have demanded the integration of GIS and OLAP technologies to give rise to a new category of tools known as *Spatial Online Analytical Processing* (SOLAP). The integration of these technologies may happen through three distinct approaches: prioritizing the resources of GIS (GIS-dominant), overlapping visual and graphic resources of OLAP tools (OLAP-dominant), and the full integration approach (SOLAP) that aggregates the functionalities of GIS with graphs, tables and maps [2].

The need for integrating GIS and OLAP technologies has driven a set of new SOLAP academic solutions. Nonetheless, no consensus was reached as to the best way to achieve this integration. The solutions differ in several respects, particularly for the data model. Without a consensus on the data model, it is very hard to provide spatial cubes on the Web.

The lack of consensus for GIS - OLAP integration and standards for the provision of spatial and multidimensional data hinders the use of different data sources at the same time. Hence, the extraction of useful information to improve the decision making process at corporations is impaired.

This paper proposes a new framework that enables the analysis of spatial cubes coming from multiple and heterogeneous multidimensional data sources. Our proposed framework can be classified as an Enterprise Application Framework, as it is concerned with OLAP domain [3]. According to Sommerville, a framework is a software that can be extended to create a more specific application [4].

The main contributions of our research is the proposal of a SOLAP framework with interfaces and abstract classes that can be implemented and extended to support new data sources, making easy the integration of heterogeneous data cubes. Hence, we offer a reusable software to interoperate spatial data warehouses from heterogeneous data sources. To the best of our knowledge, this is the first study on this question.

Furthermore, in order to validate the proposed framework, we present a case study on the accountability analysis of the TCE-AC (Court of Accounts of the State of Acre – Brazil). In the case study, we extended our SOLAP framework to access cubes coming from two different sources: *Microsoft SQL Server Analysis Services* (SSAS) and *GeoMondrian*.

The rest of this paper is organized as follows. Section II discusses related work on SOLAP. Section III addresses the architecture of the proposed framework. Section IV presents a case study involving the Court of Accounts of the State of Acre – Brazil. Finally, Section V highlights the conclusions and further work to be undertaken.

II. RELATED WORK

Spatial OLAP has been a very active research area for a long time. Surveys on SOLAP can be found in [5][6][7]. Salehi et al. propose a formal model for spatial datacubes [8]. Aguila et al. address a conceptual model for SOLAP [9]. Baltzer focuses on spatial multidimensional querying [10]. Glorio and Trujillo highlight the optimization of spatial queries [11]. Tahar Ziouel et al. propose an approach for cartographic generalization of SOLAP applications [12].

Several SOLAP tools have been developed over the last years in many contexts. Rivest et al. propose a generic SOLAP tool, called JMap Spatial OLAP, comprising two modules: an administrative one, allowing for the setup of the connection with a multidimensional spatial database; and a visualization module, allowing for the interactive exploration of data through charts and maps [2]. The proposed tools are based on Relational OLAP (ROLAP) architecture and support the three types of spatial dimensions: geometric, non-geometric and mixed.

Bimonte et al. developed the GeWOlap SOLAP tool, highlighting the synchronization of different forms of data visualization. Their architecture comprises three layers: data, SOLAP server and client layers (user interface) [13].

Escribano et al. proposed a tool called Piet, integrating GIS and OLAP technologies and executing the precomputation of the map layers [14]. The Piet architecture also comprises three layers: data, SOLAP server and client layers.

Another challenge faced in SOLAP solutions is the issue of aggregation performance when queries involve considerable amounts of spatial data. Jiyuan Li et al. combine SOLAP approach with the Map-Reduce model for processing large amounts of data in parallel [15]. Saida Aissi et al. propose a multidimensional query recommendation system aiming to help users to retrieve relevant information through SOLAP, improving the data exploitation process [16].

The integration of the GIS and OLAP technologies was also explored through data communication techniques, with the objective of enabling their interoperability. Silva et al. proposed a Web Service called GMLA WS, which combines XML for Analysis (XMLA), Geography Markup Language (GML) and Web Feature Service (WFS) [17].

Dubé et al. presented an XML format to supply and exchange SOLAP cubes through web services [18]. The proposed XML format does not depend on the OLAP/GIS tool and represents all the necessary data (facts and members) and metadata (scheme), besides supporting spatial dimensions and members. The advantage of exchanging data through Web Services is that the communication is not limited to traditional client-server platforms, but also supports ubiquitous mobile computing environments.

These solutions differ in the data model and there is no standard for provision and analysis of spatial data. In this perspective, the Open Geospatial Consortium (OGC) published in 2012 a report (white paper) containing an evaluation of the ways that the OGC standards (e.g., WMS - Web Map Service, WFS, WPS - Web Processing Service, etc.) could be extended, in order to promote the use of geospatial information and the interoperability of GeoBI applications. However, neither extension nor standard for this purpose has been published by OGC yet.

We observed that the key features to compare SOLAP solutions are: (I) to enable the creation of queries through a visual query language; (II) to provide an integrated view of both multi-dimensional and spatial data; (III) to support spatial operators allowing for more comprehensive analysis; (IV) to give access to various multidimensional data sources (cube servers); (V) to enable geocoding data to provide spatial analysis in pure OLAP sources; (VI) to use open technologies to reduce costs; (VII) to be extensible so that new features can be added; and (VIII) to enable data visualization through maps, tables and graphs. Table I presents a comparison among the related work in which the cells that contain an X mean that a given solution implements a given feature and those that contain a – mean that a given solution does not implement a particular feature.

This paper presents a SOLAP framework, known as SOLAP_Frame that enables the connection to multiple and heterogeneous data sources. The framework was developed using open technologies. Furthermore, the proposed framework presents an integrated visualization of multidimensional and spatial data, allowing for the creation of queries by means of a visual specification language with support to spatial operators and data visualization through maps, tables and charts. Finally, SOLAP_Frame also enables the geocodification of the data and is extensible, providing support for addition of new functionalities, such as new operators or data visualization methods. To the best of our knowledge, this is the first work to propose a SOLAP framework that is able to interoperate with new heterogeneous data sources.

TABLE I. RELATED WORK COMPARISON

Solutions	Features							
	I	*II*	*III*	*IV*	*V*	*VI*	*VII*	*VIII*
JMAP [2]	X	X	-	-	-	X	X	X
GeoWOLAP [11]	X	X	-	-	-	-	-	-
Piet [12]	-	-	-	-	-	-	-	-
SOLAP_Frame	X	X	X	X	X	X	X	X

III. SOLAP_FRAME: A FRAMEWORK FOR SPATIAL ANALYSIS USING HETEORGENEOUS DATA SOURCES

This section presents the SOLAP_Frame architecture. In the next subsections, we describe the architecture and the extension points of the proposed SOLAP framework.

A. Architecture

The framework architecture comprises three layers: client, application and data layers, as shown in Figure 1.

The client layer comprises a set of graphical Web interfaces in which the user can connect to a multidimensional spatial cube, geocode members, compose queries by means of a visual specification and visualize the data.

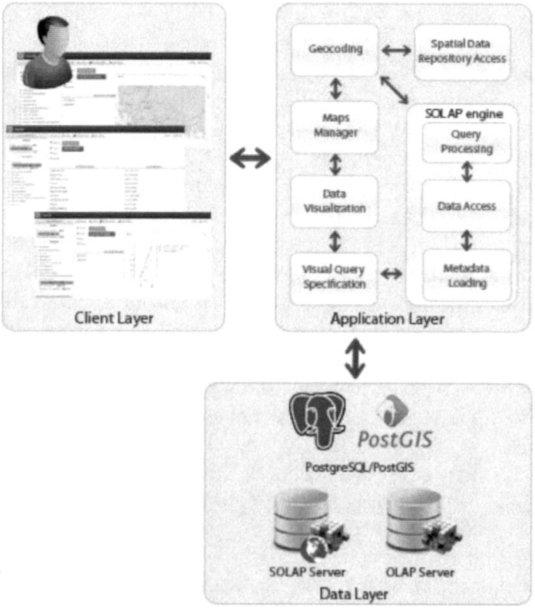

Figure 1. The SOLAP_Frame architecture.

The data layer comprises the multidimensional data sources to be analyzed and the spatial data repository. The framework is capable of accessing several multidimensional servers (cube servers), employing different technologies and manufacturers. The framework also supports the geocoding of cube members, enabling the spatial analysis of non spatial OLAP cubes. The application data repository is stored in the *PostgreSQL* DBMS, with the PostGIS spatial extension. The spatial data resulting from the geocoding of members of the cube are stored in this repository, characterizing a Data Warehouse federated approach. Any spatial DBMS can be used for this purpose, by simply extending the proposed solution through its extension points.

The application layer is responsible for the implementation of the whole application logic. This layer has six modules: visual query specification, data visualization, map manager, spatial data repository access and the SOLAP engine. We highlight the SOLAP engine as the main module of this layer, providing communication between the application and the multidimensional servers (OLAP or SOLAP) attached to the data layer.

The visual query specification module controls the query execution and results visualization, turning the interactions between users and the graphical interface into objects that compose the query visual specification. After receiving the result of a visual query, the visual specification module forwards the result with its markup to the data visualization module so that the data can be transformed and presented in the specified format. Depending on the markup type, data may have to be transformed, for example, grouped to compose the map layers or graph axes. After being transformed, data are forwarded to the most appropriate component of the interface for visualization (e.g., tables, maps, charts, text and caption).

The map management module is responsible for displaying data in maps. As such, this module receives, from the data visualization module, a set of spatial and numerical data, query results, and the markup. The repository access module, in turn, was implemented to retrieve the metadata and the data from the spatial tables stored in the spatial data repository. The metadata and the data on the spatial tables are used by the geocoding module, which is accessed using the Java Database Connectivity (JDBC) driver for the PostgreSQL database management system (DBMS) with the PostGIS spatial extension.

The SOLAP engine comprises of three sub-modules: data access, metadata loading and query processing. This engine is responsible for: implementing the connection to a given multidimensional data source; loading of metadata from cubes to be analyzed; translating the visual specification to the destination query language; submitting the translated query; and receiving the result data.

The implementation of the SOLAP engine depends on the manufacturer of the SOLAP server to be accessed.

The data access module enables the connection to the multidimensional data source and the choice of the cube to be analyzed. This module is also in charge of executing queries in the language of the accessed technology and returning the results of these queries. To accomplish the data access module, it is necessary to have information on the connection properties that match both source and cube properties. The source properties state where and how to connect to the multidimensional data source, while the cube properties address which cube belonging to a given source should be accessed. The data access module knows how to handle heterogeneous sources.

The metadata loading and the query processing modules of the SOLAP engine interact with the data access module, which is specialized, that is, its implementation depends on the adopted technology.

The metadata loading module is responsible for retrieving cube metadata. In order to connect to a multidimensional data source, the *ConnectionProperties* and *DataSource* objects must be informed. The metadata coming from this connection will be turned into Cube objects, which will be loaded into memory for subsequent use by other modules of the proposed framework.

The query processing module is responsible for translating the visual queries to queries in the target technology native language; executing them using the data access module; and returning the query. In order to retrieve the data, besides the connection properties, a visual query is passed as parameter to the processing query module.

B. SOLAP_Frame: Extension Points

SOLAP_Frame contains extension points that enable to connect it to heterogeneous data sources. In this section, we provide details of the communication interfaces.

B.1 SOLAP Engine Facade

The main extension point of the proposed framework is the implementation of the SOLAP engine facade. This facade is responsible for the communication between the proposed solution and the multidimensional data source. The message exchanges between the solution and the SOLAP engine consist of requesting metadata and data from a cube. The facade standardizes this message exchange. To request metadata from a cube to the SOLAP engine, the facade contains the *loadCube* method that receives as parameters the connection properties of both the data source and cube and returns an object that represents the cube.

To request data from a cube to the SOLAP engine, the facade contains three methods: *processQuery*, *getLevelMembers* and *filterLevelMembers*. The *processQuery* method receives as parameter an object that models the query, which is part of the visual specification defined by the user. This visual query should be translated into the query language of the source technology; and then executed. The query result must be modeled in a return object called *VisualQueryResult*.

The *getLevelMembers* method receives as parameter an object that represents a hierarchical level. This level is used to retrieve the members of the cube. Finally, the *filterLevelMembers* method receives as parameter, besides the level, a filter that can be either conventional or spatial. This filter will be used to select the members to be retrieved.

The solution will automatically identify the implementation of the front end by means of the Contexts and *Dependency Injection services* (CDI) present in the Java Enterprise Edition platform, and will register it for use. A name and a type must be associated with the SOLAP engine in order to be presented to the user. The type is used by the BI engine manager keeping the mapping between types and implementations available in the solution.

B.2 Connection properties interface

The communication process requires that the user provides the connection properties. This information will be used every time the SOLAP engine needs to communicate with the data source. Thus, another extension point in our framework is the implementation of this user interface.

The connection properties depend on the technology to be used. Hence, the parameters that must be provided vary according to the technology.

The front end for the SOLAP engine contains a method called *getLoaderPopup*, which returns an object called *LoaderPopup*, which, in turn, contains the necessary information for the exhibition of the component. This object is used by the interface that lists all the available engines in the solution. The *LoaderPopup* object is formed by another object called *LoaderBean*, that needs to be implemented. The *LoaderBean* is the controller responsible for preparing the component for exhibition and for enabling access to the properties of connections created by the user. Figure 2 presents a class diagram for the *LoaderPopup* and *LoaderBean* object.

B.3 SOLAP engine for XMLA server

In order to provide access to several heterogeneous multidimensional data sources, we also developed a SOLAP engine for servers that provide their data through the XMLA protocol. To enable the XMLA engine to access a specific technology, it is necessary to implement the abstract classes described in the following. In the implementation of the SOLAP engine for XMLA, we used the XMLA driver supplied by the Open Java API for OLAP (olap4j), which is also an open specification for the construction of OLAP applications based on the JDBC specification. Once connected to the data source, the user chooses the cube that will be analyzed. After that, an alias is assigned to the cube. This alias will be used to identify the cube in the system.

After the selection of the cube, its metadata will be loaded. For this, it is necessary to convert the metadata from the native format into the target one. The metadata loading is carried out by the *Olap4jXMLACubeMetadataDAO* class, and the abstract class *AbstractOlap4jXMLACubeConverter* implements the basic methods necessary for the conversion of the cubes from the native format to the format used in the solution.

The methods of the *AbstractOlap4jXMLACubeConverter* abstract class are spatially related and depend on the technology used by the XMLA server. This is due to the fact that XMLA does not specify a standard format for the transportation of spatial data. Furthermore, the Multidimensional Expressions (MDX) query language, used by the XMLA server, does not specify spatial functions. Figure 2 presents a class diagram for the XMLA engine.

To load the data, the *AbstractOlap4jXMLAQueryDAO* class supplies the basic functionalities necessary for the correct operation of the solution. However, it is necessary to implement the method responsible for translating MDX filters into the language for the chosen technology. The abstract method is necessary due to the fact that the solution has filters that use spatial functions. Since these functions are not standardized for MDX, they vary according to the technology used.

IV. CASE STUDY APPLIED TO THE COURT OF ACCOUNTS OF THE STATE OF ACRE - BRAZIL

In order to evaluate the SOLAP_Frame, we ran a case study on public accountability of the Court of Accounts of the State of Acre – Brazil (TCE-AC). The aim of this case study was to help in the decision making process related to the definition of more efficient management strategies to achieve an effective control of public spending.

To run this case study, the framework was extended to connect to two multidimensional data sources: *SQL Server Analysis Services* and *GeoMondrian*, of which the first one provides access to conventional data, and the second one provides access to spatial data. Three cubes are available for the analysis: Commitment, Liquidation and Payment. The Commitment cube uses the opensource *GeoMondrian* server, while the other ones utilize the Microsoft SSAS.

Figure 2. Class Diagram for XMLA Engine.

The fact tables to be analyzed are represented by the measures: Commitment values, Liquidation values and Payment values. Besides the measures modeled in the *Spatial Data Warehouse* (SDW), the cubes have two additional measures: number and mean of the values.

In this context, we highlight two queries aiming at validating the proposed framework with respect to the simultaneous access to several multidimensional data sources. Moreover, the queries proposed enable to address the functionalities related to the spatial analysis of multidimensional data. The two queries are:

- "Display a thematic map with the sum of the Commitment values for each city in the State of Acre – Brazil"; and
- "What is the sum of the liquidated values in the neighbor cities of Rio Branco city, concerning the functions Administration, Agriculture and Legislative?"

In order to solve the first query, the Commitment cube was used. The Commitment Value measure was added to the tab "Columns", while the spatial hierarchy District Name was added to the tab "Layers". The caption, created for the Commitment Value measure, was used to visualize the data on the map. The District Name level was used as a label for the geometries. Figure 3 presents the result of the query.

The second query demonstrates the use of the spatial filters available in the framework. In this example, the cube Liquidation was used; the Liquidation Value metric was added to the tab "Columns", the hierarchy Function Description was added to the tab "Rows", and the hierarchy District Name, to the tab "Layers". The members of the Function Description level were filtered. A geographic filter was added to the districts, and the spatial operator *Touches* was used to filter neighbor cities of Rio Branco municipality. To visualize the data, we used a caption in spatial panels. Figure 4 presents the result of the query.

V. Conclusions and Future Work

Currently, there is a demand for exploring spatial data sources to improve the decision making process. However, no consensus has been reached yet regarding the best way to accomplish this integration. This lack of standard makes it hard to analyze spatial cubes from heterogeneous multidimensional data sources.

From the survey of the state of the art, it was possible to compare the strengths and weaknesses of the main existing solutions and conclude that these solutions do not provide analysis of spatial cubes from heterogeneous multidimensional data sources.

This paper presented a framework that executes spatial data analysis from heterogeneous multidimensional data sources. The framework provides interfaces and abstract classes that can be extended to incorporate new multidimensional data sources. Reusability is a main issue addressed by our framework concerning the implementation of SOLAP tools. Furthermore, by using geocoding operation, it is possible to have a SOLAP tool over a traditional OLAP one, which emphasizes the flexibility of the proposed framework.

To validate the proposed ideas, we conducted a case study in a Court of Accounts, in which where Microsoft SSAS and GeoMondrian cube servers were accessed through extensions of the proposed framework. This case study has enabled a concrete evaluation of the SOLAP functionality.

The spatial cube model of the proposed framework proved to be efficient, allowing the spatial analysis of cubes from multiple heterogeneous sources.

As a future work, the framework can be explored in various expansion points in order to make it more robust. For example, the incorporation of new SOLAP operators and support for faster data.

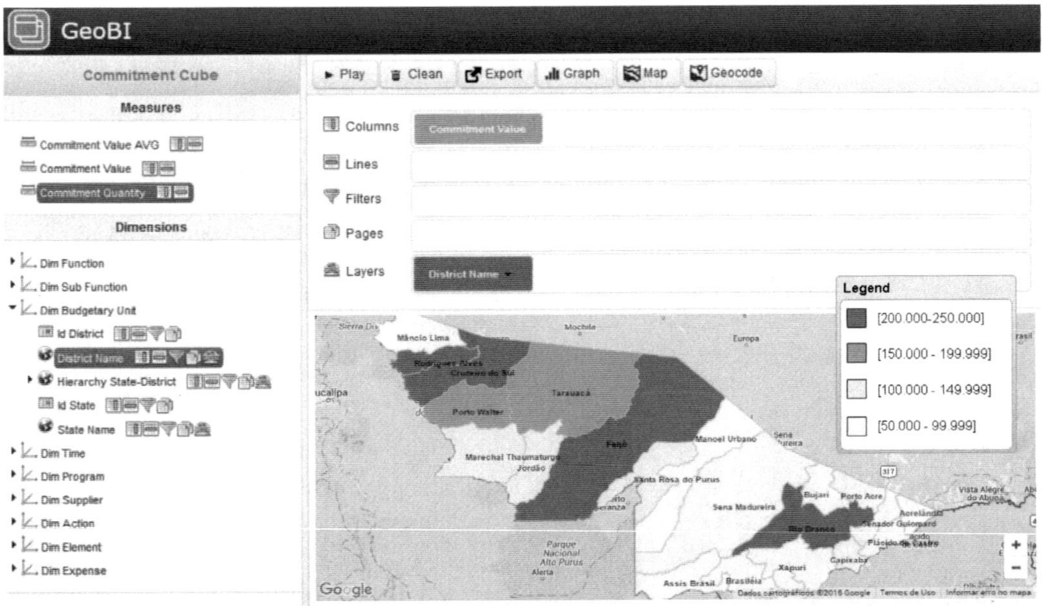

Figure 3. Query 1: result.

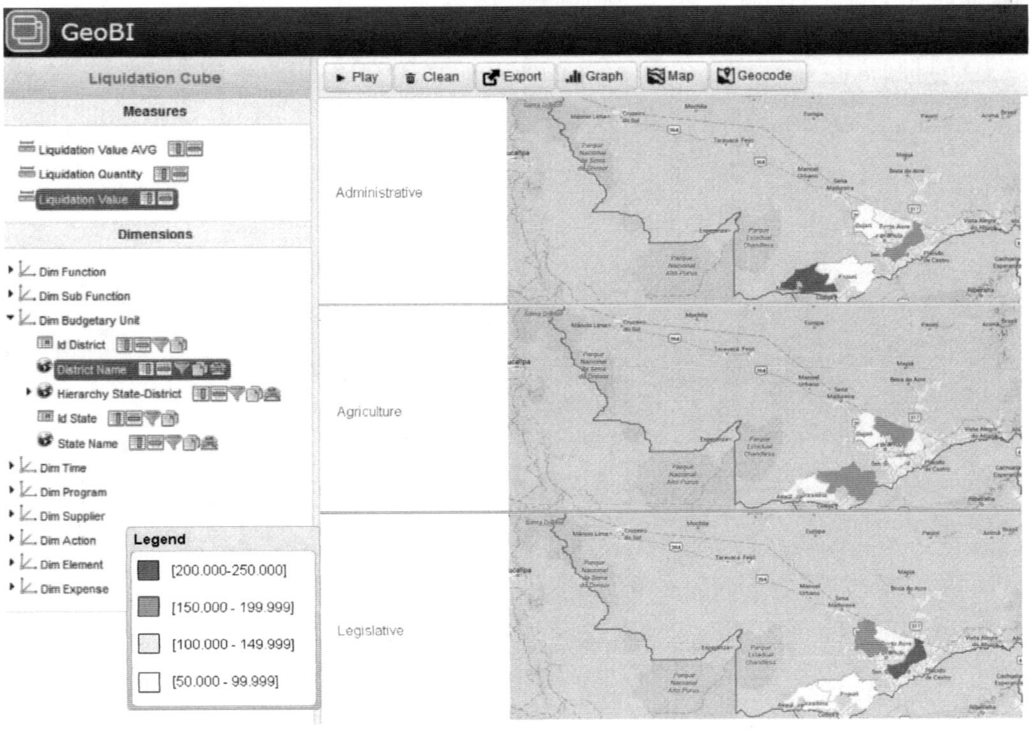

Figure 4. Query 2: result.

ACKNOWLEDGEMENTS

The authors thank the Court of Accounts of the State of Acre – Brazil (TCE-AC), the Brazilian ElectricityRegulatory Agency (ANEEL) and the National Council for Scientific and Technological Development (CNPq) for funding this research.

REFERENCES

[1] S. Chaudhuri, U. Dayal, and V. Narasayya, "An overview of business intelligence technology," Communications of the ACM, vol. 54, no. 8, 2011, pp. 88-98.

[2] S. Rivest, Y. Bédard, M. Proulx, F. Hubert, and J. Pastor, "SOLAP technology: Merging business intelligence with geospatial technology for interactive spatio-temporal exploration and analysos of data," ISPRS P&RS, vol. 60, no. 1, 2005, pp. 17-33.

[3] M. E. Fayad and D. C. Schmidt, "Object-oriented application frameworks", Communications of the ACM, vol. 40, no. 10, 1997, pp. 32-40.

[4] I. Sommerville, Software Engineering, 10th edition, 2015, Pearson.

[5] L. Gómez, B. Kuijipers, B. Moelans, and A. Vaisman, "A Survey of Spatio-Temporal Data Warehousing," JDWM, vol. 5, no. 3, 2009, pp. 28-55.

[6] S. Bimonte, A. Tchounikine, M. Miquel, and F. Pinet, "When Spatial Analysis Meets OLAP Multidimensional Model and Operator," JDWM, vol. 6, no. 4, 2010, pp. 33-60.-

[7] G. Viswanathan and M. Scheneider, "On the requirements for user-centric spatial data warehousing and SOLAP," in Proceedings of the 16th DASFAA, 2011, pp.144-155.

[8] M. Salehi, Y. Bédard, and S. Rivest, "A formal Conceptual Model and Definition Framework for Spatial Datacubes," Geomatica, vol. 64, no 3, 2010, pp. 313-326.

[9] P. Aguila, R. Fidalgo, and A. Mota, "Towards a more straightforward and more expressive metamodel for SDW modeling," in DOLAP 2011, pp. 31-36.

[10] O. Baltzer, "Computacional Methods for Spatial OLAP," Ph.D. thesis, 2011.

[11] O. Glorio and J. Trujillo, "Designing Data Warehouses for Geographic OLAP Querying by Using MDA," in Proceedings of the international Conference on Computational Science and its Applications, 2009, pp. 305-519.

[12] T. Ziouel, K. A. Derbal, and K. Boukhalfa. "SOLAP On-the-Fly Generalization Approach Based on Spatial Hierarchical Structures," CIIA 2015, pp. 279-290.

[13] S. Bimonte, A. Tchounikine, and M. Miquel, "Spatial OLAP: Open Issues and a Web Based Prototype," in M. Wachowicz & L. Bodum (Eds.), Proceedings of the 10th AGILE International Conference on Geographic Information Science, 2007.

[14] A. Escribano, L. Gomez, B. Kujipers, and A. Vaisman, "Piet: a GIS-OLAP implementation," in Proceedings of the ACM 1oth international workshop on Data Warehousing and OLAP, 2007, pp. 73-80.

[15] J. Li, L. Meng, F. Z. Wang, W. Zhang, and Y. Chai, "A Map-Reduce-enabled SOLAP cube for large-scale remotely sensed data aggregation," Computers & Geosciences, vol. 70, 2014, pp. 110-199.

[16] S. Aissi, M. S. Gouider, T. Sboui, and L. B. Said, "Personalized recommendation of SOLAP queries: theoretical framework and experimental evaluation," SAC 2015, pp. 1008-1014.

[17] J. Silva, V. Times, and A. Salgado, "An Open Source and Web Based Framework for Geographical and Multidimensional Processing," SAC 2006, pp. 63-67.

[18] E. Dubé, T. Badard, and Y. Bedard, "XML enconding and Web Services for Spatial OLAP data cube exchange: an SOA approach," CIT, vol. 17, no 4, 2009, pp. 347-358.

Geographic Metadata Searching with Semantic and Spatial Filtering Methods

Tristan W. Reed*, Elizabeth-Kate Gulland*, Geoff West*, David A. McMeekin* and Simon Moncrieff*
* Cooperative Research Centre for Spatial Information
Department of Spatial Sciences, Curtin University, Bentley, Western Australia
Email: {tristan.reed, e.gulland, g.west, d.mcmeekin, s.moncrieff}@curtin.edu.au

***Abstract*—Web search engines, such as from Google, are very good at finding relevant information in documents and web pages. However, when such tools are used to find spatial web services, the user has to be very specific in describing what they are looking for to find relevant results in high-ranked positions. To locate an Open Geospatial Consortium-compatible web service relating to soil in Australia, a query such as "getcapabilities australia soil" is required to find relevant results, as there are no spatial constraints available. Current spatial data discovery systems, such as spatial catalogue systems, generally keyword match user queries to the content of metadata catalogues. Such systems also provide basic spatial constraints, which limit the user's ability to find results. A combination of semantic and spatial search techniques are required to effectively search geospatial data, as existing systems are primarily designed to search human-readable documents. A search algorithm is presented which uses such techniques to expand text queries to find more relevant spatial datasets through spatial filtering, natural language query decomposition and the use of thesaurus graphs to expand queries. A Resource Description Framework (RDF) schema that extends the ISO 19115 specification is explored as part of the query expansion technique, including the evaluation of tools to generate these graphs from unstructured documents which allows the overcoming of restrictions to access data behind an organisation's firewall. A prototype has been written as a web application, using the Django framework and the Python programming language and the Natural Language Toolkit (NLTK) interface to WordNet. Initial tests on seperate components of the system as well as the above system has shown the feasibility of the search system as a whole.**

***Keywords*–Semantic search; Resource Description Framework; Spatial search; Metadata; Thesaurus; Graph; Ontology.**

I. Introduction

Index-based web search engines, such as Google [1], are successful at generating automated methods to build indexes of information available on publicly accessible documents and HTML pages on the Web [2]. Such search engines cannot index information that is hidden from the Web, such as that held in file systems and databases behind firewalls.

Using search tools to find spatial data that is publicly available on the Web from an Open Geospatial Consortium (OGC)-compatible web service is only possible if the user specifies their query in a very precise manner. Such an example would be "getcapabilities australia soil". This query returns links to relevant OGC-compatible web services about soil in Australia, but the user has to know that "getcapabilities" is a word used in the schema of said web services. Without the use of "getcapabilities" in the query, relevant results are ranked lowly as the content of the machine-to-machine XML-based structure is different to the content of the human-readable HTML structure the search engine is looking for.

Specific spatial information tools exist for searching catalogues of metadata, such as GeoNetwork. The commercial Google Map Engine (GME) also searches metadata catalogues, but does not take advantage of technology used in Google's own web search engine. The CKAN cataloguing system can be used for spatial data, also searching a metadata catalogue. All of these search tools are restricted by search methods that keyword-match the user's query with the content of metadata records [3].

Limitations of keyword-matching approaches include incomplete source data, such as the content of metadata records, as well as syntactic differences between user specified queries and metadata record content with similar meanings.

Typically, metadata generation is a manual process and leads to minimal metadata being supplied by spatial data custodians for many data sets. Manually generated metadata is rarely a complete description complying with the ISO 19115 metadata standard. Due to the difficulties in automatically generating metadata to fill records, such as determining the context and relevance of data [4], it is proposed to approach improvements from the other side, automatically expanding user queries instead. It is easier to find relevant metadata records by creating contextually relevant queries than attempting to create contextually-relevant metadata, due to the size of the query compared to the size of the metadata record.

To achieve this, natural language processing is applied to queries to separate the spatial and non-spatial components of a query, allowing the application of spatial operations on data sets. Graph-based query expansion is used to parse the non-spatial part of the query, which allows the discovery of more data sets that have metadata syntactically different to the user's query, but similar in meaning. The expanded queries are run over traditional metadata records, while integrating into the expansion a graph-based domain thesaurus extracted from non-structured resources, such as reports found behind firewalls within internal repositories. Queries such as "Parks in Perth" identify an object ('Park'), a spatial operator ('in') and a location ('Perth') and can look for data sets of interest inside bounding boxes or polygons describing Perth, depending on the services used.

Much development has gone into the standardisation of metadata, including ISO 19115, used by many spatial data providers. The ISO 19115 specification is explored to help determine the best methodology to automatically generate metadata discovered through searching file systems and databases, possibly behind a firewall. This allows metadata to be acquired from sources such as PDF reports hidden in a data provider's repository. This technique is used to populate a thesaurus of similar domain-specific terms also acquired from the repository. The source of the data is noted and related to other records found in the same document, as well as in other documents containing the same terms. This information is then used to expand the location and object part of the query respectively.

Such metadata must be generated by each data provider to overcome restrictions on access behind firewalls. To this end, a number of commercially available software tools have been explored to determine their capabilities including how they can generate publishable metadata for consumption by the system. On the web, RDF models can be used to store metadata. An RDF schema of ISO 19115 [5] has been explored for its suitability, as well as research being conducted for a 'domain thesaurus'.

The paper is organised as follows: Section II explores current systems used to search and manage geospatial metadata; Section III discusses the use of semantic and spatial filtering techniques to improve search results and Section IV presents the results of a prototype system implementing some of the proposed techniques.

II. Current Approaches and Systems

Three systems currently used to search and manage geospatial metadata are GME [6], CKAN [7] and GeoNetwork [8]. GME is a commercial product from Google, Inc. which extends the abilities of Google Maps to allow more complex spatial data to be overlaid upon Google's base maps. Data custodians upload data files alongside their associated metadata to Google's cloud. Each layer, or other asset, has associated metadata which can be searched through the Maps Engine API or the Google Maps Interface itself. The search is based on keyword matching, looking for occurrences of the user's exact phrase within the metadata. CKAN is an open-source cataloguing system that, whilst not designed solely for spatial data, is commonly used to catalogue and search spatial data by various jurisdictions, including many Australian government departments.

The open source GeoNetwork is a similar system, except that data is not stored within the system itself but rather is accessed through OGC-compatible web services such as the Web Feature Service (WFS), Web Map Service (WMS) and Catalogue Service for the Web (CSW) [9]. These services expose relevant metadata about geographic data sets, which GeoNetwork keyword-matches with the user's query. The function 'GetCapabilities' exposes much of the metadata accessed by GeoNetwork, which complies with some of ISO 19115 [10]. CKAN functions in the same manner as GeoNetwork for OGC-compatible services, but also allows spatial data to be uploaded in file-based formats as well. In that case, the metadata must be manually generated rather than harvested. These services also expose the data sets themselves for use in other systems.

The OGC WFS standard defines a number of possible spatial operators including 'Contains', 'Intersect' and 'Equals' which can be applied to any known spatial feature type such as polygons or points. However, as there is no requirement for a WFS dataset to implement all of these operators, the availability of these operations cannot be assumed in all cases [11]. Another complication is that the syntax used to describe these operations varies depending upon the version of WFS specified in the data request.

All three of these systems rely upon keyword matching of the user's query; if the user misspells a word or uses a synonym of a keyword within the metadata, valid results will not be included in the result set. Much like traditional web search, keyword indexing is the primary way this is achieved. Optimisation of queries and a lack of support for alternatives means that important spatial operators such as 'in', 'within' and 'near' are ignored.

These systems do not have free access behind a firewall; GeoNetwork allows only basic authentication rather than more sophisticated methods which would use permissions to expose extra data to certain groups of machines or people. Without access behind a firewall, it is possible that many data sets and their metadata cannot be interrogated, despite the fact the user may have access to the data set.

All three systems provide the ability to restrict the search set spatially with a bounding box. However, there is no ability to restrict the search based on a polygon or text term. Being able to restrict a search by a polygon is important as polygons allow the user to use a complex many-edged shape that is more representative of real-world spatial boundaries than a bounding box.

The use of a visual bounding box drawn on screen by the user in CKAN and GeoNetwork's case is difficult and time consuming. This is particularly so when the map is small. A rectangular box is not always representative of an area of interest; consider a collection of irregular islands such as Hawaii. The bounding box as implemented in these tools also only allows the use of an implicit 'within' spatial operator; others such as 'next to' are unable to be used.

Another common theme with all of these tools is the manual generation of metadata - even in the case of automatically harvesting metadata from a 'GetCapabilities' call, the data must originally be manually generated by the data custodian. This leads to issues of quality and completeness, as metadata is typically a low priority for data custodians. Such metadata includes a title and description of the data set alongside metadata tags which briefly describe the dataset. Often these fields are subject to standardisation by data providers, leading to metadata which is sufficient for some groups but less useful for other users. It is rare that more complete ISO 19115 descriptions are provided, either formally or as part of a description field.

The features of contemporary web search tools exceed those of geospatial web search tools by allowing more complex queries from users. Many of these capabilities are examples of semantic search capabilities used in a general sense. Through using linked data, web search engines are able to deliver results that are similar to but not exactly the same as the user's query. By matching components of the user's query with synonyms and correctly spelt words (in the case of misspellings), more relevant results can be returned.

In Google's web search, this can be seen through the use of their proprietary Knowledge Engine graph, which injects some semantic capabilities into Google's web search, which expands the user's query to find data that is not expressed in the same way [12]. This can be achieved through the use of a thesaurus graph, which details relationships between words and even in some cases misspelled words. The Knowledge Graph is Google's proprietary version of this. However, it is to be noted that Google's search algorithm is designed to search human-readable HTML pages, rather than machine-to-machine XML documents such as OGC-compatible web services. The methods can however be applied to a system designed to read this format of data.

Systems such as that from [13] improve upon geospatial

search to provide more modern semantic features, however they are manual in nature and designed specifically for the geospatial context in which they are used. Rather, they are good at generating a corpus of known OGC-compatible web services and how they are related, but are not specialised in providing an improved user-facing query method.

III. Approach

A. Proposed New Semantic and Spatial Filtering Techniques

These issues highlighted above are addressed through the use of semantic technologies as seen in contemporary web search tools, including ontology-based, graph-powered natural language processing and extending the use of ontologies into handling geographic phenomena to find semantic matches to a text query containing location information. Spatial filtering techniques are also employed to further filter results of user queries to return more relevant results.

The use of these technologies enable more targeted and relevant data sets to be returned to the user for a given search query, by finding metadata that is phrased differently to the user's query, alongside being able to use spatial filters to remove geographically irrelevant data sets.

To effect these techniques, the user enters a query in one of a number of forms. Currently these are restricted, but more complex formats will be considered as needed. Natural language processing is then used to classify the query as one of the following four formats understood by the algorithm, in precedence order as seen in Figure 1.

I	{ Object } { Operation } { Location } (e.g. Parks in Perth)
II	{ Location } { Object } (e.g. Melbourne forests)
III	{ Object } { Location } (e.g. Boundaries Sydney)
IV	{ Object } (e.g. Admin boundaries)

Figure 1. Natural language precedence algorithm

In these rules, a 'Location' is a geographic area in which to restrict the search, an 'Operation' is a spatial operation upon the said 'Location' and the 'Object' is the data being inquired for, related to the 'Location'. It is assumed that, for this system, the operation will be 'within' for rules III and IV.

Rule IV is a fallback for when a location cannot be determined using rules I to III - in essence, the query is treated as a standard text query that will not take advantage of the improved ability of spatial operations and geographic restriction. Logically, it is meaningless to express this in terms of an operation and location. An overall representation of the design of the system can be seen in Figure 2.

If the query is of type I, II or III, the system will attempt to find the most relevant area using a WFS call to a service providing boundary information. The design of this part of the system allows administrators to specify the WFS, layer and field types required.

The queries are decomposed using a simple 'split' method on a list of spatial operations for rule I - these operations are 'in', 'near', 'next to' and 'intersects', however WFS and the GeoDjango system used to complete the filtering can use other operations [14]. The location will be found to the right of the split, and the object to the left.

If the algorithm cannot find an operation in the query, it attempts to find the location and object based on rules II and III. For rules II and III, the system makes calls to the WFS for the beginning and end components of the query - attempting up to three words on each side. For example, a query such as "West Perth bus stops" would try the following in precedence order as locations and the balance as objects:

- West *(rule II)*
- Stops *(rule II)*
- West Perth *(rule II, polygon found)*
- Bus Stops *(rule III)*
- West Perth Bus *(rule III)*
- Perth Bus Stops *(rule III)*

The search stops as soon as at least one result is returned by the WFS. Otherwise, the query is assumed to be of type IV.

As the user's query is not subject to any restrictions on spatial operations, it is then attempted to match the geographic location (where possible) in the text query to a polygon region (step 2a in Figure 2). Depending on the data source used, this polygon may be located in a database or accessed via a web service. It is this polygon that is then used, in conjunction with the available spatial operations, to restrict the search to more relevant results.

In some cases, it may be preferable to use a point rather than a polygon, for example spatial queries using the 'near' operator. This requires a second service that returns points rather than polygons. This can be complex, as the centroid is not always an accurate indicator of a point representing regions, as such a number of results may be needed.

The Spatial Identifier Reference Framework (SIRF) [15] was investigated for determining location information, however a method was chosen independent of SIRF that allows further extensibility and modularity. SIRF is a developing repository of location information about features within Australia, linking records where they appear in more than one dataset. SIRF primarily uses the same source data for Australia, with the linked data aspect of the system not required for the purposes of this system - only a single 'ground truth' is needed, rather than cross-referencing (assuming the supplied WFS is authoritative). There are disadvantages to not using SIRF; namely that alternative boundaries can be chosen using SIRF; ideally this should not be an issue, but calls to the SIRF API did not return the alternatives during exploration.

To determine the matching boundaries, a call is made to the chosen WFS used for the boundaries. A list is returned based on features matching the below criteria specified in the GET query string (Figure 3), where `LAYER` is the layer in which the boundaries are stored, `NAMESPACE` is the namespace used by the WFS, `PROP_NAME` is the name of the property storing the boundaries and `LOCATION_TERM` is the 'Location' term from the user's query. The approach allows the WFS to search for any properties in the boundary layer matching and containing the location. This is effected through a WFS filter on the boundary data set.

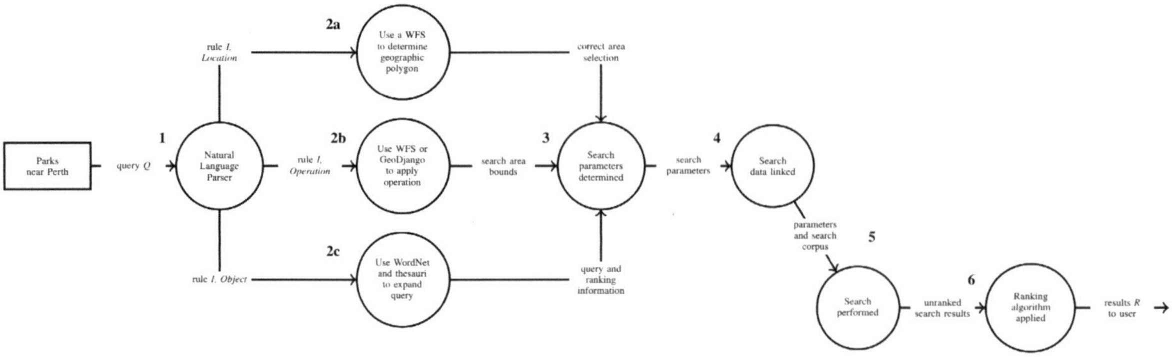

Figure 2. Diagram of the query process

```
?service=WFS&version=1.1.0&request=
GetFeature&outputFormat=json&typeName=
NAMESPACE:LAYER&propertyName=NAMESPACE:
PROP_NAME,NAMESPACE:the_geom\&filter=
<Query><Filter><PropertyIsLike wildCard=
"*" "singleChar="." matchCase="false"
escape="!"><PropertyName>NAMESPACE:
PROP_NAME</PropertyName><Literal>
*LOCATION_TERM*</Literal>
</PropertyIsLike></Filter></Query>
```

Figure 3. WFS Call for administrative boundaries

In cases where there are multiple alternatives to the boundaries, the user is able to choose from a list the boundary they intended, with the least specific area (i.e. largest sized) initially assumed. However, the query results will take into account all alternatives for the boundary, with the ranking positively influenced by the increased size of the area.

Semantic resources allow for this approach to be extended further. Using data stored in unstructured reports, including those behind firewalls, thesaurus graphs of related words can be generated alongside including information about related geographic areas. Using this information, similar areas are also searched but ranked lower based on a combination of the 'distance' of each word from the original, due to being less relevant. The information is stored as an RDF graph, which is explained in more detail in Section III-B.

If the query is of type I, a WFS call will be made to check whether the particular source supports spatial operations. If it does not, the spatial operation will be performed manually using the GeoDjango extension to Django, which allows spatial operations to be performed on data sets. In query types II or III, where this is only a geographic area specified, queries will be reduced to data falling within the boundaries of the polygon – in effect, an "in" spatial operator will be assumed. For query type IV, no processing of this type is completed.

Most current tools can only indirectly restrict results to spatial criteria via text searches - that is, to results that contain the geographic terms within the metadata records, or allowing the user to restrict their query to data sets that contain features within a bounding box. However, they cannot take advantage of spatial data sources with the ability to apply more specific spatial operations such as 'in', 'near' or 'intersects'.

Two methods are used to achieve this: either the built-in function as part of an OGC-compatible web service data set (where available), or as a manual spatial operation through GeoDjango. As this data is extracted from the user's text query, the method is hidden from the user, improving usability of the interface. These syntax and implementation specifics should not be required in a clean interface focusing on natural language queries, as they complicate the interface and are unlikely to be known by users seeking data.

As shown in Figure 4, after parsing the query to determine the spatial operation required in step 1 (of Figure 4), the system adjusts to the web services available operations by then requesting and searching its capabilities, as seen in step 2a. It can then select the most specific operation available to it, following a sequence of possible operations such as:

1) Operations include "Or", "Intersects", and "Within".
   - Return all records that are within or intersect with the boundaries of the comparison polygon(s).
2) Operations include "Within".
   - Return all records within the comparison polygon(s).
3) Operations include "BBOX".
   - Convert the comparison polygon(s) into a bounding box.
   - Return all records within the bounding box.
4) No relevant operations found.
   - Return all records.

The boundary polygon on which to apply the operation is then retrieved in step 2b (of Figure 4). Retrieving boundary polygons can go beyond simple geocoding of region names from WFS's. For example, a query for "bus stops *near* me" can make use of a user's location, obtained from their web browser, and create a buffer around that point location.

At processing step 3, the capabilities of the data source can be combined with the search areas from step 2b, depending on the system's capabilities. If the contents of a WFS contaning bus stop data were searched, for instance, the point features could be filtered by multiple polygons or, if the WFS does not have this capability, a single bounding box is created from the polygons and used as an alternative filter.

Once the geographic location and spatial operation has been determined, the rest of the query is interpreted similar

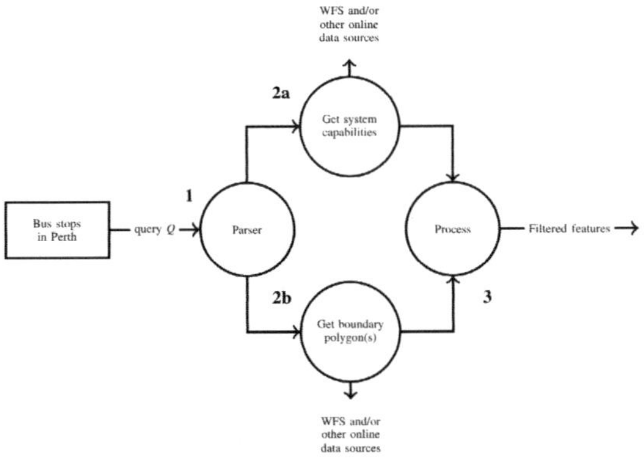

Figure 4. WFS spatial operations

to a traditional 'keyword' search, but with the query expanded using the semantic graphs. Query Expansion is performed to ensure that a broader base is used to find relevant metadata to the user's query, effectively creating and using multiple queries which are ranked depending on relevance.

The NLTK [16] toolkit is used to determine similar words and phrases to that supplied by the user. Using one of the built-in similarity ranking algorithms within NLTK, a ranking is applied to how similar an expanded query is to the original. A keyword-match of each of these queries is undertaken against the metadata records of the OGC-compatible web service data sets supplied to the system. The results are then ranked by a combination (where applicable) of the similarity of the query, the level of matching of the keywords and the geographic proximity of the results (in the case of a spatial operation such as 'near').

Through leveraging WordNet, GeoMeta is able to achieve a similar ability to Google's web search to match indirect queries, through query expansion of metadata corresponding to geospatial data sets. This consists of comparing queries using a ranking algorithm against the content of metadata records (as used in the existing approach by GeoNetwork et al.), but also through other sources of metadata. Metadata records are sourced from data sets they are attached to, located using OGC-compatible web services. Other sources of metadata are generated from unstructured reports and similar documents.

The NLTK is able to provide an interface to WordNet that links related words such as synonyms. For example, if the user entered the word 'park' as part of their query, the system would also look for 'reserve' as well. It has been previously shown [17] that it is feasible to use WordNet for query expansion. As such, investigation is ongoing as to the best way to use WordNet for this purpose. As the interface allows the user to determine and supply also the particular type of the word (noun, verb), relevant expansions can be determined. In cases where it is likely that the query has two meanings and the NLTK is unable to determine which, the user is asked to choose their intended meaning from a choice of possibilities, or the user can choose the option to rank both equally. A dictionary of domain-specific terms is also used in parallel and in the same manner, explained in more detail in Section III-B.

Investigations are ongoing to determine if any of the many similarity algorithms within WordNet are suitable. As each algorithm weights differently the relationship between two words, testing is being undertaken to determine which is most accurate in the context of GeoMeta. The similarity algorithm will then be used as part of the ranking algorithm. The system can also easily by extended by the user's own controlled vocabulary ontologies either automatically generated by an extension to the software or manually by the user.

Finally, the result set is returned to the user interface in JSON format (a lightweight data container used to store complex data), through the callback of the original AJAX request. The Google Maps API is used to visualise the data set by displaying a bounding box of the data on a map, alongside some traditional text-based metadata (such as a description) and a link to the data set being returned.

B. *Use of Semantics in Geospatial Search*

The use of semantically linked data greatly increases the relevance of returned results. Semantic graphs allow data to be linked together by meaning, and as such can be used to extend the context of the user's search query. In the context of this search system, similar locations are linked together alongside similar domain terminology in a 'thesaurus'-type format. An RDF schema is proposed for each of these, with the schema for the locations being influenced by the ISO 19115 RDF schema. This allows further expansion of queries to be matched by the algorithm.

ISO 19115 is the de facto standard for defining relevant metadata for geospatial data sets, providing a set of mandatory and optional metadata [10]. It is advantageous for use within a geospatial search system, as the standard allows for searching over the description and classification of a wide range and type of geographic metadata. Examples of this metadata includes traditional text-based descriptions alongside more detailed information about the physical, spatial and lineage aspects of the corresponding data set [18].

In practice, ISO 19115 is rarely used to its full potential, due to the burden of manually generating the required metadata and large quantity of optional fields [19]. OGC-compatible web services only require the bare minimum of the standard to be complied with, as that data is used as part of the 'GetCapabilities' function. For these reasons, often very little additional use of ISO 19115 is described in said web services.

As part of this search system, the extension of metadata available in ISO 19115 to a more flexible representation is proposed, which would allow more comprehensive coverage of data sets and hence more likely to satisfy a variety of user needs.

This would make use of the automatic generation of linked data from non-structured sources. Each of these allows the extension of queries into logically similar but syntactically different forms, therefore catching more metadata record resources for each query than keyword-matching alone. This therefore returns more complete search results. Such an approach therefore reduces the burden for both the user and the data custodian, as this information will be generated automatically.

The RDF metadata graphs are expressed through an RDF schema, which detail the kind of metadata terms of interest to be extracted from unstructured documents (such as reports or PDF files) and linked together based on lexical distance within

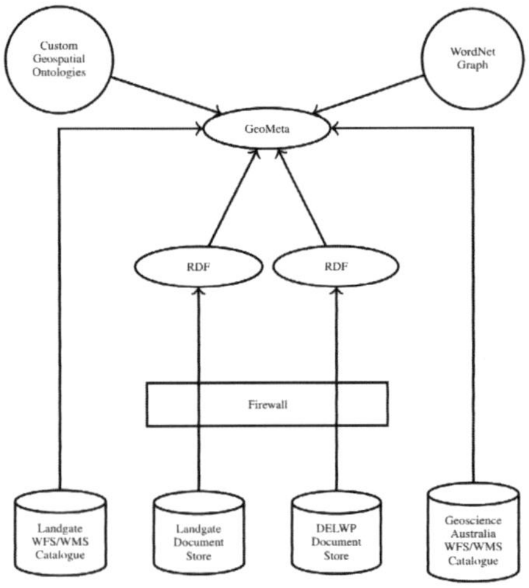

Figure 5. Data sources for GeoMeta

the text. In this way, it is similar to the WordNet graph used for query expansion, but rather than the links being determined by the similarity of the word, it is determined by both the similarity and distance within the text. This technique is used directly for the purposes of the domain terminology schema, whereas it is applied only for locations with respect to that schema.

The generation of schema-complying RDF descriptions is achieved through the use of existing tools which already search unstructured data. Voyager [20], Sintelix [21] and Omniscient [22] are being explored as part of the project. Plugins are being written to convert the underlying databases of these products to RDF descriptions compliant with the schemas. These tools allow the automated creation and updating of the RDF descriptions with minimal user input.

The design of the interoperable schema allows any tool to be used for this purpose in a plug-in modular format, as long as the tool can produce RDF that complies with the schema. The design of this architecture allows custodians to run the tool behind a firewall, only exposing the minimum data required for the schema, allowing the use of a much greater corpus of metadata that would often be left behind the firewall. This is a benefit over existing tools, which are generally restricted to searching the public Web, or private networks through basic authentication.

The sources of data for the case study of the search system can be seen in Figure 5, with the RDF metadata being the central two sources. Each data source can be used to generate data complying to either schema, depending on content is within the file. For the location schema, such data as described in the ISO 19115 standard is generated from unstructured documents.

Alongside geographic information, descriptions of more 'general' terms of relevance related to physical 'things' will be generated, which may aid finding results for both the spatial and non-spatial component of the user's query. Examples of such extra fields are 'Person', 'Organisation' and 'Currency'. These are determined by the above software tools, as they are designed to look for specific types of information. ISO 19115 has already been generated into RDF [5], and is explored, modified and expanded for this purpose.

IV. Results

A system named GeoMeta has been produced, consisting of a user facing front-end (where the user enters a text-based query) and a server side back-end. The front end, written using HTML5 and jQuery, provides a native application-like experience giving more feedback to the user. For example, as the text queries are edited, the results are updated on-the-fly, and loading screens are provided to show that the system is processing a query. This is achieved through Asynchronous JavaScript and XML (AJAX) queries from the front-end to the back-end.

The back-end system is written using the Django [23] framework in the Python programming language. Modular components were designed to allow for new features to be dropped in. This enables, for example, the WordNet interface (used in part to expand queries) to be exchanged with a comparable system. Currently, this interface is provided through the NLTK [16] library. It is entirely possible to substitute other thesauri graphs through the NLTK library (or other libraries) to use them instead. The NLTK connects locally to downloaded versions of the WordNet corpora. WordNet is able to be used as a service from a remote server [24]; however as the corpora is static this method is not being used.

The first version of the prototype is a demonstration of using a text-based query that can be split into both spatial and non-spatial components. The system used the Google Geolocation API to determine a bounding box of the location, and was fixed into only being able to search using the 'in' spatial operator. As a data source, the Shared Land Information Platform (SLIP) [25] of Landgate, the Western Australian Land Information Authority, was used. This was accessed through the Google Maps Engine API, however as the API will soon be discontinued, future versions will support different ingest mechanisms for data sources (namely OGC-compatible web services).

The GME API allowed the dataset title, description, metadata tags and bounding boxes to be extracted to create an effective metadata record for each data set. Although there are other metadata fields available for use within GME, these were rarely used. Hence, these three fields were the only ones used through the API. This data was generated manually by Landgate, and contains some gaps and repetitive template text with limited relevance.

Many of the descriptions followed a standard format consisting of generic information about the agency and contact information, which negatively influenced the search results. For example, a search for 'imagery' had many matches because the agency is described as supplying imagery, even when the dataset itself did not contain any. This data is also not fully viewable without authentication, hence a system such as proposed in this paper would solve both of the above issues of inadequate metadata and the metadata being behind a firewall.

This demonstrator proved that even the basic additions of spatial filtering proved useful. The ability to filter spatially allowed results that were not fit for purpose, by being located in other areas of Western Australia, to be excluded from the search results, leaving only more relevant candidate data sets for consideration.

A second demonstrator, currently under development, uses a more advanced natural language processing classifier as described in Section III-A. This enables the program to sort queries into four types, allowing more advanced spatial operations to be performed. The second version allows these to be performed but only through the GeoDjango method. A rudimentary version of the WordNet graph used for query expansion has been implemented, alongside the use of OGC-compatible web services as data sources due to the retirement of the Google Maps Engine API. This also allows the system to sit on top of existing systems such as FIND [26] (a GeoNetwork instance publishing many datasets with the custodian being the national Australia geographic agency) and other GeoNetwork installations (which primarily catalogue OGC-compatible data sets). This architecture reduces the work required by custodians to exploit the GeoMeta system.

Small-scale tests on parts of the second system have been undertaken on each component to determine their suitability. A 'region finder' has been implemented based on the Administrative Boundaries data set from the Australian Bureau of Statistics, available through the Australian Government's 'National Map' [27]. This component of the system works successfully, with polygons being returned for various types of boundaries that match the user's 'Location' part of the query. The polygons returned are are as expected. The 'rule classifier' has also been built, and has been successful in categorising a wide range of queries into the relevant type. Tests continue to fine-tune and improve the system for more advanced uses, such as long 'Location' terms for rules II and III that exceed three words in length.

The method to apply spatial operations built in to some OGC-compatible web services has been explored for suitability. The method has the benefit of being able to offload some of the heavy processing to remote servers, reducing response time and the need for caching (as the queries will need to be processed live on the remote server). It is possible that remote server load issues could be encountered, and as such a method will need to be determined as to when the local processing should be used instead. Indeed, this has already happened in informal tests. Formal tests are being conducted to determine the best way to work out if a OGC-compatible web service will be 'too slow', as the apparent processing speed is a function of latency, data set complexity, the processing power of the remote machine and other factors, not all of which are able to be determined ahead of time.

Examination of the three tools used to automatically generate metadata from unstructured documents show that they are all able to pick out relevant information that can then be used within an RDF schema. Comparisons of these results are being undertaken to measure the quality of results from each system. Initial results show that all three tools are suitable to generate useful metadata.

The system can be investigated at http://research.haxx.net.au/geometa where a prototype of the system resides.

V. Conclusion and Future Plans

This paper has presented a search algorithm to overcome many of the limitations of using contemporary geospatial search engines to find spatial data relevant to a user's query. The algorithm presented extends upon the traditional search technique of keyword-matching the user's query with the content of metadata records, by using natural language processing to split a query into spatial and non-spatial components.

The splitting of the spatial component of the query allows sophisticated spatial operations to be undertaken by the user to find relevant data sets that satisfy the operation, defined by bounding boxes and polygons. The use of a natural language text-based interface enables a user-friendly experience that more easily articulates the users' intentions, compared to a map image-based input system for spatial queries. This reduces effort and uncertainty, as well as providing better results than a pure keyword-matching approach.

The design of a prototype, GeoMeta, is presented as well as some results from an initial proof of concept based on a case study with Landgate, the Western Australian Land Information Authority, that has been using the Google cloud to store spatial data which is accessed using GME. This experience indicated that there are advantages in not only opening up data sets to be indexed by search engines, but also having data custodians run automatic metadata-generating software over their own internal document repositories to generate improved metadata.

Future plans include the integration of automatically generated metadata. As this metadata will be generated by third-party tools in a format complying with the proposed RDF schema, future research will need to be able to interpret these and link them in with existing data. This will be achieved by using the data within the RDF files to further facilitate query expansion, ranking results by distance in the same manner as the existing system.

The WordNet algorithm currently implemented will be fine-tuned to better articulate the users' intention to ensure that the query expansion is relevant, and will remove automatically generated 'nonsensical' queries from those being searched. This will ensure that only relevant expanded queries are used, improving response time to the end user. Future research will also allow the user to augment this with their own ontologies, allowing more domain specific terms to be used in queries.

Finally, polygon comparison will be implemented on the data sets where available. This will further enhance the accuracy of results when compared to a bounding box comparison. Further investigation is needed to determine whether it is feasible to process data files when the polygon is not exposed via a Web Feature Service. Polygons will also be displayed in the preview map image in later versions.

VI. Acknowledgement

The work has been supported by the Cooperative Research Centre for Spatial Information, whose activities are funded by the Business Cooperative Research Centres Program.

References

[1] Alphabet, Inc., "Google," https://google.com/, 2016, [Online; accessed 2016-02-10].

[2] S. Brin and L. Page, "The anatomy of a large-scale hypertextual web search engine," Computer Networks and ISDN Systems, vol. 30, no. 1/7, 1998, pp. 107–117. [Online]. Available: http://dx.doi.org/10.1016/S0169-7552(98)00110-X

[3] S. Grill and M. Schneider, "Geonetwork opensource as an application for SDI and education," in GIS Ostrava, 2009, pp. 25–28. [Online]. Available: http://gis.vsb.cz/GIS_Ostrava/GIS_Ova_2009/sbornik/Lists/Papers/039.pdf

[4] C. Jenkins, M. Jackson, P. Burden, and J. Wallis, "Automatic RDF metadata generation for resource discovery," Computer Networks, vol. 31, no. 11, 1999, pp. 1305–1320.

[5] A. Saiful, "Development of a web-based modeling system using metadata concepts and databases," Doctoral Dissertation, Drexel University, 2004.

[6] Alphabet, Inc., "Google Maps Engine," https://developers.google.com/maps-engine, 2015, [Online; accessed 2015-10-12].

[7] Open Knowledge Foundation, "ckan - The open source data portal," http://ckan.org/, 2016, [Online; accessed 2016-02-10].

[8] Open Source Geospatial Foundation, "GeoNetwork opensource," http://geonetwork-opensource.org/, 2016, [Online; accessed 2016-02-10].

[9] N. Chen, X. Wang, and X. Yang, "A direct registry service method for sensors and algorithms based on the process model," Computers Geosciences, vol. 56, jul 2013, pp. 45–55. [Online]. Available: http://linkinghub.elsevier.com/retrieve/pii/S0098300413000691

[10] K. Senkler, U. Voges, and A. Remke, "An ISO 19115/19119 Profile for OGC Catalogue Services CSW 2.0," in 10th EC GI GIS Workshop, Warsaw, Poland, 2004.

[11] A. Friis-Christensen, M. Lutz, and N. Ostländer, "Designing Service Architectures for Distributed Geoprocessing: Challenges and Future Directions," Transactions in GIS, vol. 11, no. 6, 2007, pp. 799–818.

[12] M. Nickel, K. Murphy, V. Tresp, and E. Gabrilovich, "A Review of Relational Machine Learning for Knowledge Graphs From Multi-Relational Link Prediction to Automated Knowledge Graph Construction," 2015.

[13] W. Li, "Automated data discovery, reasoning and ranking in support of building an intelligent geospatial search engine," Doctoral Dissertation, George Mason University, 2010. [Online]. Available: http://ebot.gmu.edu/handle/1920/6013

[14] Django Software Foundation, Django Documentation (Release 1.8.6), 2015. [Online]. Available: http://media.readthedocs.org/pdf/django/1.8.x/django.pdf

[15] Commonwealth Scientific and Industrial Research Organisation, "SIRF," http://portal.sirf.net/, 2015, [Online; accessed 2015-12-16].

[16] NLTK Project, "Natural Language Toolkit - NLTK 3.0 documentation," http://nltk.org/, 2016, [Online; accessed 2016-01-15].

[17] D. Buscaldi, P. Rosso, and E. Arnal, "A WordNet-based Query Expansion method for Geographical Information Retrieval," in Working notes for the CLEF workshop, 2005. [Online]. Available: http://citeseerx.ist.psu.edu/viewdoc/download?doi=10.1.1.99.8031&rep=rep1&type=pdf

[18] Standards Australia, "AS/NZS ISO 19115.1 Geographic information Metadata Part 1: Fundamentals," Tech. Rep., 2015.

[19] O. Karschnick, F. Kruse, S. Töpker, T. Riegel, M. Eichler, and S. Behrens, "The UDK and ISO 19115 Standard," in Proceedings of the 17th International Conference Informatics for Environmental Protection EnviroInfo, 2003.

[20] AAM, "AAM Group — Geospatial Excellence," http://aamgroup.com/, 2016, [Online; accessed 2016-01-21].

[21] Semantic Sciences, "Sintelix - gaining value from your corporate documents," http://sintelix.com/, 2016, [Online; accessed 2016-02-09].

[22] Omnilink, "Omniscient Spatial Metadata — OMNILINK Property and Location Data Management," http://omnilink.com.au/products/omniscient-spatial-metadata/, 2016, [Online; accessed 2016-02-09].

[23] Django Software Foundation, "The web framework for perfectionists with deadlines — Django," https://djangoproject.com/, 2016, [Online; accessed 2016-01-15].

[24] A. Fred, J. L. G. Dietz, K. Liu, and J. Filipe, Knowledge Discovery, Knowlege Engineering and Knowledge Management. Springer, 2011. [Online]. Available: http://dx.doi.org/10.1007/978-3-642-19032-2

[25] Landgate, "SLIP Home," http://slip.landgate.wa.gov.au/, 2016, [Online; accessed 2016-02-09].

[26] Department of Communications, "Find - Office of Spatial Policy," http://find.ga.gov.au/, 2016, [Online; accessed 2016-02-09].

[27] Australian Government, "NationalMap," http://nationalmap.gov.au/, 2016, [Online; accessed 2016-02-09].

A Linear Approach for Spatial Data Integration

Alexey Noskov and Yerach Doytsher
Mapping and Geo-Information Engineering
Technion – Israel Institute of Technology
Haifa, Israel
emails: {noskov, doytsher}@technion.ac.il

Abstract—**The developed method allows the user to integrate polygonal or linear datasets. Most existing approaches do not work well in the case of partial equality of polygons. The suggested method consists of two phases: searching for counterpart boundaries or polylines by triangulation, and rectifying objects without correspondent polylines by a transformation and a shortest path algorithm. At the first phase, middle points of polygon boundaries are used to implement the triangulation. In order to define correspondent boundaries, the polylines of the two datasets which are connected by triangles are compared based on the lengths of lines and the distances between the nodes. At the second phase, vertices of the polylines without counterparts are shifted with respect to the lengths of the shortest distances to the nodes of the polylines with counterpart. The method is effective for pairs of datasets with different degrees of accuracy. Less accurate datasets use precise elements of other datasets for integration and improvement of their accuracy. The resulting data are well integrated with a more accurate map. A review implemented by specialists enables us to say that the results are satisfactory.**

Keywords-Geometry fusion; triangulation; shortest path; topology.

I. INTRODUCTION

We live in the information age. Terabytes of spatial information are available today. Hundreds of sources produce thousands of maps and digital layers every day. We encounter serious problems when trying to use different maps together.

Let us list some popular data producers. Survey companies and agencies prepare accurate topographic maps and plans. Aero and satellite images act as a basis for numerous variations of derivative maps (e.g., thematic and topographic maps). A special niche is reserved for crowd sourcing maps, e.g., OpenStreetMap (OSM) [2]. Significant parts of this sort of map contain data derived from users' devices, mainly GPS devices.

It is very difficult to use all these data together. In many cases, the user decides to draw a map from scratch, despite having existing maps with most of the required elements for the user's map. One of the reasons for this situation is a low degree of integration of existing datasets even when we consider maps containing many identical elements. For instance, soil maps need to be based on topographic maps. Today, soil maps could take basic contours from different sources.

In an ideal situation, spatial datasets use the objects (polylines or polygons) from more accurate datasets. In the real world, many maps are produced by measuring/digitizing objects from satellite images. As a result, despite the fact that most of the objects on different maps are identical, they are presented with small positional discrepancies. The problem is compounded by the fact that different objects in a Geographic Information System (GIS) environment could be depicted by the same geometries (e.g., square or circle). Thus, specific tools and algorithms need to be developed. This makes it difficult to detect identical objects on different maps. The obvious advantage of integrated databases is efficiency of data storing. Equal elements from different maps link to the same object in the storage memory. We do not need to take up extra storage on a disk. Additionally, editing of objects will be reflected on all maps, which contain them.

The benefits of data integration are demonstrated in this paper by using the city planning and cadastral datasets. A cadastral map is a comprehensive register of the real estate boundaries of a country. Cadastral data are produced using quality large-scale surveying with total station, Differential Global Positioning System devices or other surveying systems with a centimeters-level precision. Normally, the precision of maps based on non-survey large-scale data (e.g., satellite images) is lower. City planning data contain proposals for developing urban areas. Most city planning maps are developed by digitizing handmade maps, using space images. Almost all boundaries have small discrepancies in comparison to cadastral maps. We need to integrate these datasets, where the identical elements in the datasets have to be linked to the same geometries. All the non-identical elements have to be coherent with shared geometries.

The approach we suggest enables the user to resolve the described problems. It consists of two main stages: defining correspondent boundaries using triangulation technique, and rectification of the remaining polylines by transformation and the shortest path algorithm. The suggested approach could be applied to polygonal and linear datasets.

This paper is structured as follows: the related work is considered in Section II. The initial processing of the source datasets is described in Section III. Section IV focuses on correspondent boundary definition. The problem of resolving line pair conflicts is described in Section V. The shortest path approach for fusion boundaries with and without counterpart is discussed in Section VI. The results are discussed in Section VII. The conclusion is presented in Section VIII.

II. Related Work

The main groups of approaches for data matching and data fusion are considered in this section

The wide spread of databases is the reason for developing attribute-based matching methods. Schema-based [10] and Ontology-based types of attribute matching could be selected. In [13], an approach based on both types is presented. Attribute-based matching could be effective when data with sustainable and meaningful structure and content of attribute database is processed.

The map conflation approaches [11] are based on data fusion algorithms; the aim of the process is to prepare a map, which is a combination of two or more [6]. The merging and fusion of heterogeneous databases has been extensively studied, both spatially [9] and non-spatially [14].

Geometry, size, or area is used in feature-based matching. These allow us to estimate the degree of compatibility of objects. The process is carried out by the structural analysis of a set of objects and analysis of the result to see whether similar structural analysis of the candidates fits the objects of the other data set [1]. In [12], comparison of objects is based on the analysis of a contour distribution histogram. A polar coordinates approach for calculating the histogram is used. A method based on the Wasserstein distance was published by Schmitzer et al. [5]. A special shape descriptor for defined correspondent objects on raster images was developed by Ma and Longin [17]. Focusing on single shapes does not allow us to apply these algorithms in our task.

In [4], topological and spatial neighborly relations between two datasets, preserved even after running operations such as rotation or scale, were discovered. In relational matching, the comparison of the object is implemented with respect to a neighboring object. We can verify the similarity of two objects by considering neighboring objects. The problem of non-rigid shape recognition is studied by Bronstein et al. [3]; the applicability of diffusion distances within the Gromov-Hausdorff framework and the presence of topological changes have been explored in this paper.

We have concluded that the mentioned approaches could not be applied to resolve the considered problem. That leads from the fact that the mentioned approaches have been developed for specific conditions. For instance, the feature-based matching is effective for detecting separate outstanding objects; attribute-based matching is effective for definite and well-designed databases. Thus, a new approach should be developed.

III. Data preparation

Spatial data sets covering a part of Yokne'am (a town in the northern part of Israel) have been used. They are depicted in Figure 1. Land-use city planning and cadastre polygons are displayed as color areas and as black boundaries, correspondingly. As can be seen in the figure, in most cases the boundaries of two datasets are the same. Some boundaries are presented in the first dataset and are not presented in the second. The white background of the cadastre polygons means that this area is not covered by the city planning dataset. It is mainly presented in the upper part of the figure. The case where black cadastre boundaries cross an area with a similar background color means that these boundaries are not presented in the city planning datasets.

The city planning data have sensitive positional irregular discrepancies. Because of the small scale, they cannot be observed in Figure 1; hence, the problem is illustrated in Figure 2. The figure shows that the problem could not be resolved by transformation only, and that a more sophisticated technique is required. The figure leads us to an approach based on defining corresponding objects and further modification of the remaining objects with respect to found pairs.

Figure 1. Source data: land-use city planning (colored background) and cadastre (black outline) maps.

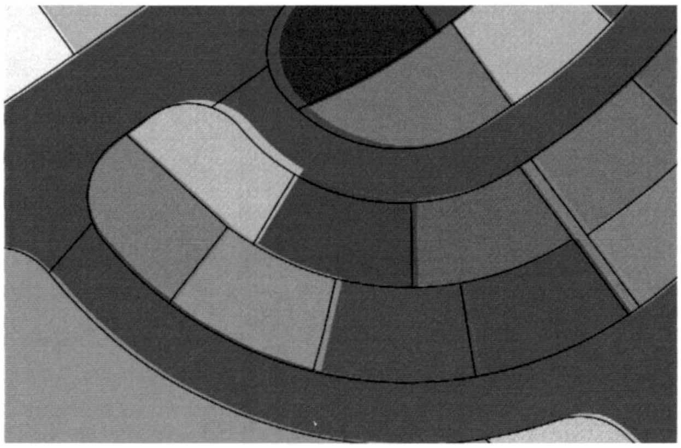

Figure 2. Positional discrepancies of city planning (colored areas) and cadastre (black lines) datasets.

In the previous approach [8], we defined correspondences between polygons. We encountered two problems. Because whole polygons are processed, it is difficult to precisely define the points connecting polygons with and without counterparts. Considering a polygon as a separate object does not allow us to unambiguously detect polygons' shared nodes. As a result, in some cases, it is difficult to correctly eliminate gaps between objects. Using centroids in the polygon triangulation approach is the reason for the second problem. For non-compact polygons, even small changes in the polygon's boundary lead to significant changes in the centroid position. It could negatively impact the results.

In this paper, we propose a technique, which is based on defining line pairs by triangulation. In most cases spatial data are found in non-topological data format (e.g., ESRI's Shape Files, GeoJSON, MapInfo Tab Files). This means, that the boundaries of neighboring objects are repeated for each polygon. This fact leads us to the possibility of modifying the boundary of neighbor polygons independently. In the most cases, it is a source of many difficulties, e.g., small gaps between boundaries or the necessity of repeating the same action for each polygon separately. Because of the problems mentioned we use topological data format provided by GRASS GIS 7 [16]. The source shape files have been converted to this format. A sample part of the city planning dataset found in a topological format is presented in Figure 3. Polygon data comprise 3 types of elements: boundary, node, and centroids. Nodes separate boundary polylines. Each group of closed boundaries could be considered as an area. Centroids link polygon to certain raw in attribute table by a category number. Each raw in the attribute table starts with a "cat" field, which could be connected to a centroid with a given "cat" value.

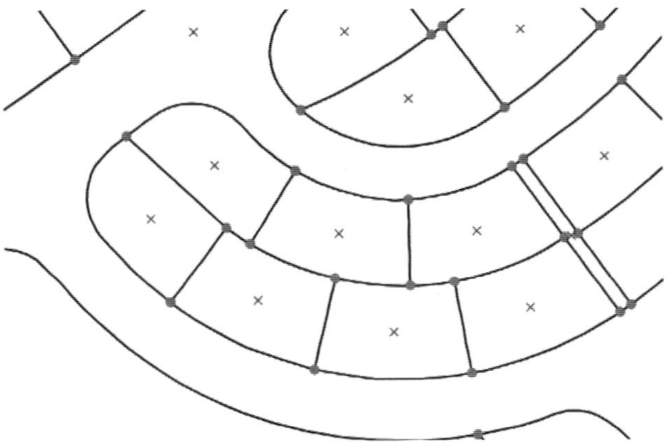

Figure 3. A sample of the city planning dataset residing in GRASS GIS's topologycal format. Nodes – red circles, centroids – blue croses, and boundaries – black lines.

We can conclude from the first two figures, that most of the counterpart polygon boundaries of the datasets are located close to each other and present the same objects. It is efficient to define a measure for detecting the fact that two objects certainly could not be defined as counterparts. In other words, we can use it as a filter. Maximal distance parameter could fulfill this role.

In addition, it is quite popular to use buffers for detecting this fact. For instance, in [15] the authors have applied a buffer with a certain buffer size, where all objects outside the buffer could not be considered as counterparts. We have found that a segmentation technique could be more sensitive and flexible in this context. Segmentation means dividing polygon boundaries (or any other sort of polyline) into equidistant segments. Point delimiters are used to calculate distances between the considered datasets. An example of segmentation is depicted in Figure 4.

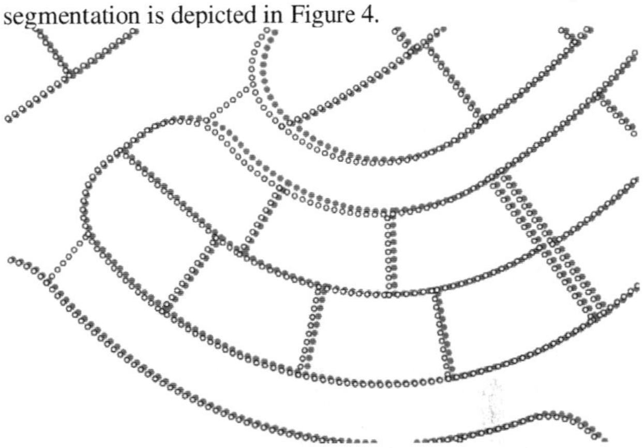

Figure 4. Point delimiter of equidistant segments. City planning – red, cadastre - black points.

Maximal distance (D_{max}) is calculated as follows. For each point in the first dataset, a distance to the closest point belonging to the second dataset is assigned. Then we apply a loop from the first to the last percentile (from the percentile with maximal number and minimal distance to that with minimal number and maximal distance) on the list of 100 percentiles of the calculated distances. D_{max} equals percentile i if the standard deviation of distances between percentiles i-1 and i is more then 1. D_{max} is used mainly to filter considered objects. In our case, the distances between the nearest equidistant points of the cadastre and the city planning data sets' boundaries are in an interval from 0 to 92.7 meters. The boundaries of the percentiles number (i.e., i decrement) 6, 5, 4, and 3 are 2.09, 4.97, 7.88, and 17.75 meters, correspondingly. Standard deviations for distances in intervals between percentiles 6-5, 5-4, and 4-3 are as follows: 0.78, 0.89, 1.46, and 2.77. Hence, D_{max} equals 7.88, because 7.88 belongs to percentile number 4 (the first with a standard deviation of more than 1). Objects residing further than D_{max} are excluded from the processing. For Yokne'am datasets, D_{max} equals 7.9 meters. A 2-meter distance between nearest points has been assigned for our test.

IV. DEFINING CORRESPONDING LINES OF DATASETS BY TRIANGULATION

In this section, the main process is described. It is based on identifying correspondent triples of polygon boundaries of the considered datasets. Delaunay triangulation enables us to easily connect points by triangles. We use it to divide boundaries into triples. Figure 5 illustrates the triangulation process. The triangulation is based on the middle points of boundaries' polylines. In the figure, the boundaries' middle points are depicted as gray circles; the boundaries are colored lines; and the triangulation layer is presented a colored background.

Now, we have grouped middle points into triples boundaries of cadastre and city planning datasets. The next step is searching for correspondent triple candidates, and it is implemented as follows.

First, the lengths of all boundary polylines are calculated. Sorted lengths of correspondent boundaries are stored into "A", "B" and "C" fields of attribute table for each triple. "A" stores the shortest length, "C" stores the longest. Then, we compare all possible pairs of triples.

To reduce the number of comparisons we consider only the nearest triples. These are defined by comparing the coordinates of the start and end nodes of their boundaries. For further consideration, all start and end nodes of the second triple boundaries have to be inside the extent of the first triple's nodes (defined by an enlarged buffer). Buffer size is equal to the square root of the median polygon area. In our case it is 32 meters. The areas of both datasets are sorted into one list to find a median value.

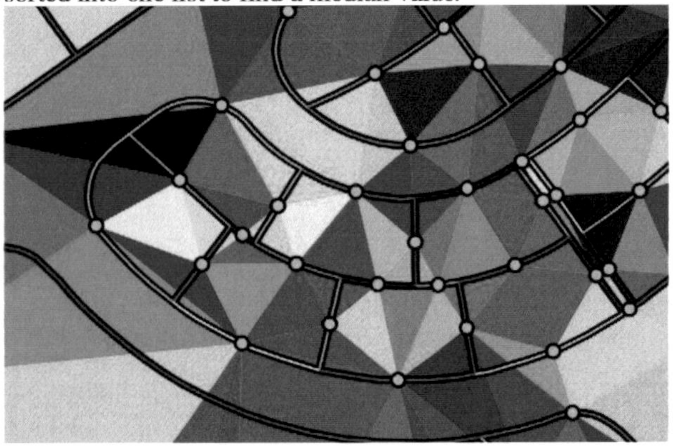

Figure 5. The triangulation of boundaries' middle points of a city planning dataset.

In the next step, we compare boundary lengths. As mentioned above, ordered lengths are stored in an attribute table ("A", "B" and "C" fields). Triple pairs are added into a list for further processing if a correspondent length (A-A, B-B, or C-C) resident in the second triple is within an interval of between 80% to 120% of a length resident in the first triple, and are considered as triple pair candidates. This two-step initial filter by extents and lengths comparison is illustrated in Figure 6. In the figure, blue lines are city planning boundaries; black lines are cadastre boundaries, grey and green triangles are candidate cadastre boundaries obtained by an extent (red rectangle) and by length comparisons, correspondingly. Candidates are defined for a triple of city planning boundaries marked by a red triangle.

At this point, we have a few candidates. In order to define the "winner" candidate, we calculate distances between nodes of the correspondent boundaries. We need to determine pair boundaries belonging to a considered triple candidate. The brute force process is implemented; all possible combinations are considered. The most acceptable combination is a combination with a minimal sum of distances between correspondent points. The brute force process is not time sensitive, because it is implemented only for a few filtered candidates. A candidate is marked as a triple pair if the maximal distance between correspondent nodes is less than D_{max}, as defined in Section III.

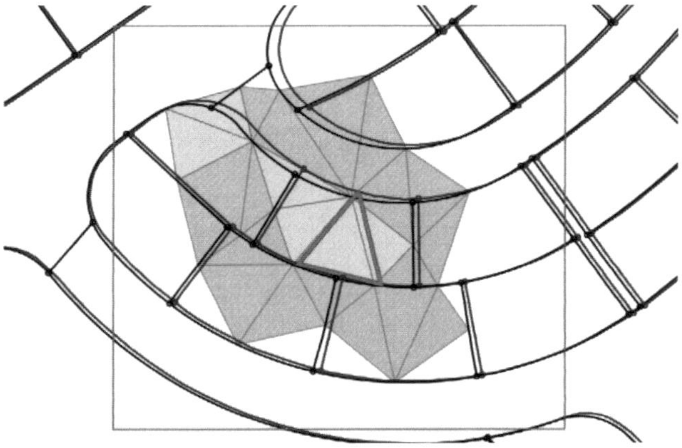

Figure 6. Filtering possible triple pairs.

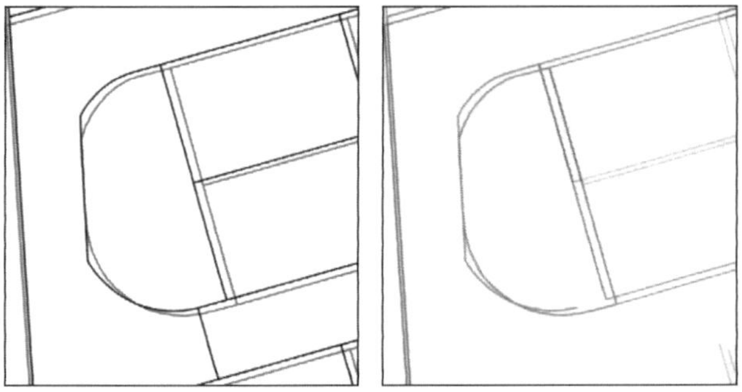

Figure 7. An example of an incorrectly found line pair. Left – original boundaries of the city planning (red lines) and cadastre (black lines) datasets. Right – detected linepairs.

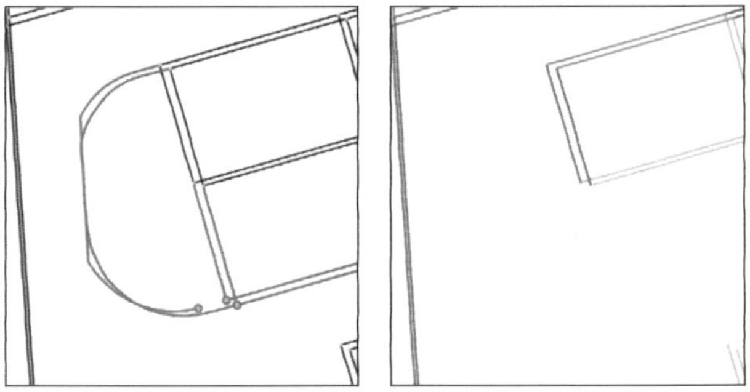

Figure 8. Detecting incorrect pairs. Left – incorrect nodes and line pairs are marked in red. Rigth – final line pairs.

In this section, correspondent boundaries have been defined. The candidate triples have been filtered by extent and lengths comparison, then line pairs have been defined by distances between nodes.

V. Resolving line pairs' conflicts

In this section, we describe the process of searching for wrongly defined boundary pairs and resolving these situations.

First of all, in many cases line pairs are repeated in neighboring triples. The participation of a line in different pairs is marked as a problem. It is quite obvious that a boundary from the first dataset could have only one counterpart boundary in the second dataset. In order to resolve conflicts, we compare the number of times they participate in triples. For instance, we have two line pairs A1-B1 and A1-B2. If A1-B1 pair is encountered in 2 triples and A1-B2 in 1, then the combination A1-B2 is eliminated and A1-B1 remains. If both are encountered simultaneously, both candidates are eliminated.

Additionally, we need to consider the situation illustrated in Figure 7. The curved purple line pair is detected incorrectly. This line is composed of two lines in the cadastre dataset, because of the line, which is connected to the bottom part. The connected line does not exist in the city planning dataset.

These types of errors could be detected by analyzing the line junctions. Each node is identified by a set of ids of lines connected to the node. The required conditions for the remaining line pairs are as follows. First, node values (set of ids of lines) have to be unique. Second, each node has to have a node of equal value, and vise versa. If one of the conditions is false, all lines connecting with the incorrect node are eliminated on both datasets. The process is illustrated in Figure 8.

VI. A shortest path approach for boundaries fusion

At this point, we have the pairs of corresponding boundaries. As mentioned in Section I, cadastre datasets are produced using quality large-scale data. They are more accurate than city planning datasets. Hence, replacing the city planning boundaries with their cadastre counterparts will significantly improve the accuracy of the resulting map. This was done in the previous step. In this section, we consider how to integrate boundaries without counterparts with pair boundaries. This is implemented in two steps.

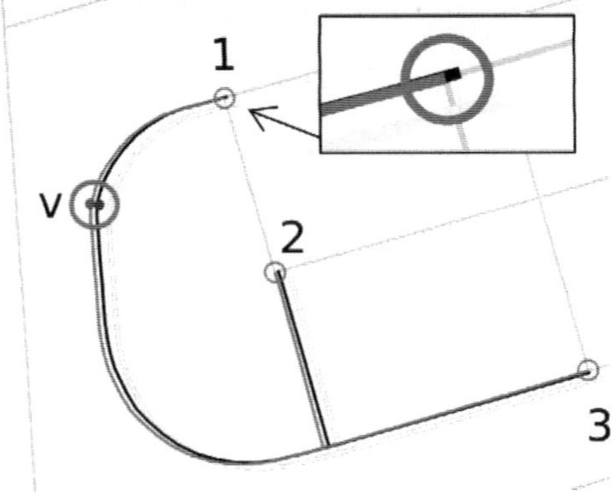

Figure 9. A vertex moved with respect to the shortest paths to bridge nodes.

In the first step we use coordinates of correspondent pair nodes as Ground Control Points for second-order affine transformation. We transform the boundaries without counterpart to make them closer to the cadastre dataset. We shall henceforth call it "transformed boundaries or dataset".

The transformed boundaries still have gaps between them and the remaining boundaries. A shortest path approach has been developed to integrate both types of boundaries.

The idea of the approach is quite simple. Each vertex (including nodes) of the transformed boundaries is processed. We calculate the shortest path from a vertex to each bridge node. Bridge nodes connect a nest (group of lines joined without gaps) of transformed boundaries to boundaries with counterparts. In figure 9, the described elements are presented.

The figure explains the algorithm. Green lines are cadastre counterparts. Black lines are transformed city planning boundaries without pairs. They still have small gaps with cadastre counterparts. Red lines are the result of applying the shortest path approach to each vertex. Vertex v is the considered vertex and 1, 2, and 3 are the bridge nodes. Bridge nodes of a transformed dataset differ from the other nodes by having a counterpart node in the cadastre pair boundaries. Thus, we can precisely say how to move bridge nodes in order to locate them exactly on the node of cadastre boundaries with pairs. It is not correct to only move a bridge node; we need to move other vertices too.

To define new coordinates we use shortest paths. Three nodes are impacted for the vertex "v". Thus, three shortest paths are calculated: v-1, v-2, and v-3. v-2 and v-3 are partially overlapped paths. We need to note an important condition. If a path touches more then 1 bridge node, the path is eliminated from further consideration. Only paths intersected by one bridge node are considered. The new coordinates of a vertex are calculated as follows.

$$c_2 = c_1 + \sum_{0}^{n} (c_{oi} - c_{ti}) \cdot (1 - l_i / l_{sum}) \qquad (1)$$

In (1), c denotes x or y coordinate; c_1 is the source coordinate; c_2 is the target. n is number of bridge nodes, i is index of the current bridge node. c_o and c_t are x or y coordinates of pair bridge nodes resident in cadastre counterpart and transformed (without pair) city planning boundaries, correspondingly. l_i is the length of the shortest path to be considered as a bridge node. l_{sum} is the sum of lengths of the shortest paths to bridge nodes from the vertex.

Let us consider an example of calculating new coordinates by the shortest path method. We have 3 paths from vertex v to bridge nodes 1, 2 and 3. The paths' lengths are 19.8, 66.8, and 76.3. c_o - c_t values are (x y) -0.39 -0.14, -0.34 -0.24, and -0.23 0.16. For such parameters we need to add -0.67 -0.18 to the x y coordinates of the vertex.

VII. Results

In order to acquire a final result, cadastre pairs of the boundaries are merged with the rectified boundaries without counterparts. Since pair boundaries have the same id and the rectified boundaries of the city planning dataset without cadastre pairs inherit the original ids, the correspondences between original and final polygons could be established by comparing ids of boundaries comprising a polygon. It is derived from the fact that each polygon could be identified by a unique set of ids of boundaries.

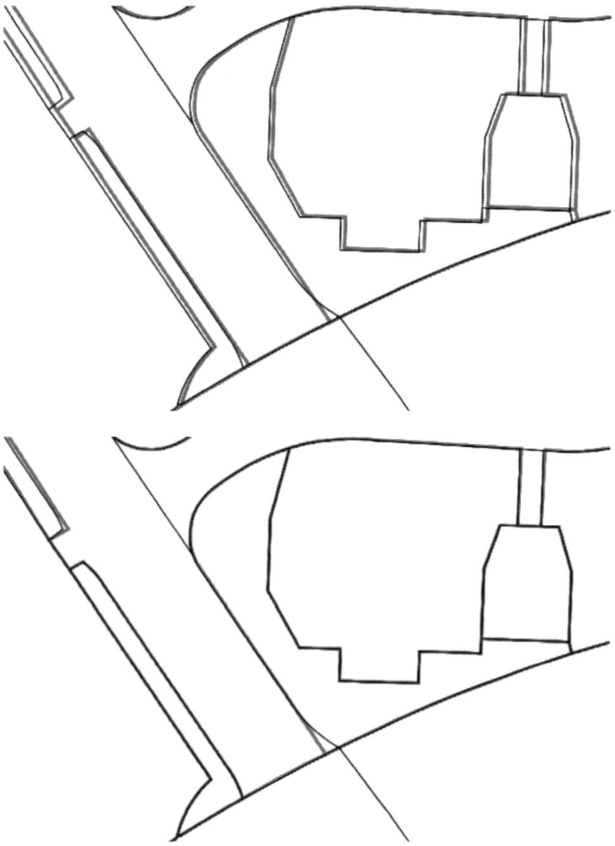

Figure 10. Zoomed-in extent 1. Bountaries of original (upper) and result (lower) datasets: city planning – red, cadastre - black.

TABLE I. AVERAGE DISTANCES AND STANDARD DEVIATIONS

Parameter	Dataset compared with cadastral layer	
	Original city planning	*Result city planning*
Average distance, m	1.15	0.24
Standard deviation, m	0.64	0.41

The result datasets are presented in Figure 10 and Figure 11. We can conclude that most boundaries have been taken from the cadastral dataset; others have been rectified to connect boundaries without corresponding pairs and boundaries with pairs. The result looks satisfactory; the final map is holistic and does not contain significant deficiencies. A review implemented by specialists enables us to state that the results are satisfactory.

In order to estimate the results quantitatively, we use distances between the closest equidistant points of the cadastral and the city planning data sets' boundaries. The distances have been calculated between original city planning and cadastral datasets, as well as, the result and cadastral datasets. Only distances less than D_{max} have been taken into account. In Table I, average distances and standard deviations are presented.

Figure 11. Zoomed-in extent 2. Bountaries of original (upper) and result (lower) datasets: city planning – red, cadastre - black.

According to the table, the average distance has been reduced by five times; standard deviation has been reduced by a factor of three. We can conclude from the table that the accuracy of the original dataset has been significantly improved.

To implement the approach, we used Python 2.7 programming language, GRASS GIS 7.1, and Debian GNU/Linux 8 operating system.

VIII. CONCLUSION AND FUTURE WORK

An approach for improving linear and polygonal spatial datasets is presented. Land-use city planning dataset locations have been corrected according to the cadastral dataset.

The outline of the approach is as follows. The conventional polygon data have been converted to topological data format. Boundaries have been split into equidistant segments to calculate D_{max}. Then, correspondent boundaries have been defined using triangulation technique. Rectification of the remaining polylines by transformation and the shortest path algorithm has been implemented.

In the future, we need to test the approach with more datasets and different parameters, to compare it with other approaches. In order to improve the presented approach by also defining correspondences between parts of boundaries (not only whole boundaries), we would like to combine this approach with the segmentation-based algorithm published in [7]. This will allow us to apply the method to other types of datasets. For instance, OSM datasets are usually complete, updated, and relatively non-accurate. In order to produce updated and precise layers, it could be useful to integrate OSM data with an accurate dataset.

ACKNOWLEDGEMENT

This research was supported by the Survey of Israel as a part of Project 2019317. The authors would like to thank the Survey of Israel for providing the financial support and data for the purpose of this research.

REFERENCES

[1] S. Belongie, J. Malik, and J. Puzicha, "Shape matching and object recognition using shape contexts," IEEE Trans. on Pattern Analysis and Machine Intelligence, 24(4), 2002, pp. 509—522.

[2] J. Bennett, "OpenStreetMap - Be your own cartographer," ISBN: 978-1-84719-750-4, Packt Publishing, 2011.

[3] A. Bronstein, R. Kimmel, M. Mahmoudi, and G. Sapiro, "A Gromov-Hausdorff framework with diffusion geometry for topologically-robust non-rigid shape matching," International Journal of Computer Vision, vol. 89(2-3), 2010, pp. 266-286.

[4] X. Chen, "Spatial relation between uncertain sets," International archives of Photogrammetry and remote sensing, vol. 31(B3), Vienna, 1996, pp. 105-110.

[5] B. Schmitzer and C. Schnorr, "Object segmentation by shape matching with Wasserstein modes," Energy Minimization Methods in Computer Vision and Pattern Recognition, Springer Berlin Heidelberg, 2013.

[6] S. Filin and Y. Doytsher, "The detection of corresponding objects in a linear-based map conflation," Surveying and land information systems, vol. 60(2), 2000, pp. 117-127.

[7] A. Noskov and Y. Doytsher, "A Segmentation-based Approach for Improving the Accuracy of Polygon Data," GEOProcessing 2015, Portugal, 2015, pp. 69-74.

[8] A. Noskov and Y. Doytsher, "Triangulation and Segmentation-based Approach for Improving the Accuracy of Polygon Data," International Journal on Advances in Software, vol. 9 (1-2), 2016, accepted, in progress.

[9] C. Parent and S. Spaccapietra, "Database integration: the key to data interoperability," Advances in Object-Oriented Data Modeling, M. P. Papazoglou, S. Spaccapietra, Z. Tari (Eds.), The MIT Press, 2000.

[10] E. Rahm and P. Bernstein, "A survey of approaches to automatic schema matching," The International Journal on Very Large Data Bases (VLDB), vol. 10(4), 2001, pp. 334–350.

[11] A. Saalfeld, "Conflation-automated map compilation," International Journal of Geographical Information Science (IJGIS), vol. 2 (3), 1988, pp. 217–228.

[12] X. Shu and X. Wu. "A novel contour descriptor for 2D shape matching and its application to image retrieval", Image and vision Computing, vol. 29.4, 2011, pp. 286-294.

[13] P. Shvaiko and J. Euzenat, "A survey of schema-based matching approaches," Journal on Data Semantics IV, Springer Berlin Heidelberg, 2005, pp. 146-171.

[14] G. Wiederhold, "Mediation to deal with heterogeneous data sources," Interoperating Geographic Information System, 1999, pp. 1–16.

[15] S. Zheng and J. Zheng, "Assessing the completeness and positional accuracy of OpenStreetMap in China," Thematic Cartography for the Society, Springer International Publishing, 2014, pp. 171-189

[16] M. Landa, "GRASS GIS 7.0: Interoperability improvements," GIS Ostrava, Jan. 2013, pp.21-23.

[17] T. Ma and J. Longin, "From partial shape matching through local deformation to robust global shape similarity for object detection," Computer Vision and Pattern Recognition (CVPR), IEEE Conference on. IEEE, 2011, pp. 1441-1448.

BIM/GIS-based Data Integration Framework for Facility Management

Tae-Wook Kang†, Seung-Hwa Park‡ and Chang-Hee Hong*
ICT Convergence & Integration Research Institute
Korea Institute of Construction Technology
Seoul, South Korea
e-mail: laputa99999@gmail.com, {parkseunghwa, chhong}@kict.re.kr
† 1st Author, ‡ 2nd Author, * Corresponding Author

Abstract— **In this study, we propose a software architecture for the effective integration of building information modeling (BIM) into a geographic information system (GIS)-based facilities management (FM) system related to smart city service. This requires the acquisition of data from various sources followed by the transformation of this data into an appropriate format. The integration and representation of heterogeneous data in a GIS are very important for use cases involving both BIM and GIS data, such as in the management of municipal facilities. We propose a BIM/GIS-based Data Integration Framework (DI) that separates geometrical information from that related to the relevant properties. For a GIS based facility management, the property information is extracted from BIM models and transformed using the ETL (extract, transform, and load) concept. In consideration of the viewpoint of geometry, the surface model for several representations in the GIS, we designed BG (BIM/GIS)-DI, developed a prototype, and verified the results through interviews. The results show that BG-DI and BIM/GIS integration has benefits such as reusability and extensibility.**

Keywords-BIM; GIS; FM; integration; surface model; BG-DI

I. INTRODUCTION

The integration of data through geographic information system (GIS)-based building information modeling (BIM) has recently emerged as an important area of research related to developing smart city services. Several studies have investigated the benefits of the effective integration of BIM and GIS [1][2].

This integration process generally involves the extraction and transformation of information required by each domain in a single project. GIS and BIM are similar in that both are used to model spatial information — the former is used for outdoor modeling and the latter for indoor modeling —and have common use cases, such as location-based municipal facilities information queries and management, etc. In order to implement some use cases based on BIM and GIS, effective interoperability between GIS and BIM should be supported by an appropriate platform.

To facilitate information interoperability in the construction sector, the buildingSMART, which is the worldwide authority driving transformation of the built environment through creation & adoption of open, international standards, has developed and standardized the Industry Foundation Classes (IFC) data model in a significant effort to accommodate industry requirements. IFC is an integrated model schema that describes construction information. It uses an object-oriented method to integrate information required by the relevant stakeholders in a project.

Even though the IFC based information model integration was researched using various methods, there were practical issues in solving the integration problem. For example, during information exchange among heterogeneous systems, which means more than two different software, or commercial modeling software using IFCs, some loss or change of information has been reported [3]. Furthermore, a related study has pointed out various issues associated with the integration of BIM and GIS [4].

In particular, a facilities management system that has accumulated data over a long period tends to lack IFC compatibility and fails to support the creation of an IFC model. In general, the formats that a facility management (FM) system supports for import and export are text, spreadsheet, and relational database file, and the data in these are heterogeneous [5]. Managing a BIM-based facility initially requires the integration of heterogeneous data with a BIM object. By integrating GIS and BIM, the BIM model including the facility management data can be utilized effectively based on GIS data. The manual integration of heterogeneous data can incur a substantial cost and cause incorrect data entries, thus hindering correct decision making.

For these reasons, integration among heterogeneous datasets related to FM, and the BIM and GIS models should be automated. Moreover, the data integration process should be adjustable according to the use case at hand, and each phase of integration should be testable.

This study approaches problems related to BIM-based data integration from a practical perspective. To integrate heterogeneous data, such as BIM, GIS, and FM data, we propose a method to map FM data from BIM to GIS by using a BIM/GIS-based information Extract, Transform, and Load (BG-DI) method. To verify the effectiveness of the proposed the architecture, we developed a prototype system and conducted interviews with experts.

The rest of the paper is structured as follows. In Section II, we present the research objective of this study. Section III describes other conventional approaches from related literature. In Section IV, we describe our BG data integration framework and, in Section V, we discuss some case studies. We conclude in Section VI.

II. RESEARCH OBJECTIVE

Our purpose in this study is to propose BG-DI architecture for the effective integration of BIM, GIS, and FM data. We design BG-DI workflow in order to define each phase to map BIM objects to GIS objects in a CityGML model after integrating external, heterogeneous data into the attributes of the BIM object. For BIM object shaping, LODs are obtained through an LOD extraction algorithm and mapped to the LOD of the relevant GIS object. Accordingly, we also propose an architecture that effectively integrates the properties and shapes of the BIM object into GIS and displays them, as this is required by the user.

In this study, we design the ETL concept for the proposed BG-DI architecture to effectively integrate BIM, GIS, and FM data, and provide object mapping that transforms the BIM model —IFC —into the GIS model, CityGML. BG-DI includes heterogeneous data extraction, data integration with the BIM object, a workflow transforming the BIM object to the GIS object, and mapping rules.

To test the usefulness of the proposed architecture, we implement simple facility management use cases. We extracted and processed information stored in the BIM facility management database of the Korea Institute of Construction Technology to check the information using the GIS through the BIM model. The model uploaded to the GIS is a surface-based model that simplifies the BIM model, which has a large capacity, and contains information of a degree of detail between LOD1 and LOD2. We can upload the BIM model to an additional viewer in order to check details beyond LOD3. When a facility object included in the BIM model is selected, the FM information can be viewed. Through this architecture, the information required according to each use-case perspective is defined, processed, and extracted through BG-DI. Thus, heterogeneous systems are cost-effectively interrelated to form a data warehouse that can be utilized for information mining. BG-DI provides various data sources and facilitates function expansion. The object geometry information of BIM can be quickly visualized by the simplified surface model.

III. CONVENTIONAL APPROACHES

Hijazi et al [6] proposed a mapping methodology to extract utility information through CityGML application domain extensions. The integration of BIM into GIS was considered in a study on GeoBIM to extend GIS data using CityGML and an open source-based BIM server [7].

Sebastian et al. proposed a method that expands BIM using an application domain extension to support interoperability between BIM and a GIS in relation to a bridge construction plan [8]. Furthermore, in order to implement a GIS-based use case, such as land selection or fire management in the construction industry, Isikdag et al. proposed a method to integrate BIM information into GIS [9]. The relevant study developed a persistent schema-level model view schema in order to convert IFC data into an Environmental Systems Research Institute (ESRI) schema. It also proposed a method that converts this data into transient temporary object model data, integrates them with a GIS geographic data model, and saves the final data in ESRI's shape file and geodatabase structures [10][11]. Expert interviews were used to confirm the results of this study in terms of quality according to ISO 9126-1.

Moreover, a few researchers attempted to solve problems arising from the difference between BIM and GIS schemas by developing a new common schema. This unified building model (UBM) analyzed IFC and CityGML schema structures and proposed a new schema [12][13].

BIM data can be converted into another schema model depending on its structure. Such conversion is considered a mapping-based process that, in general, consists of several mapping conditions and rules.

In relation to this approach, Nour partially utilized a model in order to use IFC in a cooperative team work environment [14]. He pointed out that schema-based modeling tools, such as Standard for the Exchange of Product Model Data (STEP) tools and Express Data Management (EDM), are complex and inconvenient for the user.

LaPierre and Cote [15] proposed another approach to integrate data that considers a web service-based solution for city data management using CityGML, Web Feature Service (WFS), and 3D Viewer. Döllner and Hagedorn [16] researched the integration of city information from GIS, computer-aided design (CAD), and BIM using a web service supported by the Onuma system, and Akinci et al. [17] proposed an ontology structure and a navigation method to resolve the CAD and GIS use cases.

A Spatial Data Warehouse (SDW) is a system that adds a 3D model of the required information from the use-case perspective to the existing data warehouse system. SDW has long been studied in the field of GIS. It supports analysis and decision making by storing non-volatile data that have integrative and temporal properties according to the relevant topic-centered spatial and non-spatial information regarding properties [18]. SDWs are constructed based on spatial data extracted from heterogeneous systems, such as a GIS and asset management systems. BIM also focuses on the re-utilization of object information over a certain space and can be effectively utilized from the perspective of BIM interoperability. An SDW can be created and renewed in a topic-oriented manner through ETL.

For data managed using FM, ETL supports effective extraction, transformation, and loading processes from heterogeneous systems. Even if numeric data are the same at the time of loading, they can have a different meaning or representation depending on the perspective of the project stakeholder or the user.

During the construction of a data warehouse (DW), it is important to load only the required information by extracting source data from the heterogeneous database management system (DBMS) or the software used by the relevant project's stakeholders. The extracted data may include geometrical spatial data, as well as non-spatial data, such as properties. From the BIM perspective, spatial data extracted and loaded through an ETL process should have a structure that assists the data analysis requirement with respect to the DW [19].

A recent study by Gökçe and Gökçe on ETL for buildings centered on a case study of the energy management system used to perform extraction, transformation, and data loading [20]. They showed that a single integration information model is not suitable for an environment where each project stakeholder uses a different database. They proposed an architecture where building information was extracted from data sources and sensor data, including multi-dimensional data, to avoid the above-mentioned problem. Information was extracted using a wireless sensor network or CAD and was managed in the DW.

As described above, SDWs are the focus of most DW- and ETL-related studies. Few studies have considered data integration in conjunction with a GIS.

Of the aforementioned studies on heterogeneous data integration, the service- and system-based approaches may be more effective than others in terms of system performance, but are disadvantageous because the programming code for system development can limit the element mapping process, in which heterogeneous data are integrated, of the relevant models. These approaches hinder the flexibility and extendibility of a system, and require extensively specialized problem-solving methods.

IV. BG-Data Integration Framework

A. *Overview for semantic data integration between*

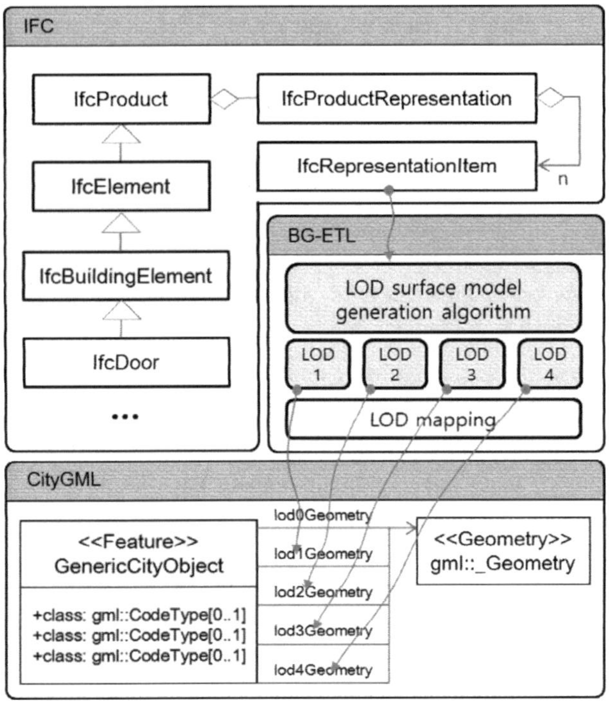

Figure 1. Data Integration Process between IFC and CityGML

For data integration between heterogeneous data models, mapping rules appropriate for a use-case are needed. Further, if an element is missing when mapping from the data source to the target information model, an algorithm that generates the necessary element using the data source needs to be executed prior to executing the mapping rules. Figure 1 shows such a process.

B. *Data Integration Framework Design*

The BG-DI architecture should consider the scalability and flexibility of the data integration method in order to support interoperability. Taking this into consideration, the architecture for supporting BIM/GIS-based FM is designed as shown in Figure 2. The BIM/GIS middleware consists of an IFC converter to represent GIS model, which is suggested, BG-DI to extract the external data related to FM. Following its extraction from the heterogeneous system, such as the excel file, external data is stored in the DW DB. The DW storage phase normalizes heterogeneous data in the form of a table. Following this, the datasets stored in each table of the DW are connected to the BIM object of IFC. To link the BIM object and external data, such as maintenance records stored in the excel file, the Primary Key (PK), such as the Globally Unique Identifier (GUID), is used as the primary key in the data schema of our database. The BIM objects are mapped in the form of CityGML to be represented in GIS format. Moreover, for prompt visualization, building facility objects in the GIS are converted to a lightweight surface format. The facility objects are represented as LOD1, LOD2, etc., in GIS, and the shape of an object at a high level, such as LOD4, is verified through a separate viewer.

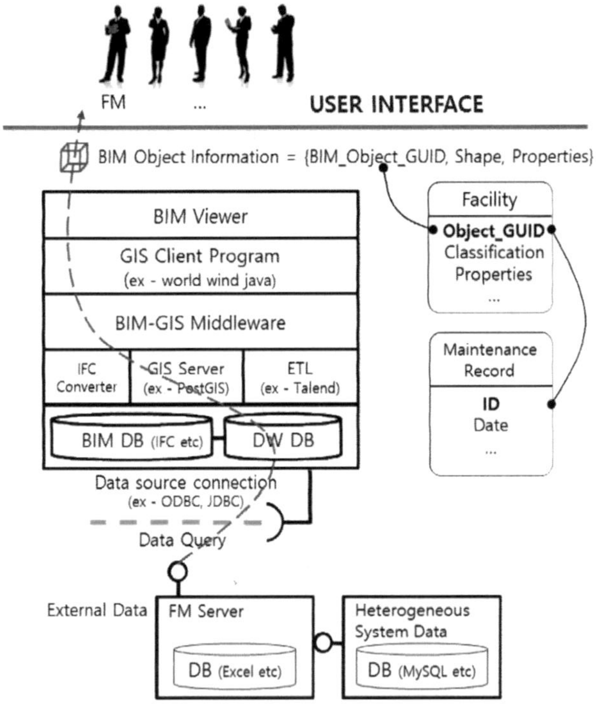

Figure 2. BG-Data Integration Workflow

C. BG-DI workflow and data mapping rule set

If connecting heterogeneous facility data and the BIM object, and representing the data and object in the GIS are not automated, the data modelers and FM staff need to perform additional tasks for data integration and manual conversion. Such manual tasks can cause side effects, including data entry errors, and the integrity of the integrated data can be compromised, thus generating unproductive results. Therefore, if possible, the workflow where the elements between set of heterogeneous data models are mapped needs to be automated, and a standard method is needed to verify the integrity of the data during mapping. A BG-DI workflow is defined to standardize such data integration. The BG-DI workflow is applied using the ETL concept and, as shown in Figure 3, mapping the elements of the data model between two heterogeneous systems is completely automated.

The definition of the BG-DI workflow consists of the following elements:

1. Extraction: Heterogeneous datasets are extracted and stored in the form of a relational database. The necessary data from the perspective of each use case are extracted; the data structure is similar to a star schema, which is built in the DW. The relationship of the table of each dataset forming the schema should be set around the PK, as with the object GUID, in order to connect it to the BIM object.

2. Transform: This process comprises two steps:

1) D2B_Binding: The dataset is integrated by linking it to the relevant BIM object. The dataset is extracted from the table stored in the DW during the extraction phase. The parameter for setting the PK needed for the data source stored in the DW and binding is defined as the "DataRecord" element. The parameter is used to bind the BIM object and the dataset of the DW through the PK field, which is designated in the "Object" element. The "category" parameter is used to distinguish the attribute categories, and any name that is designated here is saved as "("+"categoryname" + ")" + "attributename" during CityGML's attribute mapping.

2) B2G_MappingRuleset: The mapping rules are defined to semantically map the integrated BIM object to the GIS object. The mapping rules are classified into "Object," which is an object attribute mapping rule, and "Geometry," which is a shape mapping rule. Mapping a shape requires the execution of a separate algorithm to generate the LODs. Thus, an algorithm implementation module can be set to the "algorithm" parameter. The mapping source is designated as "source" and the mapping target as "destination."

3. Load: The format mapped to a GIS object is represented in CityGML and, for effective visualization of the actual building facility objects when they are loaded in the GIS, data integration post-processing may be needed. That is, the data can be converted to a lightweight format optimized for visualization. Such post-processing requires an additional algorithm.

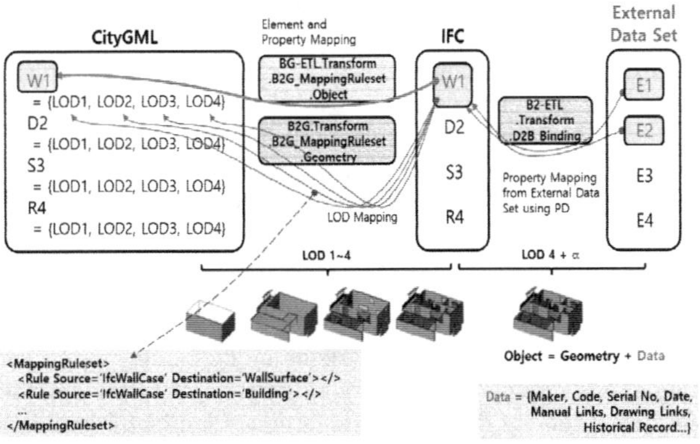

• In example, W1 = Wall#1, D2 = Door#2 S3 = Slab#3, R4 = Roof#4.
• External Data Set – External data set related to the specific use-cases such as FM

Figure 3. BG-Data Integration Workflow

BG-DI defines a mapping rule table for semantic mapping from the BIM model to the GIS model. The mapping tables are classified into two types, attribute mapping and shape mapping, and is defined. Attribute mapping is defined by naming the source and target object, where the attribute tag "TYPE" is used to store the type of the source object in the corresponding target object because 1:1 mapping is difficult. For shape mapping, the LOD unavailable in the BIM model needs to be obtained through an LOD generation algorithm and then properly mapped to the LOD of an object, as shown in Figure 4.

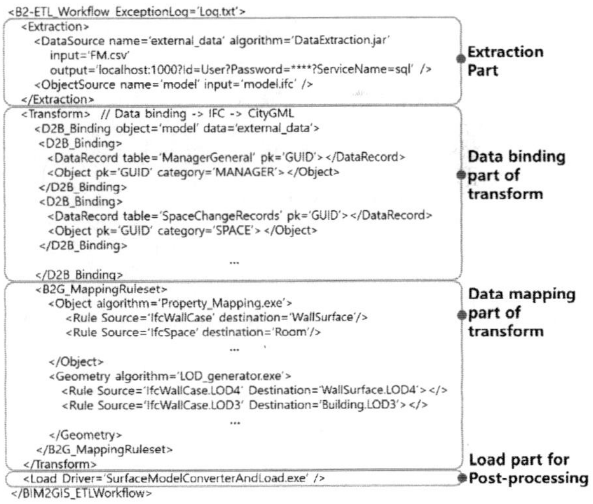

Figure 4. BG-Data Integration Workflow

V. CASE STUDY

We developed a prototype system by utilizing the BIM/GIS-based FM software architecture described in Section 4. The databases, which were integrated with the current system for information interoperability, were Excel-based structures constructed for the BIM-based FM of the main building at the Korea Institute of Construction

Technology (KICT), and the BIM objects were modeled using the Revit software. The databases were constructed according to existing managed documents. Thus, its maintenance history was managed through documents and drawings. Many items of information were hence unavailable because of illegible handwriting. Furthermore, it was difficult to obtain maintenance history data in terms of BIM objects. Therefore, the facility maintenance historical data was constructed on the basis of space.

The FM database for KICT was constructed in only two months. Therefore, it was primarily divided into structural data and maintenance history data for the space, and was managed in Excel files. The classification code system for the facility object information was defined according to the construction information classification system published in 2006 by the Ministry of Land, Infrastructure, and Transport (Table 1).

TABLE I. FACILITY DATA ITEM (KICT).

No	Item	Description
1	Information classification code	Space classification based on facility and configured as two-digit numbers
2	Actual space name	Actual space name
3	Space ID	Revit's zone object ID
4	Manager	Name of manager
5	Space number	To manage room space, facility managers are assigned this additional number
6	Space modification history information	Maintenance history information, such as space modified date, space area, space perimeter, space volume, space ceiling height, and the number of occupants
7	Floor maintenance history information	Maintenance history information, such as space floor finish, partial repair, repair rate, total repair, and final repair date
8	Wall maintenance history information	Maintenance history information, such as space wall finish, partial repair, repair rate, total repair, and final repair date
9	Ceiling maintenance history information	Maintenance history information, such as space ceiling finish, partial repair, repair rate, total repair, and final repair date

This table was extracted from the excel file and included space, floor, wall, and ceiling management information developed to manage KICT building and facilities

The FM data were extracted, transformed, and loaded into the DW by the ETL process, and each property was represented from the user's perspective. From the viewpoint of geometrical representation performance, the Surface Model format was about 13.6 times faster than IFC format in terms of the data loading time for sample data with 643,279,768 vertices. Figure 5 shows our prototype system.

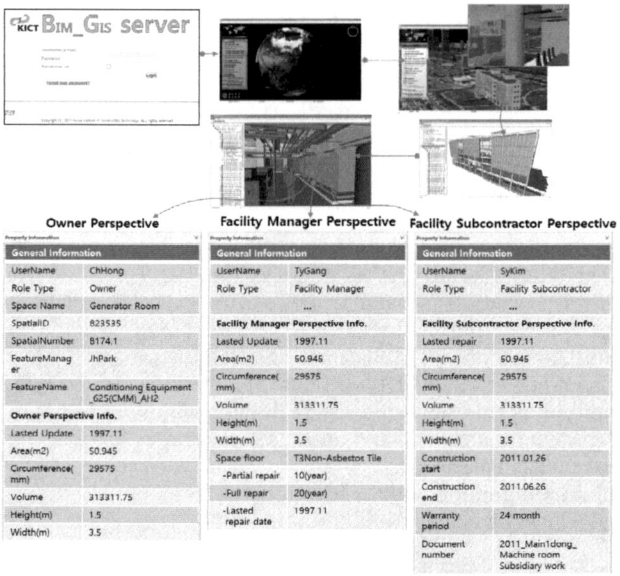

Figure 5. Prototype System for BIM/GIS-based FM software

VI. CONCLUSIONS

In this paper, we proposed a BG-DI architecture for the effective integration of data from heterogeneous systems of BIM, GIS, and FM.

From the practical viewpoint of data integration, data were divided according to their geometry and property information, allowing the problem of GIS- and BIM-based information interoperability to be addressed. Property information was extracted and transformed to obtain the required information from a use-case perspective by utilizing BG-DI. Applying BG-DI, we designed an effective architecture for the support of information interoperability between heterogeneous BIM, GIS, and FM systems, and developed a prototype that implemented FM use cases. Thus, we verified the effective integration of the required data from the project stakeholder's perspective.

In future work, we intend to analyze the spatial data of a topic on the basis of the proposed architecture, and study the effect on datasets using linkage analysis between the previously analyzed spatial data and other spatial data. We also intend to obtain query information required for decision making through data mining based on BIM.

ACKNOWLEDGMENT

This research was supported by a grant from the Strategic Research Project (Development of BIM/GIS Interoperability Open-Platform 2015) funded by the Korea Institute of Construction Technology.

REFERENCES

[1] W. Wu, X. Yang, and Q. Fan, GIS-BIM Based Virtual Facility Energy Assessment (VFEA)–Framework, Development and Use Case of California State University, Fresno. In Computing in Civil and Building Engineering, pp. 339-346, 2014.

[2] I. Hijazi, M. Ehlers, S. Zlatanova, and U. Isikdag, IFC to CityGML Transformation Framework for Geo-analysis: A Water Utility Network Case, in 3D GeoInfo, Proceedings of the 4th International Workshop on 3D Geo-Information, Ghent: Ghent University, pp. 123-127, 2009.

[3] J. I. Lim et al., IFC Test between Commercial 3D CAD Application using IFC, Korea Institute of Construction Engineering and Management 9, pp. 85-94, 2008.

[4] R. E. Meouche, M. Rezoug, and I. Hijazi, Integrating and Managing BIM in GIS, Software Review, in Proceedings of the International Archives of the Photogrammetry, Remote Sensing and Spatial Information Sciences, Vol. XL-2/W2, pp. 31-34, 2013.

[5] S. H. Park, "openBIM-based Operation and Management of Architectural Design Information according to LOI (Level of Information) for Integrated Design Process." Ph.D. diss., Kyung Hee University, 2015

[6] I. Hijazi, M. Ehlers, S. Zlatanova, and T. Becker, Initial Investigations for Modeling Interior Utilities within 3D Geo-context: Transforming IFC-Interior Utility to CityGML/UtilityNetworkADE, Advances in 3D Geo-Information Sciences, pp. 95-113, 2011.

[7] R. D. Laat and L. V. Berlo, Integration of BIM and GIS: The Development of the CityGML GeoBIM Extension, Advances in 3D Geo-Information Sciences, pp. 211-225, 2010.

[8] R. Sebastian, M. Böhms, and P. van den Helm, BIM and GIS for Low-Disturbance Construction, in Proceedings of the 13th International Conference on Construction Applications of Virtual Reality, 2013.

[9] U. Isikdag, J. Underwood, and G. Aouad, An Investigation into the Applicability of Building Information Models in Geospatial Environment in Support of Site Selection and Fire Response Management Processes, Advanced engineering informatics 22, no. 4, pp. 504-519, 2008.

[10] R. Pickard, ArcView Shape Files: A Read/Write OCX, Help File. [Online]. Available from: http://arcscripts.esri.com/details.asp?dbid=11810/ 2015.12.21.

[11] M. Zeiler, Modelling Our World: The ESRI Guide to Geodatabase Design, ESRI Press, Redlands, CA, US, 2001.

[12] M. El-Mekawy, A. Östman, and I. Hijazi, An Evaluation of IFC-CityGML Unidirectional Conversion, International Journal of Advanced Computer Science & Applications 3, no. 5, pp. 159-171, 2012.

[13] M. El-Mekawy, Integrating BIM and GIS for 3D City Modeling, Royal Institute of Technology (KTH), 2010.

[14] M. Nour, Manipulating IFC Sub-models in Collaborative Teamwork Environments, in Proceedings of the 24th CIB W-78 Conference on Information Technology in Construction, pp. 111-117, 2007.

[15] A. Lapierre and P. Cote, Using Open Web Services for City Data Management: A Test Bed Resulting from an OGC Initiative for Offering Standard CAD/GIS/BIM Services, City and Regional Data Management, pp. 381-391, 2007.

[16] J. Döllner and B. Hagedorn, Integrating City GIS, CAD, and BIM Data by Service based Virtual 3D City Models, City and Regional Data Management, pp. 629-635, 2007.

[17] B. Akinci, H. Karimi, A. Pradhan, C. C. Wu, and G. Fichtl, CAD and GIS Interoperability through Semantic Web Services, CAD and GIS Integration, ITcon Vol. 13, pp. 39-55, 2008.

[18] S. Chaudhuri and U. Dayal, An Overview of Data Warehousing and OLAP Technology, in Proceedings of ACM International Conference on Management of Data, ACM SIGMOD 26, pp. 65-74, 1997.

[19] K. Krivoruchko, C. A. Golway, and A. Zhigimont, Statistical Tools for Regional Data Analysis using GIS, in Proceedings of the ACM International Symposium on Advances in GIS, pp. 41-48, 2003.

[20] H. U. Gökçe and K. U. Gökçe, Multi-Dimensional Monitoring, Analysis and Optimization System for Energy Efficient Building Operations, in Proceedings of ECPPM 2012, 2012, pp. 139-144.J. I. Lim et al., IFC Test between Commercial 3D CAD Application using IFC, Korea Institute of Construction Engineering and Management 9, pp. 85-94, 2008.

Context-aware Indoor-Outdoor Detection for Seamless Smartphone Positioning

Niklas Kroll, Michael Jäger, Sebastian Süß
Institute of Software Architecture
Technische Hochschule Mittelhessen – University of Applied Sciences
Gießen, Germany
Email: {niklas.kroll, michael.jaeger, sebastian.suess}@mni.thm.de

***Abstract*—Localization for smartphones typically relies on distinct approaches for indoor and outdoor contexts, since the Global Positioning System (GPS), which is typically used outdoors, does not perform well within buildings. A localization system supporting both needs to detect indoor-outdoor transitions automatically in order to provide seamless operation across the different contexts. This paper proposes a transition detection method that combines GPS signal evaluation with a GPS-less sensor-based machine-learning scheme in order to provide maximal accuracy, reliability and adaptability to new environments without unneccesary power consumption.**

***Keywords*–IO Detection; Smartphone Positioning; Indoor Positioning; Seamless Transition; Context-aware Computing**

I. Introduction

Location-aware mobile applications need capabilities for determining the current position of a mobile device. Smartphone positioning in outdoor areas typically relies on Global Navigation Satellite Systems (GNSS) such as the Global Positioning System (GPS). Numerous indoor solutions have been proposed in the last few years, e.g., hybrid methods that fuse WiFi fingerprinting with sensor-based Pedestrian Dead Reckoning (PDR).

Accurate indoor localization methods typically rely on some infrastructure. WiFi fingerprinting, e.g., leverages a radio map that contains a large set of locations with associated *Received Signal Strength Indication* (RSSI) values for a set of WiFi access points. In order to determine the current position, the radio map is scanned for a location with a signal strength profile similar to that of the current location. Other methods utilize the information contained in building models, e.g., the positions of doors, walls, stairs, etc.

Whereas most recent publications propose indoor-only solutions, the problem of suitably combining indoor and outdoor positioning, e.g., for pedestrian navigation, has deserved far less attention. Larger areas comprising outdoor ranges as well as several buildings, e.g., company premises, will typically be heterogeneous in the sense that a single indoor positioning method which is suitable for one building might not be applicable in another that lacks the required infrastructure. As a consequence, an appropriate localization system has to incorporate multiple indoor positioning methods. In [1], a multi-scheme approach was presented that supports multiple outdoor and indoor positioning methods with seamless transitions. A crucial problem is to detect automatically that a localization method switch is necessary. Moreover, detection should be energy-efficient and without considerable delay.

An important prerequisite for multiple scheme support is the reliable recognition of indoor-outdoor transitions, which is referred to as Indoor-Outdoor (IO) detection in this paper.

The rest of this paper is organized as follows. After presenting related work in Section II, the proposed IO detection system is explained in Section III. In Section IV, we describe the current state of an implementation and the remaining tasks. Finally, Section V reviews some benefits and shortcomings of the presented approach, open problems, and future research plans.

II. Related Work

Using GPS signal strength changes as an indicator for IO transitions is proposed in [2]. Alternatively, the signal-to-noise ratio (SNR) of the GPS signal can be observed, as proposed in [3]. However, continuously searching for GPS signals within buildings will drain the battery quickly. Moreover, the method might be unreliable and inaccurate if the signal is weak, which can occur outdoors as well as indoors, e.g., near a building entrance.

In order to save energy, other approaches try to avoid GPS usage and rely on a restricted set of less power-consuming smartphone sensors only, e.g., for ambient light, cell signal, or magnetic field. Whereas IO detection according to Zhou et al. is based on checking the sensor values cross empirically determined fixed thresholds [4], Radu et al. show that a semi-supervised machine learning approach [5] provides a much better adaptability to different environments.

III. Proposed IO Detection System

This chapter describes an advanced IO detection system that is expected to provide fast and reliable context detection without unneccessary power consumption. In the multi-scheme approach proposed by Jäger et al., a three-level positioning architecture is described, where the top-level algorithm, called Coarse Positioning System (CPS), is responsible for context transition detection and appropriate selection of lower-level localization schemes, e.g., GPS- or WiFi-based. An important property of this system is its context-awareness. Except in the inititialization phase, positioning always uses exactly one scheme, which is the most appropriate for the current location. For example, in outdoors mode, a GPS-based hybrid scheme is selected, also utilizing PDR for better accuracy.

Supposed that at some location GPS is switched on anyway, e.g., for navigation, leveraging the available signals additionally for IO detection will not impact power consumption. On the other hand, if WiFi fingerprinting is used in a building, IO detection reliability can benefit from considering changes of RSSI values without extra battery drain.

Extending the machine-learning approach of [5], context-aware IO detection is not confined to some basic standard sensors, but also incorporates and extends the GPS signal evaluation approaches of [2] and [3] in order to provide the highest possible accuracy at no additional cost with respect to battery life.

Two independent sets of sensor values are used as classifiers in a co-training scheme. After an initial supervised offline training phase, unsupervised learning supplies each classifier with further training data consisting of the labels from the other classifier. The basic data sources include light intensity, cell signal strength, battery temperature, sound amplitude, time, proximity sensor and magnetic field.

In the outdoor context, the classifier contains also the number of GPS satellites in reach, the signal-to-noise ratio (SNR) of the GPS signal, and the angles of visible GPS satellites. If a sufficiently strong signal can be received indoors, the probability that it is received from a near horizon satellite through a door or window is expected to be considerably higher compared to an origin from a vertical one. The SNR and the number of GPS satellites in reach are indicators for the quality of the signal.

Moreover, depending on the available hardware features, non-standard smartphone sensors are also considered. These include ambient temperature, atmospheric pressure, and relative humidity. The atmospheric pressure fluctuates when a door is opened or closed [6]. Additionally, it can be used as a pressure altimeter. The ambient temperature measured indoors in rooms with air conditioning or heating will often be different to the outdoor temperature. Though ambient temperature sensors are not too widespread in smartphones, the value can often be inferred from battery temperature [7].

IV. Current State and Remaining Tasks

Context-aware IO detection has been implemented in a reusable library, which loads and runs two classifiers previously trained in an offline phase. The training data includes both, the sensor data and the ground truth as given by a user.

Once the context is detected, the system can react to the transition by selecting the appropriate positioning method. This allows for turning off GPS when the user enters a building and turning it on again on leaving it. This is crucial for an effective power management.

A mobile application has been developed for ground truth acquisition and persistent storage. Each training data record consists of measurements for each of the classifier's attributes and an associated user-supplied context classification. The mobile application as well as a prototype featuring seamless positioning have been implemented for the Android platform. It supports positioning with GPS outdoors and NFC in combination with PDR indoors.

There are several open tasks left. First, a thorough evaluation is needed to measure the gain in reliability of IO detection resulting from the evaluation of additional data sources. Particularly, the influence of using the GPS signal for detecting outdoor-to-indoor transitions has to be investigated as well as the impact of leveraging WiFi RSSI values for detecting indoor-to-outdoor transitions in a WiFi indoor context. Furthermore, the machine-learning approach can not only be used to detect that a user has entered a building and to switch to another positioning scheme. It can also be applied to determine which building has been entered, e.g., by using building classifiers based on RSSI values. For indoor navigation purposes, the same approach is expected to enable reliable determination of the current floor level within a multi-storey building, particularly, if the classifiers utilize the atmospheric pressure in addition to RSSI measurements.

It can be seen as a drawback that the proposed machine-learning algorithm is to some extend tailored to a specific non-standard sensor equipment, i.e., the IO classifiers exploit athmospheric pressure and temperature measurements. The impact of these sensor values on the IO detection results needs to be evaluated. While both are not expected to be crucial for IO detection, an extended usage of the classifiers for floor level determination will probably benefit considerably from a barometer. However, on a smartphone without one, WiFi RSSI measurements might also allow reasonably reliable floor level classifications.

V. Conclusion

The context-aware IO detection presented in this paper can be used to switch seamlessly between several indoor and outdoor positioning methods. It combines the advantages of the GPS-using and the GPS-less IO detection approaches presented in Section II. Thus, adaptability to unknown contexts by semi-supervised learning is preserved, while classification reliability is expected to increase considerably by context-dependent usage of GPS or WiFi signal information and additional smartphone sensors for classification. The approach can be extended in a straightforward manner to determine also which building is entered or which is the current floor level and, thus, offers multiple new possibilities with regard to context-aware computing.

The system architecture is extensible and expected to work with arbitrary positioning methods in addition to those used in the prototype.

References

[1] M. Jäger, S. Süß, and N. Becker, "Multi-scheme smartphone localization with auto-adaptive dead reckoning," International Journal on Advances in Systems and Measurements, vol. 8, no. 3 / 4, pp. 255–267, 2015.

[2] R. Hansen, R. Wind, C. S. Jensen, and B. Thomsen, "Seamless indoor/outdoor positioning handover for location-based services in streamspin," in 2009 Tenth International Conference on Mobile Data Management: Systems, Services and Middleware. IEEE, 2009, pp. 267–272.

[3] Y. Kim, S. Lee, S. Lee, and H. Cha, "A GPS sensing strategy for accurate and energy-efficient outdoor-to-indoor handover in seamless localization systems," Mobile Information Systems, vol. 8, no. 4, pp. 315–332, 2012.

[4] P. Zhou, Y. Zheng, Z. Li, M. Li, and G. Shen, "Iodetector: A generic service for indoor outdoor detection," in Proceedings of the 10th ACM Conference on Embedded Network Sensor Systems. New York, NY, USA: ACM, 2012, pp. 113–126.

[5] V. Radu, P. Katsikouli, R. Sarkar, and M. K. Marina, "A semi-supervised learning approach for robust indoor-outdoor detection with smartphones," in Proceedings of the 12th ACM Conference on Embedded Network Sensor Systems. New York, NY, USA: ACM, 2014, pp. 280–294.

[6] M. Wu, P. H. Pathak, and P. Mohapatra, "Monitoring building door events using barometer sensor in smartphones," in Proceedings of the 2015 ACM International Joint Conference on Pervasive and Ubiquitous Computing. New York, NY, USA: ACM, 2015, pp. 319–323.

[7] A. Overeem et al., "Crowdsourcing urban air temperatures from smartphone battery temperatures," Geophysical Research Letters, vol. 40, no. 15, pp. 4081–4085, 2013.

A Raster SOLAP for the Visualization of Crime Data Fields

Kasprzyk Jean-Paul
e-mail: jp.kasprzyk@ulg.ac.be

Donnay Jean-Paul
e-mail: jp.donnay@ulg.ac.be

University of Liege
Department of Geography (Geomatics Unit)
Liege, Belgium

Abstract—**In order to effectively extract synthetic information from large spatial data sets, Spatial OnLine Analytical Processing (SOLAP) combines Geographic Information Systems (GIS) with Business Intelligence (BI) to query data warehouses through interactive vector maps. On the other hand, crime strategical analysis is usually based on raster maps computed by Kernel Density Estimation (KDE), then independent of any artificial boundary. This paper introduces an alternative vision of SOLAP which uses the raster model (instead of the vector one) in order to integrate crime data fields computed by KDE. It allows a continuous visualization of spatial data which, until now, has not been compatible with other SOLAP tools. The original geo-model is validated by a prototype adapted to the police needs.**

Keywords-Data Warehouse; Kernel Density Estimation; GIS; Business Intelligence; Crime Hotspots Analysis.

I. INTRODUCTION

In big cities like Seattle or London, crimes are prevented by allocating the available police forces in the field. In order to optimize this prevention, the identification of risk areas (crime hotspots) is based on the analysis of large amounts of data. Indeed, in London for instance, more than one million crimes are listed each year [33].

On the one hand, GIS and BI [8] techniques include efficient tools like SOLAP to easily represent aggregated information (through tables, charts and maps) out of data warehouses [19][3]. On the other hand, police analysts usually use a KDE technique to identify crime hotspots in raster maps [12]. Indeed, these maps are efficient to visualize the distribution of crimes in a continuous space which does not depend on artificial boundaries (everyone knows the expression "crime has no boundaries"). Until now, this raster representation of data (used in crime mapping and other fields like ecology) has not been compatible with current SOLAP tools (based on the vector model).

In this paper, an original SOLAP model is suggested [18]. It considers a continuous space for the exploration of a raster data warehouse. The paper is organised as follows. In Section II, a state of the art is developed about crime mapping and SOLAP. In Section III, the raster SOLAP model is presented as well as its exploitation of KDE maps. In Section IV, the model is validated by an operational prototype including a crime data set from the Seattle police.

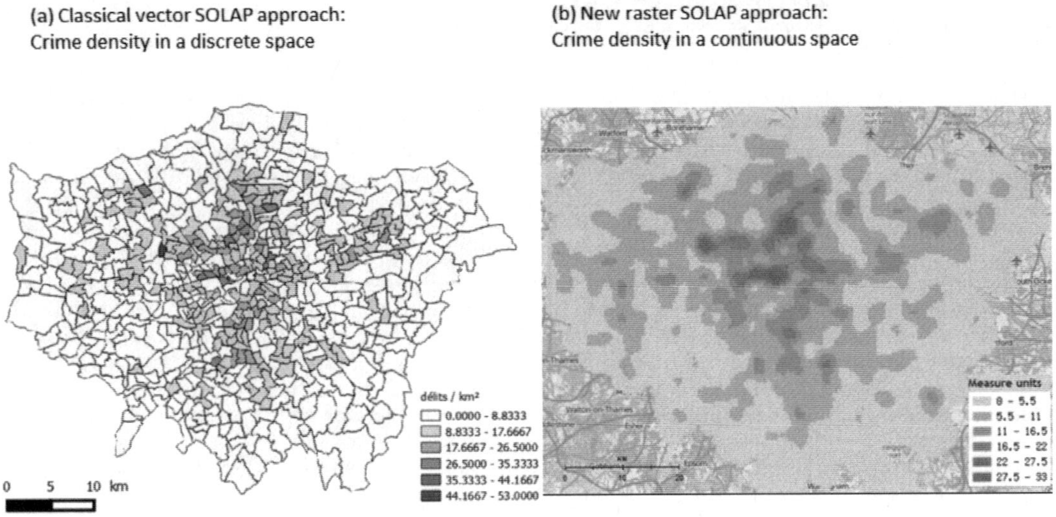

Figure 1. Spatial discrete aggregation (a) and continuous aggregation (b).

II. STATE OF THE ART

In this section, a state of the art about crime mapping and SOLAP is presented. A hypothesis is then drawn from this related work.

A. Crime mapping and Kernel Density Estimation

In crime mapping, crimes are basically modelled as georeferenced points [32]. However, when large clouds of points are analysed (involving thousands of data), point maps are not efficient anymore. Large data sets have to be spatially aggregated. In cartography, classical choropleth maps show data aggregations depending on pre-defined spatial entities like police sectors (Figure 1a). In this case, data are represented in a discrete space. However, hotspots identified with this method are too much influenced by the shape of artificial boundaries which are independent of analysed data (crimes). This well-known issue of geography is called the Modifiable Areal Unit Problem or MAUP [25]. To minimize MAUP, police analysts use KDE to aggregate data in a continuous space (Figure 1b). KDE generates a field where pixel values only depend on the number of crimes and their proximity in a neighbourhood area determined by a parameter called "bandwidth". The bandwidth and the KDE function itself are the only parameters that significantly influence the field smoothing and so the shapes of the hotspots [7].

The problem with KDE is that heavy computations are needed to generate a map (especially for large data sets) and several tests are generally performed in order to determine the best parameters for the KDE function [6]. Moreover, crimes are not only characterized by space but also by other dimensions like time (for instance, months, days of the week or hours of the day) or crime type. It is thus interesting for the analyst to generate several KDE maps depending on these non-spatial dimensions (for instances, a KDE for each crime type, a KDE for each month, etc.).

B. SOLAP

An OLAP server allows decision makers to quickly query data hypercubes which are models of pre-aggregated data depending on several dimensions [9]. Users can easily navigate into data hypercubes through interactive tables (called "pivot tables") or charts. When a data warehouse (which manages hypercubes) is spatialized, a SOLAP server can handle spatial operations through interactive maps [2]. Until now, popular SOLAP tools, like GeoMondrian [31] or Map4Decision [16], have only been able to represent spatially discrete aggregations (Figure 1a) which can be exposed to MAUP. Indeed, a classical SOLAP tool only considers vector data which are mostly efficient for spatially discrete representations. Yet in some recent researches, SOLAP models can spatially interpolate vector data on the fly [1][34] [5][35].

In GIS, the raster model is an efficient alternative to the vector one for continuous space representation [11]. This is useful for continuous phenomena analysis (temperature, precipitation, pollution, etc.) but also for continuous results of treatments applied on discrete data like KDE. In other SOLAP researches, vector grids have been considered, which are very close to the raster model [24][22][14][20][28]. The potential of raster in SOLAP, especially for continuous phenomena represented by fields, was demonstrated. Nevertheless, raster results of treatments like KDE were not considered and most of the time, validation prototypes were still implemented with the vector model (pixels stored like square polygons) whereas real raster data, stored as arrays, offer better performance for specific treatments like data aggregations on rows and columns. Moreover, data arrays are already used in Multidimensional OLAP, or MOLAP [13], which has never been considered in SOLAP. In [18], it was worked out that vector SOLAP is closer to Relational OLAP, or ROLAP [13], which uses relational database management systems like Oracle [27].

C. Hypothesis

According to SOLAP literature, an original raster SOLAP model could be developed to analyse fields resulting from KDE. Based on this model, a SOLAP prototype would allow police analysts to easily explore crime data warehouses through interactive and continuous KDE maps, but also tables and charts.

III. MODEL

In this section, OLAP basics are first explained in order to introduce our original model: raster SOLAP. Then, details are given about the way KDE fields are integrated (KDE SOLAP).

A. OLAP basics

Most of OLAP models are based on multidimensional data warehouses which are conceptually described by a simple star schema [19]. An example is given in Figure 2. A dimension (branch of the star) is a finite set of members. For instance, "burglary" and "robbery" are members of the "crime type" dimension. These members can possibly be organized by a hierarchy made of children and parents belonging to different dimension levels. For instance, in the "time" dimension, "January 2012" member (belonging to the "month" level) is a child of "2012" member (belonging to the "year" level). Each possible combination of dimension members is called "fact" and a fact is always associated to a measure. For instance, "burglaries of January 2012 in Liege province" is a fact associated to a number of crimes (the measure). When it is not stored in the data warehouse, a measure (most of the time, a numerical value) is always the aggregation of fact measures belonging to a more detailed level of dimensions (more detailed facts). A dimension can possibly be entirely aggregated. It would be the case of the "crime type" dimension if the fact was "crimes of January 2012 in Liege province" (including every crime type).

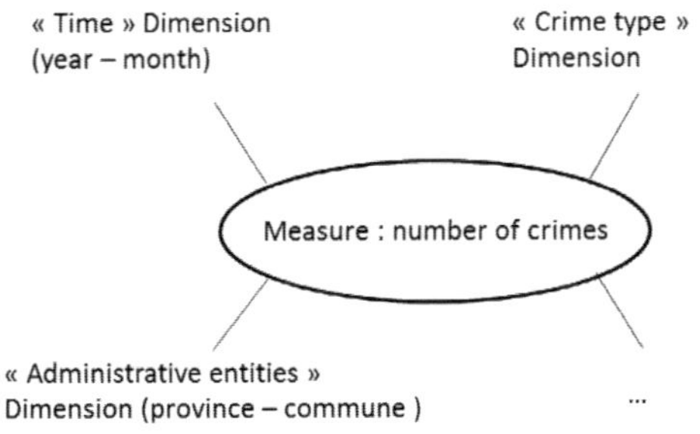

Figure 2. Example of a star schema.

An instance of a star schema is called "data hypercube" (or simply "data cube"). A cell of the cube is a fact, its value is the measure, and its coordinates in the multidimensional space are members which describe the fact. Physically, the data warehouse at least stores the "base data cube" (Figure 3). It is the cartesian product of all dimensions at the most detailed level (the set of all detailed facts). This base data cube is the minimum set of measures from which the measure of every possible fact can be computed by aggregation. In ROLAP, detailed facts are stored in a "fact table" of a relational datawarehouse. In MOLAP, detailed facts are stored in multidimensional arrays. The main advantage of ROLAP is that null facts (for which no crime is associated) do not have to be stored. In MOLAP on the other hand, even though all possible detailed facts (including unneccessary null facts) are stored, performances of aggregation operations (on rows and columns) are usually better [4][13].

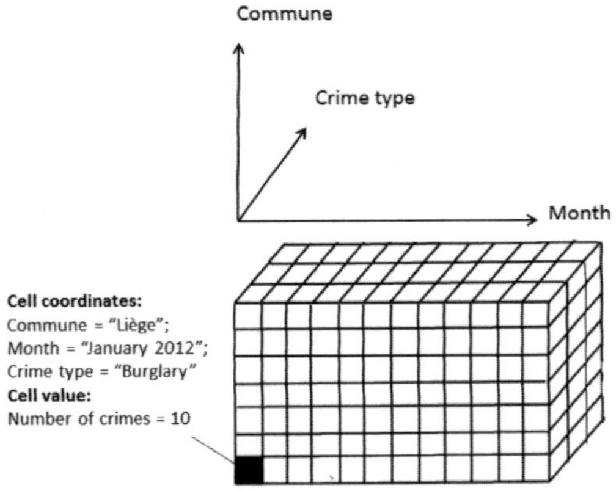

Figure 3. Example of a base data cube.

Slice on "administrative entities" dimension:

Province = Liège

Type \ Month	2012
Robbery	265
Shoplift-ing	305
Burglary	156

Roll up on time dimension ⇧ ⇩ Drill down on time dimension

Type \ Month	Jan 12	Feb 12	Mar 12	Apr 12	May 12	Jun 12	Jul 12	Aug 12	Sep 12	Oct 12	Nov 12	Dec 12
Robbery	40	30	35	10	42	10	5	0	10	13	30	40
Shoplift-ing	10	12	13	12	50	40	30	10	42	65	10	11
Burglary	20	21	22	18	19	18	20	0	0	5	10	3

Figure 4. Example of drilling and slice operations on a data cube.

Typical OLAP operations are drillings and slices on dimensions. Figure 4 illustrates an example of drill down / roll up on the hierarchy of a time dimension. In this way, users can easily switch from an aggregation level of dimension to another one. The slice operations allow users to consider only one portion of the data cube. In Figure 4, the slice isolates facts linked to "Liege" member of "Province" level. By isolating a smaller data cube, slices reduce the number of aggregation operations necessary to compute the measures. On the other hand, a roll up, which shows less detailed facts, can imply heavy treatments. For this reason, in addition to the base data cube, several cuboids (data cubes at less detailed levels) can be stored in the data warehouse to speed up heavy aggregations computations [4]. Slices and drillings can be applied directly on table interfaces, charts but also maps in the case of SOLAP.

B. *Raster SOLAP*

In a classical SOLAP approach (vector SOLAP), data cubes handle spatialized dimensions [2][4]. Spatialized members are associated to one vector geometry (point, line or polygon) in order to represent spatialized facts on a map interface (like the one in Figure 1a). Sometimes, SOLAP involves spatialized measures which can be numeric values (distances, for instance) or geometries [15]. In this case, measures can be aggregated with spatial operations like union, intersection, etc.

In order to represent the spatial continuity, our original model considers geographical space in a very different way. Indeed, space is not a property of spatialized members anymore but it is directly included as *X* and *Y* dimensions in the star schema (Figure 5a). *X* and *Y*, simply called "spatial dimensions", are cartographic coordinates of any point in the study area coverage [17].

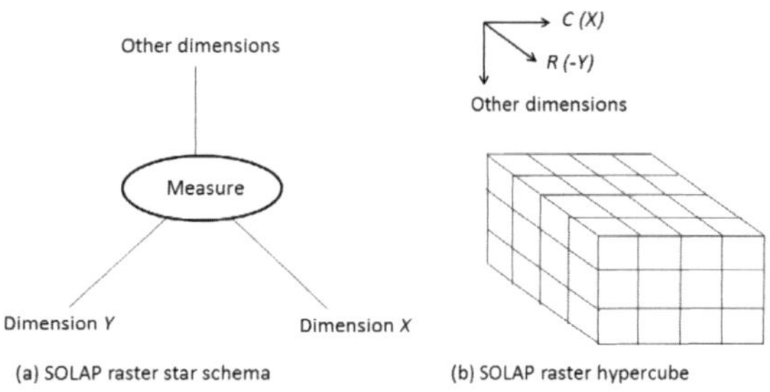

Figure 5. Raster star schema and raster hypercube.

The instance of this star schema is a data hypercube where *X* and *Y* dimensions are respectively deducted from columns (*C*) and rows (*R*) of raster data (georeferenced images). It is simply called "raster hypercube" (Figure 5b). Details of the affine transformations between (*X*, -*Y*) and (*C*, *R*) are described in [18]. In addition to *X* and *Y*, "other dimensions" are non-spatial attributes that can be modelled in any classical OLAP environment: time, customer, product, etc.

X and *Y* are very particular dimensions. Spatial members are actually pixels and they are always described by both *X* and *Y*. Like raster pyramids in GIS [26], several raster cuboids can be stored with different resolutions to represent spatial continuity at different detail levels (Figure 6). For the user, switching from a raster cube to another one is a roll up/drill down operation on spatial dimensions *X*, *Y*.

Including *X* and *Y* directly in the star schema also brings an important spatial flexibility in SOLAP. Since space is defined at the pixel level, any geographical member (Liege province, for instance) can be imported on the fly during the user's analysis. A simple GIS operation, including a raster layer of geographical entities, identifies the space members (a set of pixels in the raster cube) which geometrically describe the new imported geographical members [18]. Then, slice and drilling operations can involve these new "geo-members" in maps, tables or charts.

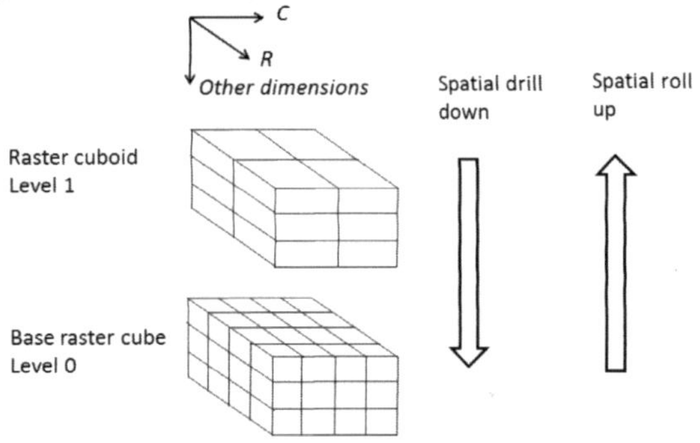

Figure 6. Precomputed raster cuboids and spatial drillings.

Another original aspect of raster SOLAP comes from its physical implementation: the hybrid management of spatial and non-spatial dimensions. Indeed, spatial database management systems like PostGIS [29] or Spatial Oracle [26] allow the storage of raster arrays as attributes of relations. Therefore in our model, non-spatial dimensions can be implemented according to a classical ROLAP approach managed in SQL (ROLAP dimensions). Spatial dimensions are included in the raster attribute, considered as a measure in the ROLAP model. Aggregations of these measures are defined by map algebra [23] like in many continuous SOLAP researches: local map algebra for non-spatial aggregations and zonal map algebra for spatial aggregations [22][34][5][14][28]. Details of raster SOLAP operations in our model are given in [18]. Thereby, raster SOLAP inherits from advantages of ROLAP for non-spatial dimensions (null ROLAP facts do not need to be stored) but also from MOLAP advantages for raster spatial dimensions (fast spatial aggregations on rows and columns of arrays).

Let us note that by considering only the *X* and *Y* dimensions of the raster SOLAP model, a raster hypercube becomes a simple raster. As discussed in [18], there is almost no difference between a 2D MOLAP cube and a raster since they both are arrays whose coordinates refer to dimension members. In raster, the transformation from the raster space (*C*, *R*) to the geographical space (*X*, -*Y*) is always an affine transformation while other indexing techniques can be used in a MOLAP cube (depending on the modelled phenomena which can be ordered or not, continuous or discrete, etc.). As already mentioned, the raster SOLAP is implemented as a ROLAP in this research. Nevertheless, the implementation of a raster cube in a pure MOLAP environment would be an interesting perspective.

C. *KDE SOLAP*

The previous section introduced our original SOLAP model able to integrate raster data (raster hypercube). Since crime mapping uses KDE raster data for crime prevention, the following section presents the way KDE raster fields are integrated into a raster hypercube for SOLAP.

A KDE is a raster field where each pixel has a value depending on the number of points (like crimes) in its neighbourhood. Moreover, the distance of each neighbour point to the pixel centre also influences the KDE value. Close points have more influence than distant points.

The whole KDE raster represents a continuous surface in the geographical space (defined by *X* and *Y*). An example was given in Figure 1b. The KDE surface smoothing first depends on the KDE function itself: quartic, normal, triangular, uniform, etc. [10]. For most of KDE functions, the value of a pixel p_j can be expressed by the following general formula:

$$p_j = \sum_{i=1}^{n} v_i * K(r, d_{ij}) \tag{1}$$

In (1), K is a KDE function (e.g., quartic), v_i is the value of a point i, d_{ij} is the distance of a point i to the pixel centre j, r is a constant bandwidth and n is the number of points for which $d_{ij} < r$. Most of the time in crime mapping, $v_i = 1$.

When KDE is applied on a large amount of points, the treatment can be quite heavy because, for every pixel of the resulting raster, d_{ij} has to be calculated for every point inside the bandwidth (a circle of radius r centred on the pixel). A raster SOLAP with KDE would imply lighter treatments during the analysis because every possible KDE raster measure (associated to a non-spatial fact) would be the result of an aggregation between pre-computed KDE raster measures (associated to detailed non-spatial facts). In this way, the number of operations to calculate a pixel value only depends on the number of detailed facts to aggregate (for instance, one fact for each month of the year). It does not depend on large amounts of points anymore. Nevertheless the total number of detailed facts in a hypercube exponentially grows with the number of dimensions [4]. For this reason, a raster hypercube cannot involve too many non-spatial dimensions to deliver rapid responses to the user.

To integrate KDE fields in the raster SOLAP, the following KDE condition has to be fulfilled. Let A and B two disjoints sets of points, $K(X)$ the raster result of a KDE function K on a set of point X:

$$K(A) + K(B) = K(A \cup B) \qquad (2)$$

In (2), the sum of $K(A)$ and $K(B)$ is actually a local map algebra operation where the result is the sum of every homologous pixel (geometrically defined on the same geographical space). In the raster SOLAP, this formula means that the sum result of two KDE raster measures respectively associated to two facts (for instance, a fact for January and a fact for February) is the same as a KDE field computed with the points involved in both KDE facts (January and February). This condition is very important because all aggregation results given by the SOLAP have to be similar to the ones given by classical KDE computations, even if only the detailed non-spatial facts (or ROLAP facts) are pre-computed with KDE.

In [18], it was demonstrated that the KDE condition is fulfilled if:

- The parameters of the function K are constant: the KDE function itself (quartic, for instance) and bandwidth r.
- The spatial metadata of raster fields $K(A)$, $K(B)$ and $K(A \cup B)$ are constant: coordinates reference system, number of columns (C members) and rows (R members), affine transformation between raster space (defined by C, R) and geographical space (defined by X, Y) including the raster resolution.

Physically, all these parameters are defined as constraints in a metadata table of the data warehouse.

In order to adapt spatial drillings of raster SOLAP to KDE, the pre-computed raster cuboids (Figure 6) are associated to different resolutions but also to different bandwidths. For one same KDE function, bandwidth r influences the KDE surface smoothing [10][7]. In other words, the higher the KDE bandwidth is, the larger the hotspots will be. Therefore, a spatial roll up implies a more global analysis (larger hotspots) and a spatial drill down implies a more local analysis (smaller hotspots). An example is given by Figure 7 (only the first quintile of the KDE is shown to isolate hotspots).

Contrary to classical SOLAP tools where the spatial drilling is defined by semantic levels (e.g., street level, commune level, province level, etc.), this alternative spatial drilling, defined by resolution/bandwidth, is not influenced by artificial boundaries. However, the choice of the bandwidth in a KDE leads to another MAUP issue as it significantly changes the shape of the hotspots. As already mentioned in Section IIa, in classical KDE, the police analysts empirically determine an optimal bandwidth using several tests (computations with different bandwidth values). With KDE SOLAP, analysts can easily and quickly test the bandwidth values by simply drilling space (until they find the best scale analysis amongst the pre-computed cuboids). In the case study of the following section, four drillings levels are arbitrary determined in order to fit to the user's needs and to the extent of the study area. A detailed methodology for the determination of these scale levels could be the subject of a future research about KDE SOLAP.

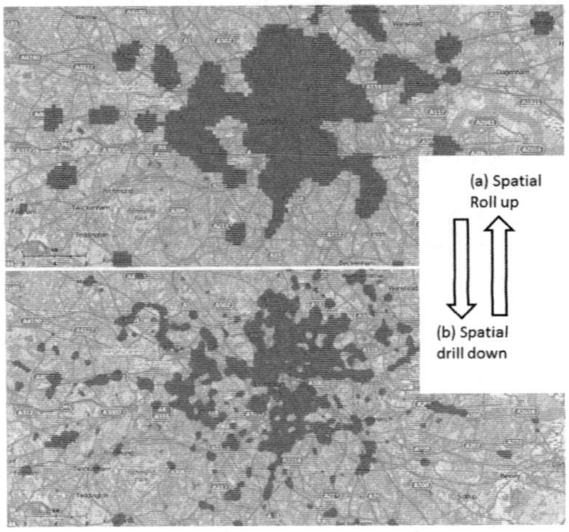

Figure 7. Spatial drilling on KDE SOLAP.

IV. VALIDATION

In this section, a case study about Seattle crime data is first introduced. Then, our research is validated by a raster SOLAP prototype (Raster Cube) which integrates Seattle crime fields.

A. *Case study*

The SOLAP raster model is validated by a prototype including several data sets. Amongst them, a raster hypercube is built with Seattle crime data. These data come from 911 calls received during the year 2013 [30]. They represent around 800 000 crimes. The star schema and the description of dimension hierarchies are shown in Figure 8. Four non-spatial dimensions are modelled as ROLAP in a PostGIS data warehouse: three time dimensions (month, day of the week and range of three hours) and a crime type dimension. According to the number of detailed members for the four non-spatial dimensions, the theoretical number of detailed non-spatial facts is equal to the following: 12 (month) * 7 (day of the week) * 8 (hour range) * 25 (crime type) = 16 800. Actually, the ROLAP management of these dimensions allowed to store only 11 304 facts because 33% of the theoretical detailed facts were null (no crime).

The two spatial dimensions (*X*, *Y*) are included in the raster measures of the data warehouse. These raster measures have two distinct values: the number of crimes (used for tables and charts) and the crime density resulting from KDE (used for maps). The spatial dimensions have four pre-computed levels depending on raster resolution (30 m for the base cube, 100 m, 300 m and 1000 m for the different cuboids). As advised by [6], a quartic KDE function was used with a bandwidth r equal to the raster resolution multiplied by 5. The users can then drill the spatial dimensions until they find the best scale for their analysis.

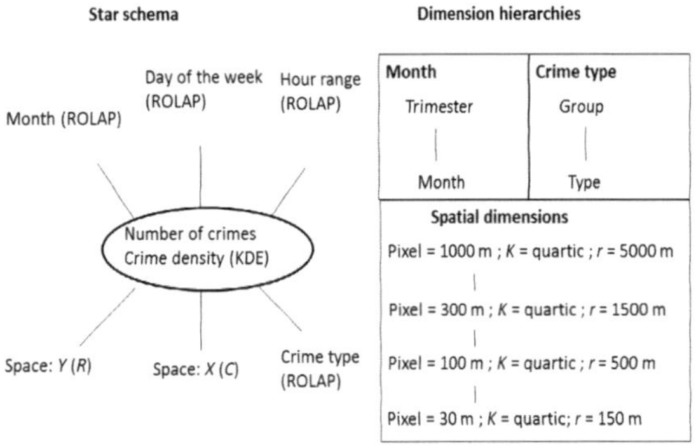

Figure 8. Star schema and dimension hierarchies for Seattle crime data.

B. *Raster Cube prototype*

The prototype, called "Raster Cube", is based on a web architecture. On the server side, the SOLAP, written in php, interrogates the PostGIS data warehouse and delivers results to the client side in HTML/Javascript. MapServer [21] was also used as a spatial data server. Raster Cube is accessible for demonstration at the following URL: http://nolap01.ulg.ac.be/rastercube .

Figure 9 is a screenshot of the user interface. The dimension tree on the left shows non-spatial dimension members organized by hierarchies. The users can check all the members they choose to slice the raster hypercube before its aggregation. For instance, the slices defined in Figure 9 implies the aggregation of all crimes of the "bike" type that happened in the "winter" trimester for the whole study area (Seattle). When the "aggregation" button is pressed, "crime density" measures (KDE) are shown on the map and "number of crimes" measures are shown on charts. The map always shows the two spatial dimensions (*X*, *Y*) and charts show non-spatial dimensions. In Figure 9, one non-spatial dimension is shown by the chart: the hour range. With an additional dimension, the interface could also show, for instance, one "hour range" chart for each day of the week (still with the slices on "winter" and "bike").

As already mentioned, in addition to the continuous map representation, raster SOLAP is also able to import new geographical members on the fly. Indeed, the users can interactively digitize spatial entities like police sectors and so add them to the dimension tree. These new "geo-members" can then be involved in spatial slices (to reduce the study area) or they can be considered as a dimension shown in charts. For instance, the evolution of "bike" crimes in the day could be easily compared with a chart for each police sectors imported on the fly.

As shown in Figure 7, space can be drilled by choosing the right cuboid and so the KDE map can be adjusted to the best analysis scale. It is also possible to generate an evolution map that for instance shows the density difference between trimesters "winter" and "spring". This is very useful to quickly see where criminality increased and where it decreased. A few options also allow users to change the classification method of the map (linear, quantiles, etc.).
Finally, a module was made to automatically build the raster hypercube from a vector data warehouse (where crimes are geometrically defined as points) and from an XML file defining the star schema. This includes the KDE computation of all non-spatial detailed facts that have to be stored in the raster data warehouse. Indeed, it is worth recalling that even if maps resulting from KDE SOLAP are identic to the one resulting from classical KDE, the way KDE are computed during the analysis is very different in this research. In a classical KDE, the field computation is based on a cloud of point (1). In KDE SOLAP, the field computation is an aggregation of pre-computed KDE raster fields which are stored in the data warehouse.

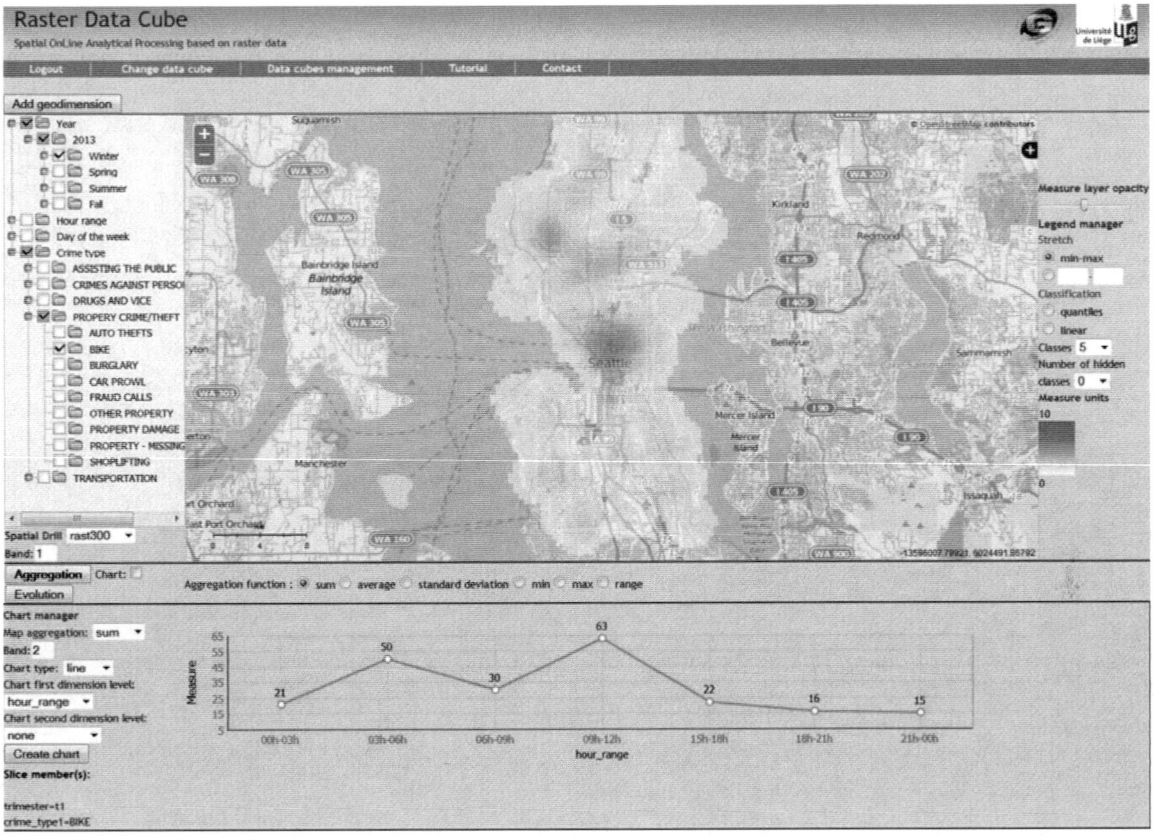

Figure 9. Raster cube interface with Seattle crime data.

V. CONCLUSION AND FUTURE WORK

The aim of this paper was the presentation of a raster SOLAP model adapted to the needs of the police for the exploration of large crime data sets. A quick state of the art about crime mapping showed that police analysts generally use a KDE raster technique to represent crimes in continuous maps. At the same time, a state of the art about SOLAP showed that current tools are vector and so not compatible with KDE raster maps (despite a few researches that demonstrated the potential of raster SOLAP data cubes).

After a brief description of the main OLAP concepts (multidimensional analysis, star schema, data hypercube, drillings and slices operations), an original raster SOLAP model [18] was introduced. Opposite to vector SOLAP tools (where space is modelled with geometries associated to semantic dimension members), raster SOLAP directly includes space inside the star schema (*X* and *Y* dimensions). It offers two main advantages. First, the SOLAP can generate continuous raster maps resulting from classical drilling and slice operations (on space or other dimensions). Then, geographical members can easily be imported on the fly during the user's analysis.

By adapting the raster SOLAP model to KDE, the tool became able to generate continuous KDE maps like the ones usually computed by police analysts. For the police, it offers an intuitive interface to explore data warehouses, including tables, charts and continuous KDE maps. Therefore, any combination of dimension members (through OLAP operations) leads to a KDE map which is usually computed case by case by crime analysts. Moreover, drillings and slices on space respectively allow the interactive adjustment of the scale and the study area shown by the map. All these concepts were validated by an operational prototype integrating crime data from Seattle.

In conclusion, this paper explained an alternative way to integrate space in an OLAP. Compared to vector, raster SOLAP can be useful for any domain which implies a continuous representation of space (fields): crime mapping, but also ecology, agriculture, epidemiology, climatology, etc. Moreover, when a SOLAP involves fields like KDE, it suffers less from the influence of artificial boundaries (MAUP) which can bias the analysis done with a traditional vector SOLAP. Nevertheless, raster and vector can be seen as two complementary approaches. Raster SOLAP, close to MOLAP, can handle dense data hypercubes described by few dimensions (global approach) and vector SOLAP, close to ROLAP, can handle less dense data hypercubes characterized by more dimensions (detailed approach). A hybrid (raster/vector) SOLAP, an interesting perspective of this research, could be a powerful tool allowing an efficient analysis of every type of spatial phenomena (discrete or continuous).

REFERENCES

[1] T. O. Ahmed and M. Miquel, "Multidimensional structures dedicated to continuous spatiotemporal phenomena", Proceedings of the Twenty-Second British National Conference on Databases: Enterprise, skills and innovation, Sunderland : Springer-Verlag, 2005, pp. 29-40.

[2] Y. Bédard, "Spatial OLAP", Forum annuel sur la R-D, Géomatique VI: Un monde accessible, Montréal, November 1997.

[3] Y. Bédard, "Beyond GIS: Spatial Online Analytical Processing and Big Data", The 2014 Dangermond Lecture, Santa-Barbara, 2014.

[4] S. Bimonte, Integration of geographic information in data warehouses and online analysis: from modeling to visualization (Intégration de l'information géographique dans les entrepôts de données et l'analyse en ligne: de la modélisation à la visualisation), PhD, Institut National des Sciences Appliquées de Lyon, Lyon, 2007.

[5] S. Bimonte and M. A. Kang, "Towards a model for the multidimensional analysis of field data, in Proceedings of the Fourteenth east European conference on advances in databases and information systems", B. Catania, M. Ivanovic and B. Thalheim, Eds. Berlin : Springer-Verlag, 2010, pp. 58-72.

[6] S. P. Chainey, L. Tompson and S. Uhlig, "The utility of hotspot mapping for predicting spatial patterns of crime", Security Journal, vol. 21, 2008, pp. 4-28.

[7] S. Chainey, "Examining the influence of cell size and bandwidth size on kernel density estimation crime hotspot maps for predicting spatial patterns of crime", Bulletin of the Geographical Society of Liege, vol. 60, 2013, pp. 7-19.

[8] T. Chee, L. K. Chan, M. H. Chuah, C. S. Tan, S. F. Wong and W. Yeoh, "Business Intelligence systems: state-of-the-art review and contemporary applications, Symposium on Progress in Information and Communication Technology", Kuala Lumpur, 2009.

[9] E. F. Codd, S. B Codd and C. T. Salley, Providing OLAP (On-line Analytical Processing) to User-Analysts: An IT Mandate, E.F Codd and Associates, 1993.

[10] M. Di Salvo, M. Gadais and G. Roche-Woillez, The kernel density estimation : methods and tools (L'estimation de la densité par la méthode du noyau: méthodes et outils), Lyon : Certu, 2005.

[11] J. P. Donnay, "Formalization of geographic information in raster". Revue internationale de géomatique, vol. 15, no. 4, 2005, pp. 415-438.

[12] J. E. Eck, S. P. Chainey, J. G. Cameron, M. Leitner and R. E. Wilson, "Mapping crime: Understanding hot spots", Washington: National Institute of Justice, 2005.

[13] B. Espinasse, Data warehouses: OLAP systems: ROLAP, MOLAP and HOLAP (Entrepôt de données: Systèmes OLAP: ROLAP, MOLAP et HOLAP). Ecole Polytechnique Universitaire de Marseille, 2013.

[14] L. I. Gomez, S. A. Gomez and A. Vaisman, "A generic data model and query language for spatiotemporal OLAP cube analysis". Proceedings of the Fifteenth International Conference on Extending Database Technology, Berlin : ACM, 2012, pp. 300-311.

[15] J. Han, N. Stefanovic and K. Koperski, "Selective materialization: an efficient method for spatial data cube construction", Proceedings of the Second Pacific-Asia Conference on Knowledge Discovery and Data Mining, Melbourne : Springer-Verlag, 1998, pp. 144-158.

[16] Intelli³, available at <http://www.intelli3.com>, [retrieved : march, 2016].

[17] ISO 19123. Geographic information - Schema for coverage geometry and functions, 2005.

[18] J. P. Kasprzyk, Integration of spatial continuity in the multi-dimensional structure of a data warehouse – raster SOLAP (Intégration de la continuité spatiale dans la structure multidimensionnelle d'un entrepôt de données – SOLAP raster). PhD. Université de Liège, 2015, Available at <http://hdl.handle.net/2268/182360>.

[19] R. Kimball and M. Ross, The Data Warehouse Toolkit: The Definitive Guide to Dimensional Modeling, Third Edition, New York : John Wiley and Sons, 2013.

[20] J. Li, L. Meng, F. Z. Wang, W. Zhang and W. Cai. "A Map-Reduce-enabled SOLAP cube for large-scale remotely sensed data aggregation", Computers and Geosciences, vol. 70, 2014, pp. 110-119.

[21] Mapserver 7.0.0 beta1 documentation, available at <http://mapserver.org/>, [retrieved: march, 2016].

[22] R. McHugh, Integration of the matrix structure in spatial cubes (Intégration de la structure matricielle dans les cubes spatiaux), Master thesis. Université Laval, Québec, 2008.

[23] J, Mennis, R. Viger and C. D. Tomlin, "Cubic Map Algebra functions for spatio-temporal analysis, Cartography and Geographic Information Systems", vol. 30, no. 1, 2005, pp. 17–30.

[24] M. Miquel, Y. Bédard and A. Brisebois, "Conception of geospatial data warehouses from heterogeneous sources: application example in forestry" (Conception d'entrepôts de données géospatiales à partir de sources hétérogènes : Exemple d'application en foresterie), Ingénierie des Systèmes d'Information, vol. 7, no. 3, 2002, pp. 89-111.

[25] S. Openshaw, The modifiable areal unit problem. Norwick (Norfolk), Geo Books, 1983.

[26] Oracle, Georaster overview and concepts, available at <https://docs.oracle.com/html/B10827_01/geor_intro.htm>, [retrieved: march, 2016].

[27] Oracle, Business intelligence, available at <http://www.oracle.com/us/solutions/business-analytics/businessintelligence/overview/index.htm>, [retrieved : march, 2016].

[28] M. Plante, Towards matrix cubes supporting on the fly spatial analysis in decision support (Vers des cubes matriciels supportant l'analyse spatiale à la volée dans un contexte décisionnel), Master thesis, Université Laval, Québec, 2014.

[29] P. Racine, and S. Cumming, Store, manipulate and analyze raster data within the PostgreSQL/PostGIS spatial database. FOSS4G, Denver, September 2011.

[30] Seattle Police, Seattle Open data, available at <https://data.seattle.gov/>, [retrieved: march, 2016].

[31] Spatialytics, Open Source GeoBI, available at <http://www.spatialytics.org/>, [retrieved: march, 2016].

[32] M. Trotta, C. Deprez and J. P. Donnay, "Impact of the environmental anisotropy in geographic profiling studies", SAGEO 2014, Grenoble, october 2014.

[33] UK Police, Crime Map, available at <http://www.police.uk>, [retrieved: march, 2016].

[34] A. Vaisman and E. Zimanyi, "A multidimensional model representing continuous fields in spatial data warehouses", Proceedings of the 17th ACM SIGSPATIAL International Conference on Advances in Geographic Information Systems, Seattle : ACM, 2009, pp. 169-177.

[35] M. Zaamoune, S. Bimonte, F. Pinet and P. Beaune, "Integration of incomplete continuous data fields in OLAP: from conceptual modeling to implementation" (Intégration des données champs continus incomplets dans l'OLAP : de la modélisation conceptuelle à l'implémentation), 9e journées francophones sur les Entrepôts de Données et l'Analyse en ligne, Blois, 2013.

Comparative Evaluation of Alternative Addressing Schemes

Konstantin Clemens
Technische Universität Berlin
Service-centric Networking
konstantin.clemens@campus.tu-berlin.de

Abstract—**Alternative addressing schemes are developed to be flexible and user friendly, while at the same time unambiguous and processable in an automated way. In this paper, four schemes are compared to WGS84 latitude and longitude coordinates - an addressing scheme for itself. An experiment with human users checks how user friendly the various schemes are and what classes of errors the users make. The results show that comprehensible and recognizable address elements are contributing towards a user friendly address scheme.**

Keywords–Geocoding; Address Schemes; Geohash; GIS.

I. Introduction

Nowadays, when referencing a location, most often postal addresses are used. That is because postal addresses are especially easy to use: A postal address is a compound of entities such as city, district, or street names. Obviously, addresses make more sense with more knowledge about the area of a specified destination: The destination can be located more accurately. At the same time, if an address refers to a location someone has little knowledge about, only the rough area is identifiable. Interestingly, this correlates with the usage pattern of an address: Users would want to recognize nearby addresses, while they would not care about the precise location of an address in a distant and unfamiliar city. Thus, humans can resolve addresses to a level that that matches their need and knowledge about the destination. Generally, the ways postal addresses are put together in various countries are specified by multiple organizations [1]. It is a special challenge to process postal addresses automatically, because in various countries different, often times not compatible, schemes for postal addresses are used [2]. Also, different address elements are reused in different addresses, when, e.g., a city name is also the name of a street in another city, or when multiple cities share the same name. Multiple on-line geocoding [3] services like [4], [5], or [6] resolve postal addresses into their WGS84 coordinates [7]. This process, however, is complex and error-prone [8], [9].

For that reason, *alternative addressing schemes* (AAS) are developed, that strive to provide both: Addresses that are easy to comprehend and to remember for a human, while also unambiguous and simple to process for a computer. However, while postal addresses grew naturally as needed, AAS are designed by hand. Key differences worth pointing out are:

1) Instead of points on the globe, postal addresses reference abstract entities, like groups of buildings, single houses, or specific entrances. A postal address of a house would remain valid, if the house is rebuild so that the main entrance moves along the street. AAS reference specific points or areas on the globe instead, so that a rebuild house could require a different alternative address.
2) AAS reference points or areas on the globe, which can be empty spaces or even open water. While a valid postal address can only address existing entities, perfectly valid AAS can reference any point on earth.
3) Postal addresses resolve to a variable degree of accuracy, as it is required. In urban centers, where many entities are to be addressed in a small area, multiple postal addresses address every single one naturally. In some cases additional address elements are added to, e.g., specify a lot within a mall with one house number. In rural areas, on the other hand, postal addresses may refer to areas with groups of buildings. AAS uniformly resolve to a fixed accuracy on the entire globe.
4) Postal addresses are composed from geographic area names as cities and regions. These areas usually have existed for a long time. Because of that, their names are well-known and easy to remember for humans. Elements of AAS on the other hand are, yet, mostly opaque for a user.

In this paper, five AAS are evaluated for their user experience: One AAS is provided by the service what3words [10]. Three words are used to identify a location. Mapcode [11] is a service that generates very short Geohash keys. Geo-poet [12] is another service that encodes Geohashes in rhymes of four words. Finally, Syllagloble [13] is another Geohash based system generating human-friendly Geohashes out of syllables. While the former two services are available and competing on the market, the latter two systems have been implemented for this paper solely. WGS84 coordinates can, as the four AAS, address any point on the globe. These coordinates make an addressing scheme for themselves. Therefore, WGS84 is used as the base line addressing scheme.

An experiment has been conducted revealing how well various AAS can be remembered for a short time. The experiment gives insight into the classes of mistakes done when using AAS. Next, the various schemes are described in detail. In Section III, the experiment and its outcomes are presented. Finally, the conclusions are drawn in the last section.

II. Addressing Schemes

As base line, WGS84 is used in this paper. In this scheme two orthogonal plains meeting in the center of the earth are defined. Every point on the surface is described by a vector from the center of the earth to the point. Vectors are specified by WGS84 latitude and longitude coordinates, which are the two angles between the vector to the two plains. These angles can thereby be arbitrarily precise.

Three AAS evaluated in this paper are based on Geohashes. Geohashes are keys of Quadtrees [14]. There are various ways to implement a Quadtree that all share the same basic idea: Areas or tiles, e.g., squares, rectangles, or even triangles [15], are split into a fixed number of smaller sub-tiles of the same shape. This process is repeated, until the desired size of a tile is reached. Keys of Quadtrees consist of multiple parts. Each

TABLE I. EXAMPLES OF LOCATIONS ENCODED WITH VARIOUS SCHEMES

	WGS84	what3words	Mapcode	Geo-poet	Syllagloble
Berlin	52.5167,13.4	dramatic liner common	VJMMB.60XJ	requesting emanation entitles demarcation	lay uxri mes ixsi
London	51.50642,-0.12721	crush activism proven	VHGQZ.RD3J	debenture consummation lamented dissertation	lac ekha kam etni
Paris	48.85693,2.3412	national slope delved	VHPM9.JZKN	unvarnished usurpation covalent obfuscation	lac igpi dav avba
Rome	41.90322,12.49565	shoebox inflame speaker	TJLFF.MR0Y	unfairly inspiration prepayment conflagration	sab isca poc uhvi

next part thereby specifies the next sub-tile to split. Having a Geohash on-hand (and knowing the way it has been computed), therefore, specifies the last sub-tile – an area inside the original space covered by the Quadtree. A Geohash that starts on a tile spanning the entire world can be used as an addressing scheme, with one extra property: Common prefixes of two Geohashes imply that the two areas addressed are located close to each other. Note, however, that differing prefixes do not imply that two areas are far apart.

Geo-poet is an address scheme that uses Geohashes. The system has been developed for this paper. Geo-poet tries to create human friendly and easy to remember Geohashes by using spoken language. Particularly, for every part of the Geohash, Geo-poet chooses a word from a specific set. The words for each part are thereby chosen so that a distich, i.e., a poem with two lines and four words, is formed. The corpus of words used in this system are taken from [16]. From this collection of words that are annotated with possible pronunciations, 289 rhyming and 5929 non-rhyming words with a specific metre have been picked. Beginning with the outer tile covering the entire world, for each word of the poem, the current tile is split in either 289 or 5929 sub-tiles, depending on whether the next word should rhyme or not. This way, Geo-poet is addressing tiles with an inner diagonal not longer than $26.1m$. This maximal distance between two points on the globe having the same Geo-poet address is reached along the equator.

A system similar to Geo-poet is Syllagloble. Similar to Geo-poet, it strives to provide easy-to-use Geohashes and has been developed for this paper. Instead of words, however, syllables are used in this addressing scheme. From a corpus of words, 13666 most common syllables have been picked so that they are easy to combine. The generated Geohashes are words that are fourteen characters long and assembled of four syllables. Thus, beginning with the outer tile, tiles are split four times enabling Syllagloble to address tiles with a diagonal not longer than $7.7m$ using a word that is easy to pronounce.

Mapcode is another system based on Geohashes, so called Mapcodes that are assembled from letters and numbers. The goal of this system is to provide Mapcodes that are short and easy to use. For that, next to global Mapcodes, many regions are also addressed with regional Mapcodes. Since the starting tile of the Geohash only need to span a region for regional Mapcodes, very short Geohashes may be used. E.g., for the region of Netherlands just four characters are addressing tiles with ca. $10m$ diagonal. The back side of regional Mapcodes is the required context, which is specifying the outer tile of the Geohash. Therefore, regional Mapcodes are out of scope for this evaluation. Only global Mapcodes with no need for context, nine characters, and roughly the accuracy of the regional tiles were used. However, Mapcode is flexible enough to provide more accuracy where needed: As with Gohashes in Quadtrees, longer Mapcodes address smaller tiles.

Another alternative addressing scheme is what3words. Like Geo-poet it uses words to encode tiles with ca. $4m$ diagonal length. However, what3words uses three random words for that and is not Geohash based. Therefore, unlike in the three previous systems, common words do not imply that two locations are close to each other. Also, changing the order of the words describing one location results in another unrelated location being addressed. The actual algorithm behind what3words is not public.

Overall, all AAS seem to be more user friendly than plain WGS84 coordinates. Also, their accuracy has the same order of magnitude: Although addressing areas and not points, all seem suitable to specify a navigation destination for a human user. Table I presents some locations encoded with each AAS. Note that Paris and London have a (very short) common prefix in Mapcode and Syllagloble. That means, both cities are in the same tile addressed by the first part of the Geohashes.

III. EXPERIMENT

The measurement undertaken for this paper is disguised as a memory quiz and is available at [17]. Not more than six participants knew the rational before taking the quiz; most of the participants followed a link advertised on various social media. Since the quiz is set up as simple as possible, it is not possible to map answers to specific participants as they are not required to identify themselves. The quiz consists of five parts, one for each AAS. Each part consists of eight questions: Eight times a specific point on the globe is encoded with the respective scheme and shown to the user for four seconds. After that, the participant has to pick the previously presented result from a list of eight possible answers. Besides the right answer, one incorrect choice within $50m$ distance of the correct answer is generated as well as six more distant options. This way, the experiment not only observes how often the correct answer is chosen. It also observes how often an incorrect choice that is close to the right answer is picked by the user. The quiz is laid out in a way that ensures only complete participations with answers to all 40 questions are taken into account. While the quiz is still on-line and collecting data, this paper only considers the 2600 data points of the first 65 participants.

The measurement results are visualized in Figure 1. The bars for each AAS are split into three parts: A part visualizing the portion of the correctly picked answers, a part for those answers that were not correct, but within 50m of the correct answer, and a part for incorrect and far-off results given. Note that to highlight the differences, the bars begin at 80%.

Looking at the correct answers, the AAS can be put into three groups: With 82.8% hit rate WGS84 is the least rememberable scheme. Mapcode and Geo-poet have 92.9% and 91% correct answers respectively and therefore are clearly more user friendly. Using what3words 96.5% and using Syllagloble 96.9% of the participants were able to recall the encoded position correctly.

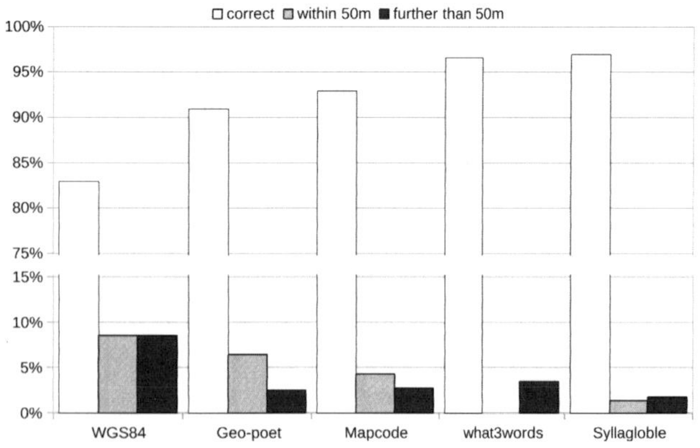

Figure 1. Error types and rates of the various alternative addressing schemes.

A closer look at the splits of the incorrect answers reveals that these are not evenly spread for the various systems. For what3words, all incorrectly picked choices were further away than $50m$ of the correct answer. Of the other AAS, most noteworthy 71.1% of incorrectly chosen answers for Geo-poet were close to the right one. Geo-poet is followed by Mapcode with 61.1%, WGS84 with 50% and Syllagloble with 43.7%. All these rates of incorrectly picked answers near the correct one are extraordinarily high: If the incorrect answers were picked at random, only 14.3% of them would have been within $50m$ radius as it is only one of the seven possible incorrect answers. Note that in total only 18 answers of what3word and 16 answers of Syllagloble encoded addresses were answered incorrect. These numbers are small enough to not represent the proper distributions of the incorrect choices.

IV. CONCLUSION

While still pretty easy to remember in the quiz, WGS84 coordinates are the least human friendly addressing scheme. This is not surprising, as WGS84 was not designed with the human use case in mind.

Geohash based AAS benefit from the common prefix property: Users often remembered parts of the address picking a wrong but similar answer. Wrong choices often become less critical therefore, as they are not too far off from the actual location. This effect is also observable with WGS84 latitude and longitude. While not exactly a Geohash, coordinates with common prefix are closer to each other too.

Interestingly, adding one single word to a scheme seems to make remembering it much harder: The ratio of incorrect answers grew from 3.4% with what3words to 8.9% with Geo-poet. At least partially, this is caused by the words chosen by what3words and Geo-poet. A look at the words in Table I strengthens this assumption: The words of what3words are all shorter. They are therefore easier to remember them selves. Potentially, Geo-poet can be made more user-friendly by tweaking the words used to encode a location. Similarly, carefully choosing the syllables used by Syllagloble, might ensure that it generates words that are even easier to recall.

A long-term experiment setup could verify how well users remember various AAS over a longer time frame. Such an experiment would also reduce the presentation bias. For example, in the experiment for this paper the AAS were always evaluated in the same order, one after the other. Moreover the choices for Geo-poet were presented using two lines each, while for every other AAS the choices only used one line. That made the result list much longer so that a participant was more likely required to scroll to the right answer. Also, a comparison to postal addresses needs to be undertaken. Such a comparison is not fair: As discussed, addresses are not covering the entire world and have a varying accuracy. Still, AAS need to gain acceptance over postal addresses if they intend to replace them in day to day use eventually.

Some of the introduced AAS work with predefined corpora of possible address elements. Geo-poet and what3words have specific sets of words; Syllagloble has definite syllables available. This property can be utilized to introduce error correction. For once, valid possibilities could be suggested as the user types. Finally, implicit error correction could be incorporated into the Geohashes. Addresses, misremembered to a certain extent, could still be resolved correctly this way.

REFERENCES

[1] S. Coetzee, A. Cooper, M. Lind, M. Wells, S. Yurman, E. Wells, N. Griffiths, and M. Nicholson, "Towards an international address standard." 10th International Conference for Spatial Data Infrastructure, 2008.

[2] K. Clemens, "Automated processing of postal addresses," in GEOProcessing 2013, The Fifth International Conference on Advanced Geographic Information Systems, Applications, and Services, 2013, pp. 155–160.

[3] D. Goldberg, J. Wilson, and C. Knoblock, "From text to geographic coordinates: The current state of geocoding," URISA-WASHINGTON DC-, vol. 19, no. 1, 2007, p. 33.

[4] "Google Developers," https://developers.google.com, Dec. 2015.

[5] "Yahoo! BOSS Geo Services," https://developer.yahoo.com/boss/geo/, Dec. 2015.

[6] "Yandex API," http://api.yandex.com, Dec. 2015.

[7] Defense Mapping Agency, "Department of defense world geodetic system 1984: Its definition and relationship with local geodetic systems." Navtech Seminars & Book and Software Store, Incorporated, 1993.

[8] K. Clemens, "Geocoding with openstreetmap data," GEOProcessing 2015, 2015, p. 10.

[9] C. Davis and F. Fonseca, "Assessing the certainty of locations produced by an address geocoding system," Geoinformatica, vol. 11, no. 1, 2007, pp. 103–129.

[10] "what3words," https://map.what3words.com/, Dec. 2015.

[11] "Mapcode," http://www.mapcode.com/, Dec. 2015.

[12] "Geo-poet," http://geo-poet.appspot.com/, Dec. 2015.

[13] "Syllagloble," http://syllagloble.appspot.com/, Dec. 2015.

[14] H. Samet, "The quadtree and related hierarchical data structures," ACM Computing Surveys (CSUR), vol. 16, no. 2, 1984, pp. 187–260.

[15] A. Szalay, J. Gray, G. Fekete, P. Kunszt, P. Kukol, and A. Thakar, "Indexing the sphere with the hierarchical triangular mesh," Microsoft Research, Tech. Rep. MSR-TR-2005-123, Sep. 2005.

[16] Carnegie Mellon University, "The CMU Pronouncing Dictionary," http://www.speech.cs.cmu.edu/cgi-bin/cmudict, Dec. 2015.

[17] "Mem-quiz," http://mem-quiz.appspot.com/, Dec. 2015.

Geospatial Content Services in the Digital Government

CASE: National Data Exchange Layer in the Finnish National Architecture for Digital Services

Lassi Lehto, Pekka Latvala, Tapani Sarjakoski
Department of Geoinformatics and Cartography
Finnish Geospatial Research Institute
National Land Survey of Finland
Masala, Finland
e-mail: lassi.lehto@nls.fi, pekka.latvala@nls.fi,
tapani.sarjakoski@nls.fi

Jari Reini
Department of SDI Services
Finnish Geospatial Research Institute
National Land Survey of Finland
Helsinki, Finland
e-mail: jari.reini@nls.fi

Abstract—**The long tradition of interoperability solutions for geospatial services appears as a challenge when new standards are taken into use in the general eGovernment development. Well-established geospatial service standards have to be adapted to the methods applied in public sector digital services. Issues related to this new integration challenge are discussed in the context of the newly introduced Finnish governmental service interoperability solution called National Data Exchange Layer. A project was initiated to investigate, how existing INSPIRE-compliant content services could be connected to this interoperability framework. Results of the pilot project indicate that in addition to the increased development burden and degraded service performance, there are also opportunities for luring new users to spatial data resources.**

Keywords-interoperability; eGovernment; data exchange; Web Services.

I. INTRODUCTION

Governments all over Europe are developing digitalized online services for their citizens [1]. The services being developed cover functionalities like taxation, applying for social benefits or a construction permit, seeking employment or checking one's retirement allowance. As more and more services are being introduced, it has become obvious that some coordination and standardization is definitely needed. Governments have initiated programs aiming at development of guidelines and policies that would improve the services interoperability. Common software modules are developed in a coordinated way to be shared among the services. Common content vocabularies and data schemas are being introduced in various sectors of public administration. Research efforts in this area include for instance the work of Dias and Rafael [2] to define open interoperability architecture for eGovernment services and Park et al. [3] to introduce a metadata standard for public Web resources on national level.

In the spatial data domain, standardization activities have already a long history. The work of Open Geospatial Consortium (OGC) and ISO Technical Committee 211, Geographic Information, are widely known and highly respected. The recent upsurge of openness, as exemplified in open standards, open data and open source software movements, has boosted the development of interoperable, network-based solutions for applications dealing with geospatial content. Spatial Data Infrastructures (SDIs) have been taken into use in several European countries. On the Pan-European level, initiatives like the INSPIRE Directive [4] are promoting the use of commonly agreed principles in the development of geospatial services.

When the new eGovernment services get widely deployed and agencies start to apply common, standardized approaches in their service development programs, a new question arises: how these developments relate to the existing, already well-established SDI platforms? Could the spatial data community affect the way the new eGovernment service standards are written? Is the spatial dimension taken into account in the generic online citizen services? If yes, is it done following the already defined spatial domain standards or do the spatial domain actors just need to adapt their existing services to the principles established in the eGovernment standardization? These are some of the new questions facing geospatial communities in many European countries. In this paper, the issue is discussed in the context of a recently introduced Finnish eGovernment interoperability framework, called the National Data Exchange Layer (NDEL) [5] and a development project with a goal to connect existing INSPIRE-compliant geodata services to this framework.

The rest of this paper is organized as follows. Section II describes some of the most important eGovernment standardization initiatives and their relation to the existing SDIs. Section III introduces the Finnish governmental program aimed at streamlining the development of digital citizen services. Section IV describes a pilot project testing the connection between the existing national SDI and the new generic approach for service development. As the conclusion, Section V details the main lessons learned in the pilot project.

II. EGOVERNMENT INTEROPERABILITY INITIATIVES

The European Interoperability Framework (EIF) is an important Pan-European initiative for facilitating the eGovernment service provision across country borders [6]. The EIF recommendations stress the importance of adopting open standards and jointly agreed dictionaries and data structures in the development of public online services. It also points out that the focus in the service development should be on ensuring security and user-friendliness. The traditional Web Services Publish-Find-Bind pattern is supported in the EIF specifications, facilitated by a Service Registry component. EIF also points out that the cornerstone for reliable eGovernment service provision is the establishment of mechanisms for signed, certified, encrypted and logged data transport over various different networks.

According to the INSPIRE Network Services Architecture [7], the EIF has similarities with the designed INSPIRE service platform. The EIF initiative aims at the development of the so-called PEGSs (Pan-European eGovernment Services) that are based on interoperable national level services. INSPIRE is based on the same kind of architectural approach. However, INSPIRE does not focus on Publish-Find-Bind pattern, nor does it define anything concerning the data transmission mechanisms. According to the INSPIRE Network Services Architecture document, INSPIRE services must be adapted to the EIF-specified communication platform in the long run.

The latest of the European eGovernment interoperability initiatives is the ISA Programme (Interoperability Solutions for European Public Administrations) [8]. The main objective of the ISA Programme is to improve cross-border and cross-sector interoperability of the national eGovernment services to create the digital single market for the EU. The new ISA^2 Programme will run from 2016 to 2020 as a follow-up of the original ISA.

In the context of ISA, there are actions that aim at bridging the gap between traditional spatial data community standards and the eGovernment interoperability solutions. They include projects like ARE3NA (A Reusable INSPIRE Reference Platform) [9] and EULF (European Union Location Framework) [10]. One of the objectives of the ARE3NA project is to facilitate reuse of INSPIRE-specified methods for interoperability outside the traditional geospatial community. ARE3NA has worked in areas like provision of geospatial data resources as Linked Data, and use of Persistent Identifiers (PIDs) to ease the use of spatial datasets as a location reference platform. The EULF project aims at increasing the use of location information in the eGovernment services and promoting the use of INSPIRE principles in new thematic areas, like transport, marine and energy. EULF will also create guidance on how to implement location enabled eGovernment services, and contribute to the further development of the EIF.

III. Finnish eGovernment Interoperability

A. Backround

The Finnish Government has started a large cross-sectorial programme, called National Architecture for Digital Services, for coordinated development of public sector online services [11]. The initiative is divided in four main areas of work: 1. Setting up a common data exchange mechanism (NDEL), 2. Developing user interfaces to government services for citizens, businesses and civil servants, 3. Enabling a common solution for secure authentication and single-sign-on in digital public services, 4. Creating a centralized solution for the management of user roles and authorization.

To establish a common data exchange platform for the public sector services, the Finnish Government made an agreement with its Estonian counterpart on the use of the Estonian X-Road platform for the purpose [12]. X-Road has been developed since 2001 in Estonia as the national solution for public services interoperability. At the moment it is used to access more than 2000 services providing access to 170 different databases. X-Road is used by half of the Estonian population, and close to 300 million requests is made over it annually.

B. XRoad Platform

X-Road is a decentralized communication and data transfer platform based on the Web Services processing model. The connections over the X-Road platform are facilitated by a special software component, called the X-Road Security Server. These components actually are running on a dedicated server hardware that is strictly defined, to ensure the highest possible level of security for the platform. Information systems always communicate with each other via two Security Servers (see Figure 1).

There is a centralized component in the system that maintains the routing information and is responsible for the centralized logging of transactions. However, Security Servers can operate independently of the Central Server, as they maintain a local copy of all the relevant information.

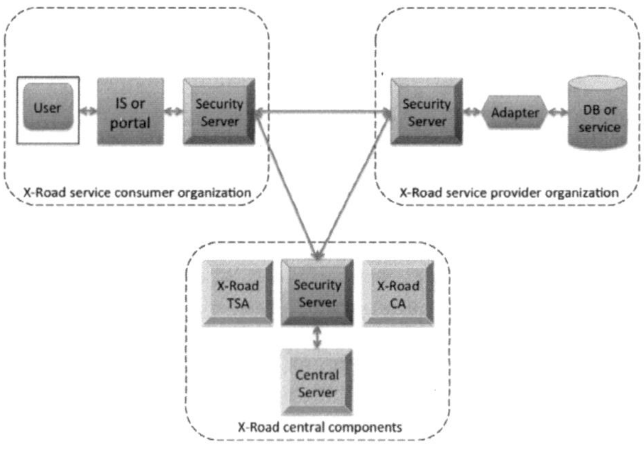

Figure 1. X-Road Architecture (CA: Certification Authority, TSA: Time Stamping Authority, IS: Information System).

The X-Road platform is based on traditional Web Services technologies. The communication is carried out using Simple Object Access Protocol (SOAP) messages [13]. The service interfaces are described using Web Services Description Language (WSDL) [14].

C. National Data Exchange Layer

In Finland, the Estonian X-Road platform has been taken into use mostly unmodified. The few Finnish additions include the so-called REST Gateway module, which facilitates the connection of traditional HTTP GET service interfaces to the X-Road's SOAP-based messaging platform. A software library has also been created to help the developers in building X-Road compatible services. The first operative version of the National Data Exchange Layer (NDEL) was launched in Nov 2015. A beta version of the first user application based on NDEL, the user interface for the citizen, was published in Dec 2015. As the first service, it provides access to the contents of the national Population Information System maintained by the Finnish Population Register Centre.

There are some plans to integrate spatial data and services to the NDEL and to the citizen's user interface. The first application would be a map user interface that displays locations of public sector service points on top of a topographic basemap, provided by the national mapping agency, National Land Survey of Finland (NLS). The second planned application would allow the citizen access to real property information from the Land Information System of Finland, maintained by the NLS. Reliable authentication of the user becomes a necessity in this context as the access will be restricted to the user's own property units.

IV. PILOT PROJECT

When the decision was made on the base technology to be used in the NDEL, it became necessary to test the connection between the already well-established National SDI and the X-Road platform. Over the recent years, the rapid expansion of the Finnish SDI has been mostly driven by the implementation efforts related to the INSPIRE Directive. Thus, a project was launched to investigate, how INSPIRE services, largely based on the OGC-specified interoperability standards, could best be connected to the NDEL [15]. The one and a half year project is funded by the Finnish Prime Minister's Office and coordinated by the Finnish Geospatial Research Institute (FGI). The project consortium includes several public sector organizations dealing with geospatial data.

A. Use Case

A use case scenario was developed, to build the pilot service development on a realistic context. The user story behind the scenario is a five-member family planning to buy or rent a cottage in the Eastern Finland North Karelia area. A mobile client application was developed for the iOS platform to demonstrate the use case (Figure 2). As a further aid to support exploring of the target location, a Differential GNSS (Global Navigation Satellite System) service, provided by the FGI and enabling sub-meter positioning accuracies, has been connected to the client application.

B. Content Services

All the organizations participating in the project provided a spatial data service to be connected to the pilot via the NDEL. The services offer various data sets that might be of interest when considering a target cottage and its neighborhoods. The services are listed below, ordered by the providing organization.

- The National Land Survey of Finland: Topographic Basemap (WMS), property information service from the Land Information System (WFS)
- Finnish Meteorological Institute: VIIRS (Visible Infrared Imaging Radiometer Suite) satellite imagery (WMS), average temperature from observation stations (WFS)
- Finnish Environmental Institute: lake water quality (WMS)
- Natural Resources Institute Finland: forest berry crop map (WMS)
- Geological Survey of Finland: surficial deposit map (WFS)

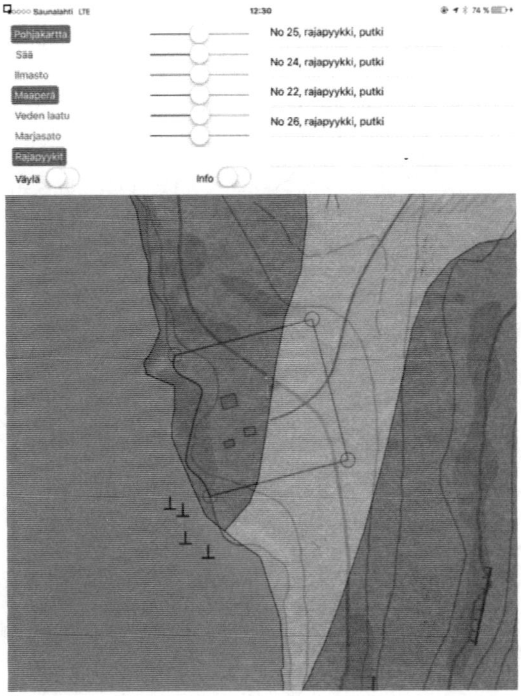

Figure 2. A map view from the mobile client application displaying a target cottage plot on top of the topographic basemap and the surficial deposit map.

C. Adapter Services

So-called Adapter Service is a crucial concept in the X-Road system architecture (see Figure 1). The task of an Adapter Service is to mediate between an existing information system and the messaging protocols of the X-Road platform. In the case of the OGC content services, like Web Map Service (WMS) and Web Feature Service (WFS), this involves unpacking the service request from inside the X-Road's SOAP messaging envelope and, subsequently, packing the resulting data set again into this envelope.

In the case of the POST WFS queries, the task of the OGC service Adapter is rather straightforward as both the POST queries and the resulting data set, in the default GML encoding, are expressed in the XML format. The Adapter Service has to simply retrieve the original XML structure from inside the X-Road SOAP message's XML envelope and, on return, embed the XML message back to the SOAP envelope.

The GET type WFS queries involve a bit more consideration. There are many different ways to encode the original query string to an XML element structure inside the SOAP envelope. The two extreme cases are: (1) The whole query string is packed inside a single XML element, (2) All query parameters are presented as individual strictly typed XML elements, possibly with carefully selected value enumerations and meaningful default values.

The corresponding POST-type query is an obvious candidate for modeling the XML structure. However, more strict schema can be specified for a given concrete service end point, thus making transactions more robust. Case (1) is easy to develop and fast to process. However, it does not reveal the details of the service interface in the WSDL service description. Case (2) is more complicated to build, but will ease the development of client side applications, as the WSDL can be used as the basis for automated code generation, and the exposed value enumerations and default values contribute to more reliable communications.

For the WMS service the situation is quite different. As only GET queries are widely supported, the model for the query encoding must be selected. However, there is no obvious candidate for this. One possible solution is to use the XML-encoded GetMap query defined in the Styled Layer Descriptor (SLD) specification.

In the case of the WMS result data set, which is normally a raster image, the processing task is more involved. Two main approaches are available: the image can be encoded and embedded inside the SOAP envelope's internal element structure or it can be sent immediately after the envelope, using mechanism called SOAP with attachments. For the embedded transmission, the image has to be encoded into text. The mostly used encoding scheme is Base64. When sent as an attachment, the image can be transmitted in binary format.

In the Finnish pilot project, the WFS messaging has so far been performed using the simplest possible approach: sending the whole query string inside a single XML element. In the case of the WMS Adapter Service, the request is sent inside a single XML element and the resulting map image as a SOAP with Attachments message.

If an existing OGC-compliant client application is used to make the request, an Adapter Service is also needed on the client side. In this case the NDEL works as secure, controlled data transfer channel, remaining completely hidden from the client and the service. This kind of architecture is shown in Figure 3. Initial tests carried out in the project show that NDEL used in this way incurs certain level of degradation in the query performance. In case of WFS, the query is 1.5 times slower compared with direct request over public Internet, whereas for WMS services this figure is approximately 2.5. The reason for slower response times is the additional processing required to encapsulate the original requests and responses into the SOAP envelope and, in case of the WMS map response, the process of transforming the map image into the Base64 encoding. Performance tests were carried out using the Apache JMeter testing tool.

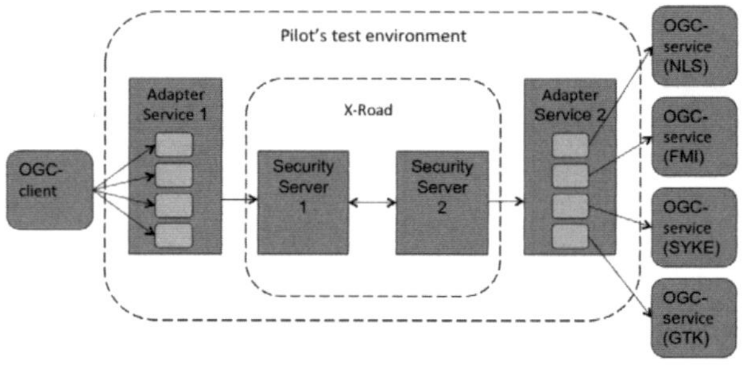

Figure 3. Service Architecture used in the pilot project.

D. New APIs

The service architecture depicted in Figure 3 represents the case, in which an OGC-compliant client application is used. However, introduction of NDEL actually opens access to traditional OGC services for a set of new client environments. These can be divided into two categories. Firstly, SOAP client code can be automatically generated based on the detailed service descriptions expressed in WSDL. Secondly, the NDEL concept of Adapter Service can be exploited to create a completely new category of service interfaces, for instance APIs adapted to the requirements of the modern Web applications programming model. These two new approaches are added to the service architecture model shown in Figure 4.

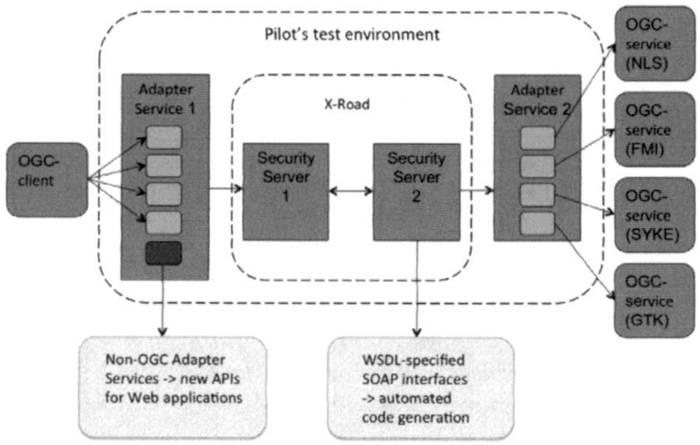

Figure 4. Service Architecture with two new access approaches added.

In the pilot project, the first implementations have been developed to test a new API for OGC services. This service interface is an attempt to provide the easiest possible access to both WMS and WFS interfaces. Geocoding functionality is embedded into the service interface, so that the client can use an address as the indicator of the queried location. The request is encoded as a REST query path. Most of the traditional query parameters can be left out. The service will use reasonable defaults for them.

An example of a query to a WFS interface could thus be expressed as follows:

http://[service domain]/spatialobject/parcel/Helsinki/Mariankatu/23

In this query all parcel objects are requested from around the point into which the given address is geocoded. For all other relevant parameters reasonable defaults are used by the service. In case the defaults are not appropriate for the client, these can be given as further path components in an order specified in the API definition.

In a similar manner, one could request a map image with the following query:

http://[service domain]/map/basemap/Helsinki/Mariankatu/23

New user groups can be given access to geospatial data resources using easy-to-use APIs like these. In Finland this opportunity is opened by the Adapter Service concept present in the NDEL platform. Thus, the approach can be seen as a positive outcome of the NSDI – eGovernment services adaptation challenge.

V. Conclusions

The Finnish NDEL platform represents an example of the new eGovernment services interoperability development programs. Connection between the already well-established INSPIRE/OGC-compliant NSDI and the new NDEL platform has been tested in a pilot project.

The need to adapt the NSDI services to the general eGovernment services interoperability mechanisms can be seen as an unnecessary burden. The tests carried out in the Finnish pilot project confirm that running a query from an OGC-compliant client to an OGC-compliant service via the NDEL platform incurs a significant performance degradation. This means from 1.5 to 2.5 times longer query times, compared with queries that go directly over open Internet.

However, connecting the NSDI with the eGovernment service platform can also open new opportunities. In the case of the Finnish NDEL, these include for instance the possibility to utilize detailed service descriptions in WSDL for automatic code generation. Another positive example is the opportunity to lure new users for spatial data sets via new easy-to-use APIs that are based on the Adapter Service concept, present in the NDEL service architecture.

The future work of the pilot project include for instance more profound testing of detailed WSDL descriptions to support automatic code generation, and performance testing of alternative methods for encoding binary information into NDEL messages. The pilot service will also be connected to the national centralized authentication service to enable single-sign-on, thus fostering more tight integration with other Finnish eGovernment services.

Acknowledgment

The pilot project described in the paper is funded by the Finnish Prime Minister's Office (contract 1940/71/2014).

References

[1] S. C. J. Palvia and S. S. Sharma, "E-Government and E-Governance: Definitions/Domain Framework and Status around the World". Foundation of e-Government. ICEG, 2007. [Online]. Available from: http://www.iceg.net/2007/books/1/1_369.pdf [accessed: 2016-01-19].

[2] G. P. Dias and J. A. Rafael, "A simple model and a distributed architecture for realizing one-stop e-government". Electronic Commerce Research and Applications 6(1), 2007 pp. 81–90.

[3] E. G. Park, M. Lamontagne, A. Perez, I Melikhova and G. Bartlett, "Running ahead toward interoperable e-government: The government of Canada metadata framework", International Journal of Information Management 29, 2009 pp. 145–150.

[4] European Commission, "INSPIRE Directive", 2007. at: http://eur-lex.europa.eu/LexUriServ/LexUriServ.do?uri=OJ:L:2007:108:0001:0014:EN:PDF [accessed: 2016-01-07].

[5] P. Kartano, "Building Digital Infrastructure, Finnish National Architecture for Digital Services". [Online]. Available from: http://www.mamk.fi/instancedata/prime_product_julkaisu/mamk/embeds/mamkwwwstructure/23377_2014-11-19_Palveluvayla_Sotu-seminaari_Mikkeli_en.pdf [accessed: 2016-01-07].

[6] EIF, Home site of European Interoperability Framework for pan-European eGovernment services. [Online]. Available from: http://ec.europa.eu/idabc/en/document/2319/5644.html [accessed: 2016-01-07].

[7] INSPIRE Network Services Drafting Team, "INSPIRE Network Services Architecture". [Online]. Available from: http://inspire.ec.europa.eu/reports/ImplementingRules/network/D3_5_INSPIRE_NS_Architecture_v3-0.pdf [accessed: 2016-01-07].

[8] ISA, Home site of ISA programme. [Online]. Available from: http://ec.europa.eu/isa/ [accessed: 2016-01-07].

[9] ARE3NA, Home site of ARE3NA. [Online]. Available from: http://ec.europa.eu/isa/actions/01-trusted-information-exchange/1-17action_en.htm [accessed: 2016-01-07].

[10] EULF, Home site of European Union Location Framework. [Online]. Available from: http://ec.europa.eu/isa/actions/02-interoperability-architecture/2-13action_en.htm [accessed: 2016-01-07].

[11] Ministry of Finance, Home site of the National Architecture for Digital Services. [Online]. Available from: http://vm.fi/en/national-architecture-for-digital-services [accessed: 2016-01-07].

[12] X-Road, Home site of X-Road. [Online]. Available from: https://e-estonia.com/component/x-road/ [accessed: 2016-01-07].

[13] W3C, SOAP Version 1.2 Part 1: Messaging Framework (Second Edition). [Online]. Available from: http://www.w3.org/TR/soap12/ [accessed 2016-01-07].

[14] W3C, Web Services Description Language (WSDL) 1.1. [Online]. Available from: http://www.w3.org/TR/wsdl [accessed 2016-01-07].

[15] FGI, Geospatial Data Sets in the National Data Exchange Layer (project Web site) [Online]. Available from: http://www.fgi.fi/fgi/research/research-projects/geospatial-datasets-national-data-exchange-layer-ptpv [accessed 2016-01-07].

Mission Exploitation Platform PROBA-V

Jeroen Dries, Erwin Goor, Dirk Daems
TAP - unit Earth Observation
VITO NV – Flemish Institute for Technological Research
Mol, Belgium
e-mail - jeroen.dries@vito.be, erwin.goor@vito.be, dirk.daems@vito.be

Abstract— **VITO and partners developed recently an end-to-end solution to drastically improve the exploitation of the PROBA-V EO (Earth Observation) data archive and derived vegetation parameters from the Copernicus Global Land Service by researchers, service providers and thematic users. The analysis of time series of data (+1PByte) is addressed, as well as the large scale on-demand processing of the complete archive, including near real-time data. Several applications will be released to the users, e.g., a time series viewer, a full resolution viewing service, pre-defined on-demand processing chains and virtual machines with powerful tools and access to the data. After an initial release in January 2016 a research platform will gradually be deployed allowing users to design, debug and test applications on the platform. From the MEP PROBA-V, access to, e.g., Landsat-7/8 and Sentinel-2/3 data, will be addressed as well.**

Keywords - MEP Mission Exploitation Platform; PROBA-V; vegetation; data analytics; on-demand processing; Web Services.

I. Objectives and Benefits

The PROBA-V MEP (Mission Exploitation Platform) builds further on the R&D (Research & Development) results, from the ESA 'ESE' project. These results consist of prototypes which were further refined in several other projects thanks to the active involvement of these projects in the ESE pilots activities. The paper is organized as follows: in this section I we discuss the objectives of the platform. In section II the technical solution is described and section III contains the conclusions and future work.

The PROBA-V MEP has the ambition to complement the PROBA-V [1] user segment by building an operational Exploitation Platform (EP) on the data, complementary data and derived products, addressing hereby the wider vegetation user community with the final aim to ease, and increase, the use of PROBA-V data. The data offering will consist of the complete archive from SPOT-VEGETATION, PROBA-V and bio-geophysical parameters from the Copernicus Global Land Service [2].

The reasons for deploying a MEP dedicated to the PROBA-V mission are numerous:

- The data and specifically the time series of daily / ten-daily data from 1998 till present is too big to be downloaded to and processed on the users' premises, at least for the majority of the users.
- On top of the Earth Observation (EO)-data mentioned above, the platform can co-locate as well complementary data in a way that it is easily accessible. Furthermore tools, libraries and applications, which can be used by the large community will be provided. This includes as well the data needed for calibration and validation activities.
- The platform can stimulate collaboration between the users, as we bring together services from various users on the same platform with a number of tools to support the publishing of and to provide feedback on these services. A further focus on documentation, knowledge sharing and user support complements this.
- The platform goes beyond offering standard products by offering in a first place applications to visualize and analyze large time series of data and pre-defined on-demand processing services, which deliver user-tailored products. In a next step we will gradually deploy a Virtual Research Environment, being a platform, which allows users to develop – debug – test an application on an infrastructure at VITO with access to the complete data archive. Successful applications from third-parties can then be offered as an operational on-demand processing service to the user community on the same platform.
- As an Exploitation Platform (EP) with a focus on open interfaces, we position the PROBA-V mission in an ecosystem of TEPs (Thematic EPs), REPs (Regional EPs) and other MEPs. In the future, the PROBA-V MEP can be integrated gradually in a federation of different platforms, including as well Sentinel Collaborative Ground Segments, in line with the current ESA strategy on the 'EO Ground Segment Evolution'.

During the PROBA-V MEP project, which will at least last till the end of the PROBA-V mission in May 2018, several third-party service projects will develop and operate applications on the operation MEP platform. We will address their user requirements to implement the shift of paradigm from "data to user" to "user to data", bridging the gap between the traditional EO ground segment and the scientist or value added industry by providing a one stop shop for access to the full PROBA-V Mission data (including derived parameters) and to external repositories of similar missions/sensors (including Landsat and Sentinel).

II. TECHNICAL SOLUTION

The PROBA-V MEP will provide scalable processing facilities with access to the complete data archive and a rich set of processing algorithms, models, open source processing libraries/toolboxes and public/collaborative software. The platform becomes the processing infrastructure hub of the mission by functioning as a powerhouse system and open access development environment.

To realize this, the platform consists of the following components:

- The existing Product Distribution Facilities [3] and [4], are serving the access to the data archive, both via a Web portal as well as standardized discovery, viewing and data access interfaces. More evolutions on these standardized machine-to-machine interfaces are planned in the near future.
- Hadoop [7], as a platform for data-intensive distributed applications, is designed to process large amounts of data by separating the data into smaller chunks and performing large numbers of small parallel operations on the data. It is applied often for processing big data and is applied in this context for the on-demand processing of EO data, as prototyped successfully in the ESE project. Oozie [5] is used as a workflow processing engine to design an EO-application as a workflow of multiple processes. Spark [6] is used intensively to allow analytics on large time series of data. The Hadoop ecosystem provides furthermore a rich and still growing set of tools, which are used to provide fast access to the data in a format needed by the specific application.
- The EO raster data is accessible via NFS (Network File System) and possibly uploaded to the Hadoop Distributed Filesystem (HDFS) using a Data Manager. This Data Manager also integrates with several catalogues implementing different protocols, so that third party-data can be ingested into the platform when needed by a specific user.
- Cloud computing technology enables dynamic resource provisioning and is therefore providing a flexible and scalable solution. OpenStack [8] is chosen as cloud middleware. Pre-configured virtual machines will be offered and can run on the OpenStack cluster at VITO, providing the environment needed for users to work with the data and develop/deploy applications on the platform, i.e., containing IDE's, a rich set of tools and access to the complete data archive.
- Interactive Web-based dashboards are designed to provide user-tailored information from the EO data archives of VITO and other providers, by combining existing components such as AngularJS, Javascript libraries and GIS components into one single solution. The combination of these different components, applied on data available in disparate data stores, offers powerful Web portals to the users in order to make vast amounts of data understandable. We can easily design user-tailored Web-based dashboards, which offer at any time near real-time information for the regional extent of interest to the user and in the format chosen by the user.
- A Web portal provides access to all applications and tools offered by the PROBA-V MEP and to the cloud consoles. Furthermore the portal provides all information on the data and components available on the platform and offers tools for e-collaboration and knowledge sharing amongst the users.
- A main concern is security since we allow users to develop and execute their applications on the platform. Their IPR shall be properly protected and the activities of individual users cannot influence the stability of the system and the work of other users. Single sign-on and proper monitoring of used resources are further requirements.

III. CONCLUSIONS AND FUTURE WORK

The platform was launched in January 2016 at the PROBA-V conference in Ghent, Belgium. Three iterations are planned to gradually expand the capabilities of the system and provide new features, in close collaboration with the first third-party projects working on the platform.

The impact of this PROBA-V MEP on the user community will be high and will completely change the way of working with the data and hence open the large time series to a larger community of users. The operational platform is based on recent R&D activities and is in line with the new ESA strategy on the 'EO Ground Segment Evolution'. Hence, as future work, the integration of the platform within a federation needs to be addressed. More applications and users will be integrated in the platform to enrich the content and enlarge the user community. Furthermore the evolutions in Big Data analytics and processing will be followed closely and integrated in the platform where relevant.

REFERENCES

[1] http://proba-v.vgt.vito.be/ [accessed: 2016-03-19].
[2] http://land.copernicus.eu/global/ [accessed: 2016-03-19].
[3] http://www.vito-eodata.be [accessed: 2016-03-19].
[4] http://land.copernicus.vgt.vito.be/PDF/ [accessed: 2016-03-19].
[5] http://oozie.apache.org/ [accessed: 2016-03-19].
[6] http://spark.apache.org/ [accessed: 2016-03-19].
[7] http://hadoop.apache.org/ [accessed: 2016-03-19].
[8] http://www.openstack.org [accessed: 2016-03-19].

On Feasibility to Detect Volcanoes Hidden under Ice of Antarctica via their "Gravitational Signal"

Jaroslav Klokočník, Aleš Bezděk
Astronomical Institute, Czech Academy of Sciences
Ondřejov, Czech Republic
e-mail: jklokocn@asu.cas.cz, bezdek@asu.cas.cz

Jan Kostelecký
Research Institute of Geodesy, Topography and Cartography, Zdiby, Czech Republic
Faculty of Mining and Geology, Ostrava, Czech Republic
e-mail: kost@fsv.cvut.cz

***Abstract*—Many not yet discovered volcanoes may be hidden under thick layers of ice in Antarctica. Discovery of two volcanoes active under the ice (from seismic network), new gravitational field models with high resolution (like EIGEN 6C4) based also on gradiometry data from satellite GOCE and progress in mapping topography of bedrock (BEDMAP 2), mostly from remote sensing by satellites, has been inspiring to seek for hypothetic volcanoes hidden under ice of Antarctica by using these data sources. Our method is novel. We do not work with direct measurements like terrestrial gravity anomalies or airborne gradiometry, but with spherical harmonic expansion for the gravitational potential. This approach is not local, but regional and global, thus it has a lower resolution than the local data. We make use of analogy with the "gravitational signal" known for volcanoes and other structures in other parts of the Earth. We utilize various functionals and functions (not only ordinary gravity anomalies) of the disturbing geopotential (being represented by harmonic coefficients in expansion of the potential to spherical harmonic series, namely by EIGEN 6C4 to degree and order 2160). We claim that our method is promising for future successful search for subglacial volcanoes, having of course in hands also other than satellite data. Our present-day attempts to discover such volcanoes hardly can be of big success, because of low resolution (mainly) of the existing gravity data and (partly) due to a low resolution of even the best bedrock topography of Antarctica now available.**

Keywords- gravity field model EIGEN 6C4; Bedmap 2; functions of disturbing potential; volcano; Antarctica.

I. Introduction

Many not yet discovered volcanoes may be hidden under thick layers of ice of Antarctica. New gravity and topography data now available (see Section III) inspired us to try to detect such objects. We had experience with studies of other objects in other parts of the world [3][4] and we utilized it here. But, there are some problems specific to Antarctica. We had to answer the following questions. Are the best present-day available gravitational and topographic data of sufficient precision and resolution? How fast is an attenuation of the "gravitational signal" of a volcano with increasing depth under the ice?

We have not found any principal problem precluding a successful detection of large volcanoes under the ice. But, there is a practical obstacle in the low resolution of the gravity (mainly) and topography data (also), even coming from the best now available data sets, so practical examples of detected volcanoes are limited to a few cases in Queen Maud Land and near the Lake Vostok.

The attenuation of the signal under the ice for the gravity anomaly is not significant; the attenuation for the second radial derivative of the disturbing potential is negligible.

II. Theory

Theory related to the present work in progress comes mainly from Pedersen and Rasmussen (1990) [1], Beiki and Pedersen (2010) [2] and from our previous research work [3][4].

III. Data

Data are of two types of data of interest: gravitational and topographic, both with significant contribution to remote sensing methods. (1) The gravitational data are the harmonic geopotential coefficients (also known as Stokes parameters) in the spherical expansion of the disturbing gravitational potential into the spherical series. European Improved Gravity model of the Earth by New techniques (EIGEN 6C4, [5]) is expanded to degree and order 2190 in spherical harmonics. It corresponds to a resolution of 5x5 arc minutes, which is ~9 km half-wavelength on the Earth's surface. But the resolution in Antarctica is lower due to the fact that there we have solely satellite data (GRACE and GOCE) available in EIGEN 6C4. (2) The topography of the ground under ice (bedrock, base of the ice sheets) in Antarctica is known as BEDMAP 2 [6], being compiled from measurements of various kinds, with penetrating radars to the ground or water under the ice, with a resolution reaching 1x1 km in some areas of Antarctica and a few kilometres in the others. A combination of both sources is also possible, but a mutual independency of EIGEN 6C4 and BEDMAP 2 is lost.

IV. COMPUTATIONS AND RESULTS

The gravity anomalies or disturbances, the Marussi tensor of the second derivatives, with the invariants and their ratios, the strike angle and with the virtual deformations are computed with our own software [7][8] in 5x5 arcmin grid everywhere where we need it. We learnt how the typical signal of volcanoes looks like outside Antarctica (e.g., [3][4]) and we now extrapolate to Antarctica under ice. We combine the gravitational data and the bedrock topography and seek for signal typical for a volcano simultaneously for more functionals or functions of the geopotential (examples in Figure 1 below). We indicate localities in the Gamburtsev Mountains, Queen Maud Land and other places (one example is in Figure 3) with candidates for volcanoes.

V. CONCLUSION

Our method, based on combining gravity and topography data, is promising for future successful search (with new forthcoming data with higher resolution) for subglacial volcanoes and other objects hidden under the ice (or elsewhere, for example under the sand of the Sahara), having of course in hand also other than satellite data only. Our present-day attempts to discover such volcanoes hardly can be of big success, because of low resolution (mainly) of the existing gravity data and (partly) due to low resolution of the best bedrock topography of Antarctica now available. But we achieved some results, of those we present here one example to show how it works in the case of known volcanoes (Figure 2) and one case of predicted, hypothetical volcano (Figure 3).

ACKNOWLEDGEMENT

For support, we are grateful to Grant Agency of the Czech Republic for the project 13-36843S.

REFERENCES

[1] B. D. Pedersen and T. M. Rasmussen, "The gradient tensor of potential field anomalies: Some implications on data collection and data processing of maps," Geophysics, vol. 55, 1990, pp. 1558-1566.

[2] M. Beiki and L. B. Pedersen, "Eigenvector analysis of gravity gradient tensor to locate geologic bodies," Geophysics, vol. 75, DOI: 10.1190/1.3484098, 2010, pp. 137-149.

[3] J. Kalvoda, J. Klokocník, J. Kostelecký, and A. Bezdek, "Mass distribution of Earth landforms determined by aspects of the geopotential as computed from the global gravity field model EGM 2008," Acta Univ. Carolinae, Geographica, XLVIII, Vol. 2, #48, Prague, 2013, pp. 17-25.

[4] Klokocník J, Kalvoda J, Kostelecký J, Eppelbaum LV, Bezdek A: 2013. Gravity Disturbances, Marussi Tensor, Invariants and Other Functions of the Geopotential Represented by EGM 2008, ESA Living Planet Symp. 9-13 Sept. 2013, Edinburgh, Scotland. J Earth Sci. Res. 2, 2014, pp. 88–101.

[5] Förste Ch., Bruinsma S., Abrykosov O., Lemoine J-M. et al.: The latest combined global gravity field model including GOCE data up to degree and order 2190 of GFZ Potsdam and GRGS Toulouse (EIGEN 6C4), 5th GOCE User Workshop, Paris 25 – 28, Nov. 2014.

[6] Fretwell, P.; Pritchard, H. D.; Vaughan, D. G.; et al.: Bedmap2: improved ice bed, surface and thickness datasets for Antarctica, The Cryosphere, 7, doi:10.5194/tc-7-375-2013, 2013, pp. 375–393.

[7] Bucha B, Janák J (2013): A MATLAB-based graphical user interface program for computing functionals of the geopotential up to ultra-high degrees and orders, Computers & Geosciences, 56, doi: 10.1016/j.cageo.2013.03.012, 2013, pp. 186-196.

[8] J. Sebera, C. A. Wagner, A. Bezdek, and J. Klokocník, "Short guide to direct gravitational field modelling with Hotine's equations," J. Geod., 87, doi: 10.1007/s00190-012-0591-2, 2013, pp. 223–238.

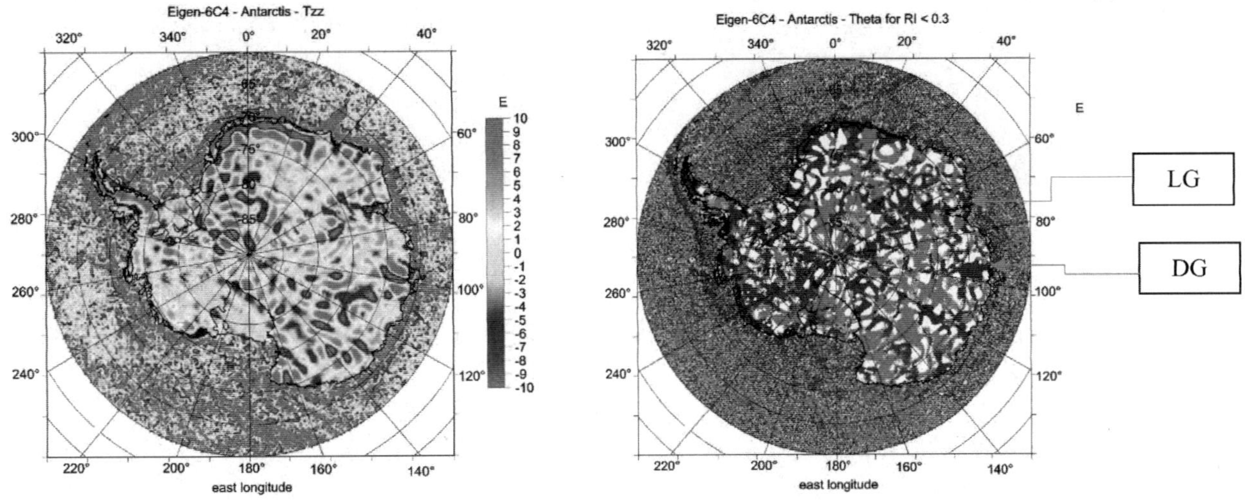

Figure 1. On the left: The radial component of the Marussi tensor (scale in Eötvös over Antarctica. On the right: The strike angle over Antarctica (in red its direction to the East, in blue to the West of the meridian). Lambert Glacier (LG), Dehmann Glacier (DG) shown by arrows.

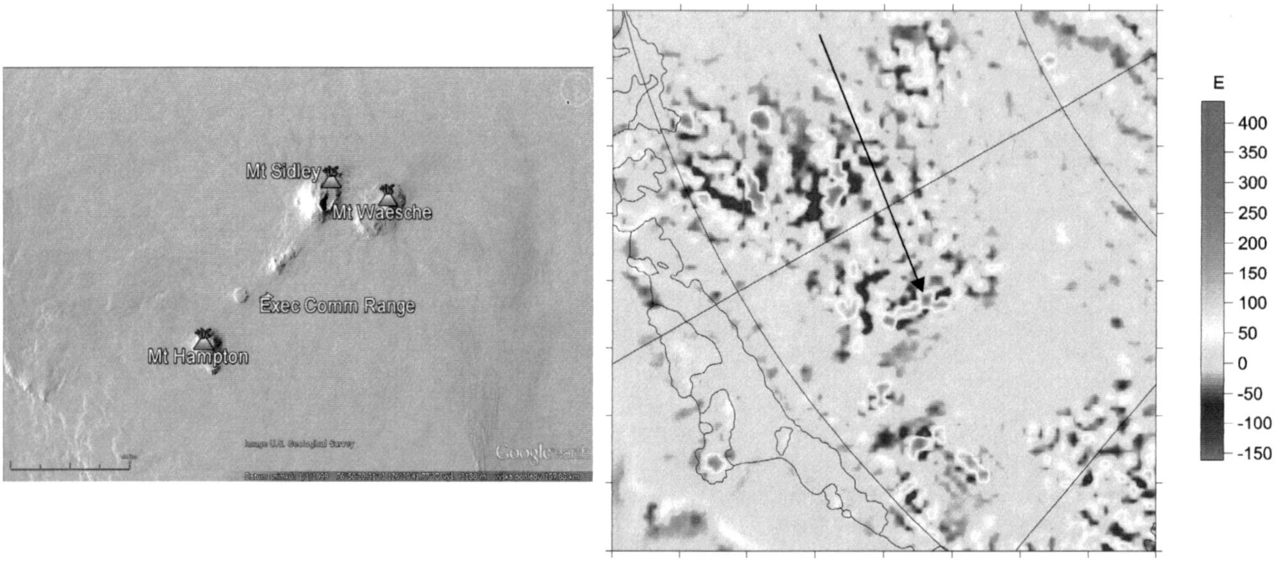

Figure 2. On the left: Topography of the area of Mt Sidley, the highest, dormant volcano in Antarctica, and the Executive Committee Range of Marie Byrd Land, from © Google Earth, example of known volcanoes, visible on the surface. On the right: The second radial component of the disturbing potential from a combination of the EIGEN 6C4 and BEDMAP 2. The arrow shows Mt Sidley.

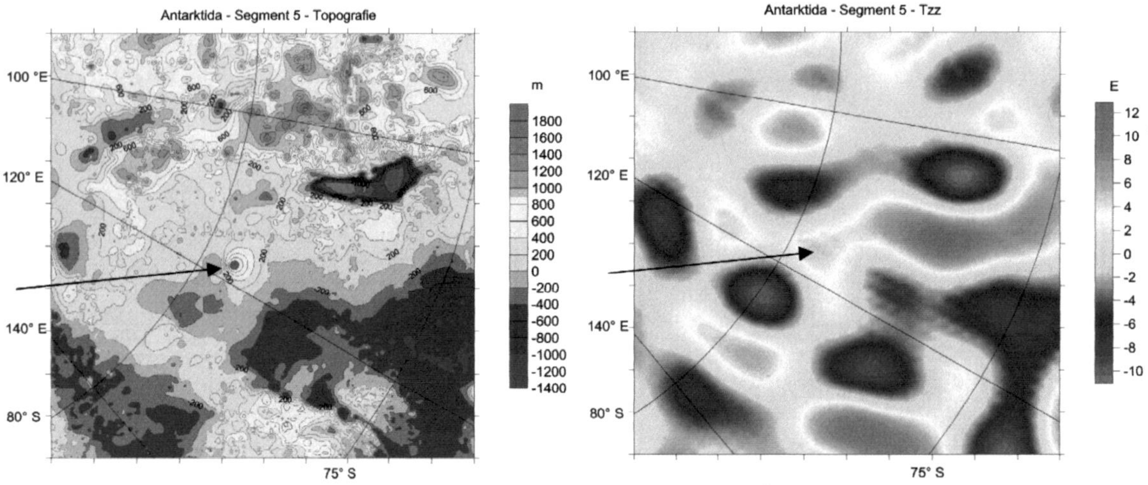

Figure 3. One example of our new results. On the left: Bedrock topography according to the BEDMAP 2 model (scale in metres) near the lake Vostok (large blue oblong depression with a depth to 1000 m, hidden under about 3 km of the ice) with suspicious cone-shaped object nearby (isolated mountain in a plain terrain, with a relative height difference of about 1200 m). On the right: The second radial derivatives of the disturbing geopotential (scale in Eötvös) according to the EIGEN 6C4 gravity field model for the same territory.

We Need to Rethink How We Describe and Organize Spatial Information

Instrumenting and Observing the Community of Users to Improve Data Description and Discovery

Benjamin Adams and Mark Gahegan
Centre for eResearch and Dept. of Computer Science
The University of Auckland
Auckland, New Zealand
e-mail: {b.adams,m.gahegan}@auckland.ac.nz

Abstract—**In Spatial Data Infrastructure or Cyber Infrastructure, the description of geographic data semantics is intended to support data discovery, reuse and integration. In the vast majority of cases the producers of these data generate descriptions based on particular understandings of what uses the data are good for. This producer-oriented perspective means that the descriptions often do not help to answer the question of whether a data set is of use for a consumer who might want to apply it in a different context. In this paper, we discuss the role geographic information observatories can play in providing an infrastructure for observing the context of data use by consumers. These observations of data pragmatics lead to operational statistical methods that will support better fitness-for-use assessment. Finally, we highlight some of the challenges to building these observatories, and briefly discuss strategies to address those challenges.**

Keywords-Data description; data discovery; data reuse; geographic information observatories; graphical model; knowledge representation.

I. INTRODUCTION

The goal of fostering data reuse and semantic integration by describing geographic data in a cyberinfrastructure has met with limited practical success, despite many years of research. One of the challenges that such efforts face is the *situated* nature of geographic knowledge—how we understand data depends strongly on our own experience and expertise, and potentially also the situation within which we intend to use it [7]. Systems that organize information are usually designed to support a set of interactions by a community of users [8]. But, if the organizational system that is used in a cyberinfrastructure does not incorporate a model of the user community—and react to what can be learned from user interactions with the infrastructure—then it will have limited utility. To date the onus for describing data has primarily fallen on the producers of the data, whether they are authoritative organizations or individual scientists. This approach means that the descriptions usually reflect what the *producers* value in the data, without consideration of whether the data will be fit-for-use by a potential *consumer* of the data. This is entirely reasonable given that the producer cannot possibly anticipate all the ways in which their data might be used. But perhaps we are approaching the problem incorrectly in thinking of it this way? To further complicate matters, the heterogeneity of geographic data, and the differing goals of its producers has lead to data being described in a multitude of incompatible ways. And despite a large number of research papers on geospatial semantics—are often not getting any easier to reconcile in practice [15]. The result is a confusion of *just-so data stories*, describing what data "means", but in a manner that provides very little help for data consumers to find the data that is suitable for their purposes.

This is not a new critique. Frank [5] identified this problem (phrased in terms of data quality) well before the era of using semantic web languages to describe geographic data. But the critiques of producer-oriented data description (see also [3][17]), have not led to improved operational approaches for organizing geospatial information, perhaps because their strategies are quite abstract in their own way, lacking a clear methodology for putting them into practice. Current efforts on geo-semantics, in contrast, make operationalization of languages (semantics) and reasoning paramount, without much consideration of fitness-for-use [11]. If, as has been argued, fitness-for-use assessment is critical for functional Spatial Data Infrastructure (SDI) (or Geo-CyberInfrastructure), then we need practical and achievable methods that allow us to understand the context of data use from the perspective of the user: that is, in contexts that may not have been foreseen by the data producers or cyberinfrastructure builders.

In this paper, we explore how some of the emerging ideas from *information observatories* could be applied in a geographical context to provide a practical solution for including the consumer in our methods of data description [1]. The approach we advocate is not without its own challenges (many of them social), but if adopted by the community, we believe will lead to statistical methods that will allow us to better support fit-for-use data assessment in cyberinfrastructure.

An Information Observatory is infrastructure designed for understanding the ecosystem of information that *observes* not just the object of study but also the conceptual structures, data, and actors involved in the process of analysis and knowledge production, from multiple perspectives. A Geographic Information Observatory (GIO) is thus an information observatory focused on geographic information AND its community of practice. Building GI observatories means building infrastructure that can observe geospatial data in the context of its use, within a community. This is quite distinct from traditional approaches to data description in cyberinfrastructure, which focus primarily on using metadata to describe data formats and the semantics of data content: in an information observatory we consider not only data but also tasks and methods performed with the data, the domain knowledge of data producers and consumers, communities-of-practice, and more; all of these facets become first-class observable artifacts (signifiers in the semiotic sense) that carry meaning and help to explain or contextualize each other.

We propose that GI Observatories can provide insight into the dynamic, geographical scientific process from multiple perspectives, from data through tasks and methods to communities-of-practice, and that this insight will help us build cyberinfrastructure that will better support these interactions [1][6]. Or to put it another way, we propose to make an empirical science out of cyberinfrastructure development. We want to observe the relationships between these different facets of geographic information, because these observables are only meaningful when brought into relation with other concepts. And by observing them, we can learn from them all kinds of practical insights that can help characterize what information is used for, by whom, using which methods, for what tasks. Over time, such observations can yield actionable intelligence that may add significant value over and above what can be achieved by formal semantics.

In the following section we review related work. In Section III we discuss the nexus of knowledge relations that are employed in geographic research. Section IV details how GI Observatories can be implemented by operationalizing the nexus as a graphical model. In Section V we discuss some considerations of the role of the community in building GI Observatories, and we conclude with a discussion of the opportunities and challenges of building these observatories going forward.

II. Related Work

The Web Observatory is a nascent idea proposed to support the notion of doing Web Science. That is, observing the web in order to understand how human activity shapes the Web, and how human activity patterns and the Web co-evolve. Web Observatories have been described as the "middle layer for broad data" meaning that, as data production has become more distributed and decentralized, the ability to perform analyses on these data has remained siloed, and the observatory serves to open up that analytic framework [19]. To date, the development of web science observatories has focused mostly on data collection / mashup tools that produce views on 'big' web data such as streaming social media content [20][21]. Recent work exploring the idea of building observatories on top of citizen science projects, such as *Zooniverse*, point to promising applications of Web Observatories to the sciences [22].

Although there are some superficial similarities between Web Observatories and the GI Observatory idea that we are proposing with respect to observing human information interaction, our focus differs in two important ways. First, the Web Observatory is a macro-scale observatory in the sense that it is designed to provide analytic infrastructure to explore properties of the web and human society, whereas the Information Observatory aims to capture observations at the granularity of data use and change, by individual researchers and within specific scientific communities [23]. Second, Web Observatories are primarily described as tools for analysis of the Web as a socio-technical system, thus there is a very specific subject of analysis, namely the Web. As noted, our motivation for building GI Observatories is in large part driven by a desire to build better cyberinfrastructure for scientific discovery, and we are interested in understanding the universe of information from a multitude of perspectives.

Personalization in information retrieval requires modeling the background context under which a user performs an information-searching task [24][25]. This context model can take the form of an explicit user model or can be based on other kinds of implicit behavioral feedback, such as search history and click-through data [26][27][28]. Group level models of personalization are based on the notion that similar users will want similar search results [29]. Recommender systems built with collaborative filtering algorithms fall within this category of personalization [30].

Several variables can play a role in creating a user model using a relevance feedback framework. For example, the temporal scope of the information can be important, so that immediate search history might be more relevant than longer-term behavior [31]. Beyond search history, measures of user interests and activities from heterogeneous sources (documents, emails, etc.) can also be useful [32]. In social search, the role of the user within a larger community becomes an important factor [33].

Personalized search based on past behavior has raised concern about the potential drawback of creating a filter bubble, where potentially relevant information is not shown because of the personalization algorithm [34]. The many personalization methodologies and algorithms that have been developed for information retrieval and web search could well be applied to observational data collected by a GI Observatory and then be used to develop new ways of searching for scientific data.

III. The Nexus of Relations

Inspired by Alfred North Whitehead's [18] writings on the intricate web of relationships that participates in knowledge representation, Gahegan and Pike [7] described a *nexus* of relations that link the many conceptual structures used in geographic research. Their nexus is shown in Figure 1. Based on this nexus the Codex system was built to capture these relations and make it possible for a user to explore resources through any and all of these relations. In practice, however, depending on the content of each of the nodes in the nexus and context that is of interest to the user, only a small subset of those relations will actually be relevant. What is missing is a way to structure the web of relations in a way that facilitates understanding without overloading the user with unnecessary information.

The ovals in Figure 1 represent the conceptual structures that the GIScience research community has expended most effort on describing, in some cases building metadata standards for describing those concepts. Semantic metadata descriptions of geographic data in cyberinfrastructure focus on "objective" aspects of geographic data shown as circles in blue. However, the nodes (purple clouds) might take on very different values depending on whether one is a producer or consumer of data (the researcher in red). The purple nodes—which are usually not captured—are important to assess fitness-for-use.

For example, we have semantic models of measurement, geographic data models, and scientific workflows [4][9][13]. Furthermore, current GIScience usually has its gaze fixed on Geographic Information (as its name suggests), so this is the subject around which other concepts are positioned. Semiotically, we could say that information (or data) is always the interpretant, with other data used to help explain it. But this misses an opportunity to focus on other key facets, such as a researcher or a method and to use data to help describe them. Meanwhile, situational knowledge is not formally captured (represented by the cloud shapes in Figure 1), other than in some cases natural language, e.g., in journal publications. Where other facets are captured, it is usually referred to as *context*, but now let's recognize that it is simply a set of observations onto a different set of objects, not typically represented in an infrastructure, but could be.

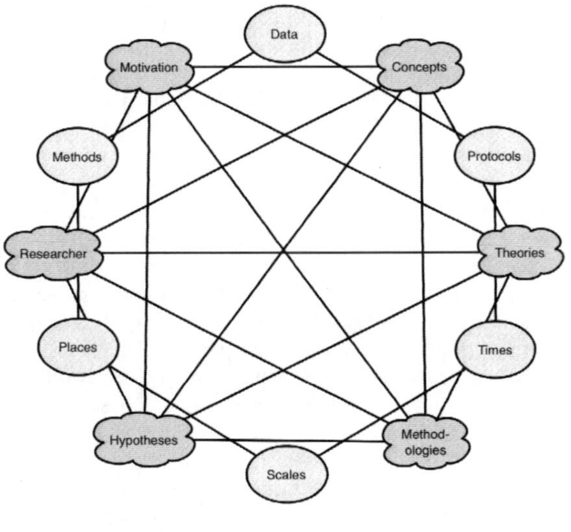

Figure 1. The nexus of relations between (some of) the conceptual structures used in geographic research, adapted from Gahegan & Pike [7].

Arguably, scientists who use these concepts in their work are more likely to be able to articulate what they want and what is important to them in terms of the conceptual structures represented by the cloud shapes—motivations, methodologies, theories, etc., rather than using the formal description languages for data that have been the target of most work on geosemantics. This in no small part might explain why—despite great effort in GIScience to advance semantics—semantic technologies are not being used by most geo-scientists in their day-to-day research. It is simply not the language that they use to think about and communicate their research.

The model of the nexus is just one example meta-model for the kinds of relationships between conceptual knowledge used in geographic research. For example, a sub-graph of these relations forms the graphical model of data production described by Gahegan and Adams [7]. In that model, each node in the nexus (*community*, *task*, *domain knowledge*, and *data*) are described by facets intended to capture the *who*, *what*, and *why* of data. The relations in that model are also directed, rather than undirected as in the original nexus meta-model.

When building cyberinfrastructure for data re-use, we are faced with a two-fold problem: first, whether the representation that we choose for each node in the nexus is a functionally useful representation (such as an ontology of conceptual structures) for scientists engaging in geographic knowledge production; and second, whether the meta-model describing the relations and directionality between the nodes in the nexus is itself useful. The GI Observatory gives us an infrastructure for addressing the first problem by collecting data on how conceptual structures relate to one another in practice, and then evaluating the information to be gained about one facet from a particular description of another facet. The second problem is encapsulated in the dynamic nature of the GI Observatory and its community-driven mandate—they are built to serve geo-scientists and the meta-model should be flexible and revised as a richer and more accurate understanding of the community's needs emerges. In this sense, Figure 1 is just a place to start, not a final recommendation.

To get us beyond the essentially meaningless result that everything is related to everything else, a GI Observatory will always need to make a commitment to a given meta-model in order to observe the nodes in the nexus. We also need to identify where to fix the GI Observatory's "telescope" and where to point it. For example, we cannot look at *tasks* from a fixed perspective without deciding first what information artifacts of *tasks* act as signifiers and then describing the object of those signs, whether they be *data*, *methods*, *communities*, etc. Most importantly, many different kinds of meta-models and signifiers and objects of enquiry can be created and could even co-exist, and the choice of those meta-models will depend on what the users of the GI Observatory want to use it for.

IV. OPERATIONALIZING THE NEXUS AS A GRAPHICAL MODEL

Relational network representations of knowledge such as the one shown in Figure 1 are found in all kinds of information systems. And such graphical models can be recast as learning problems that are solved by doing statistical inference on networked nodes that represent variables that can take on many values [12]. Because mature computational patterns have been built for statistical inference on graphical models, they are applied to stochastic problems in a wide range of fields, including speech recognition, natural language processing, statistical physics, spatial statistics, and bioinformatics. The network of conceptual structures used in practice and observed through a GI Observatory can similarly be modeled as graphical models. The variables observed can span across data (formats, semantics, geometric models), tasks, analytical methods, computational workflows and people (researchers, communities).

Graphical webs of relations (Figure 1) can be either undirected or directed. A directed graph (mathematically represented as an acyclic graph) is used to model cause-and-effect relationships, whereas an undirected graph can be used when the causal structure is undefined. The model presented in [6] is directed and explicitly represents data production as a generative model with variables that describe causal relationships between communities that perform tasks with domain knowledge, which results in data of some kind. In contrast, the nexus in [7] is undirected, nodes are connected but there is no explicit directionality to the connections. Both kinds of models can form the backbone of a GI Observatory and utilizing the methodologies of statistical inference on graphical models (such as Bayesian Belief Network Learning or more general Inductive Model Discovery) can provide unique insights into the geographic information ecosystem [2][14]—and importantly without having to choose beforehand which causes what.

However, in order to access these patterns through data-driven discovery we need sufficient data, and it is in providing access to data about people (both producers and consumers of data), methods, workflows, and intention that GI Observatories show real promise in improving our geo-infrastructure. Several strategies can be employed to observe the variables in the nexus of relations. The lowest hanging fruit is to extract pragmatic relationships connecting people, ideas and things to datasets from natural language available in data repositories and on the web. Figure 2 shows how some of this information is encoded already in abstracts and webpages for researchers and investigating organizations. The text highlighted in purple in Figure 2 represents information that can be extracted from natural language and mapped to entities that represent motivations, concepts, theories, etc. In this example the data producer describes very specific affordances that the data provides, but we know very little about how the consumers of the data have used it and whether the entities that the producer highlighted are in fact of high informational value to potential consumers.

V. Involving a Community of Users

GI Observatories will need to be built to serve the scientific community, so scientists should get real value out of what we can learn from these Observatories. Thus, ideally we involve geographers and domain scientists in their design, building, and re-building [1]. One of the most significant challenges to building GI Observatories is that, in order to observe many aspects of the scientific process, the Observatory will need access to information about what researchers do that is usually not readily available. For example, what kind of methods are scientists who are working in a specific field of research using with what kinds of data? Here we can learn from recent work by David Ribes on what he calls *scaling up ethnography*: "The object of analysis for the ethnographer ... becomes the methods, techniques and technologies used by actors to know and manage their enterprise." [16]. The instrumenting of the community to better understand data use presents several ethnographic challenges. But the fact that researchers increasingly interact with data and methods through infrastructures and workflow scripts means that at least some of the information we seek is in fact readily available. We simply need to instrument the SDIs to record it.

A GI Observatory provides the research platform for doing scaled up ethnography on not only large-scale geospatial cyberinfrastructure projects, such as spatial data infrastructures, instantiations of Digital Earth, and smart cities; but also the activities of the geographic scientific community writ large.

VI. Results and Findings

A prototype system for capturing the patterns of interaction between a community of users, tasks, methods, concepts and datasets has been created. It uses the notion of Description Spaces for: Space and Time, Domain Semantics, Processes and Community. Each space contains a smaller number of descriptive attributes. The Description Spaces allow us to compute a compound distance score between any pair of items, such as a researcher and a dataset or a method and a task. Over time, any use case that connects such items is remembered, effectively that can be used to 'bring them closer'. Bayesian inference is used on these Description Spaces and past histories to predict likelihood values that a certain researcher might be interested in a certain dataset or method, and so forth: in essence a recommender system but drawing from a much richer description of the problem domain than is usually found in conventional SDI or GIS. A complete account, with examples, is provided in [6].

In comparison to current SDI, this new approach simply broadens the focus, so everything we now consider as SDI still applies, but in addition we must (i) broaden the conceptual model used along the lines of the Descriptions Spaces describe above and (ii) 'instrument' the community of users so that we study and learn from what they do, thus it represents a much broader focus. But in return, these enhancements offer a more complete picture of how communities operate in practice, and we believe that such knowledge is extremely valuable, on a par with theory in terms of its usefulness to researchers. As an example: a recommendation such as this might be very helpful: "most climate change impacts researchers so far have used *this* interpolation method with *that* kind of dataset when working on coastal erosion problems".

Of course, these additional insights require additional effort to design and build the SDI initially and more research is yet needed to find: (i) ways to capture use patterns unobtrusively during the research process and (ii) ways to insert recommendations and insights back into the research process.

VII. Conclusions and Future Work

Easy discovery, reuse and integration of geospatial data have been the predominant motivations for decades of research on geosemantics. However, despite concerted effort by the GIScience community, a large gap continues to exist between the aspirations of the geosemantics research and applied outcomes that are used in the daily work of scientists. This stands in contrast to the very successful adoption of other GIScience research (e.g., spatial statistics), which has found wide adoption and successful implementations in a variety of geographic information systems.

We argue that the reason for this state of affairs is that the way working scientists assess the "meaning" and quality of geographic data is based on wide variety factors that are not currently brought together in a holistic way in our semantic technologies. The *situated context* of data use comprises more that just the "meaning of the data" from the perspective of the producer. It is also who has used it, how they have used it, and why they have used it. All of these properties of data use are potentially valuable pieces of information to aid discovery, reuse and integration.

The nexus of relations surrounding geographic data use represents a complex system composed of simpler interrelated parts—each of these simpler parts is possible to observe and measure. If we build our cyberinfrastructure to observe this wider context of data use, then we have an opportunity to build systems that can describe data in ways that better match how scientists themselves assess the fitness of data for their purposes. In many other fields, graphical and probabilistic methods that can learn structure in the face of complexity and uncertainty have demonstrated significant success in the last decade. We contend that if we are successful in instrumenting our community to observe this complex network of relations surrounding data use, then we too can put these powerful methods of inference to build cyberinfrastructure that better works with geoscientists, and finally begin to realize the long stated potential of geosemantic technologies.

Additional work is required to refine the Description Spaces used so that they are both useful (richly describe the problem domain) and easy to use (simple for the user to grasp). Descriptions used must be either easy for a user to enter, or easy to learn via use-cases. We make no claims that our current Description Spaces are optimal, merely that they are useful. But others could be more useful. If such Description Spaces are treated as meta-models, it should be possible to test different arrangements, to see which offer the best balance between usefulness and simplicity. As noted already, additional research is also needed to find ways to unobtrusively harvest useful information during the research process (for example from workflows) and to then make it available to future researchers as recommendations.

REFERENCES

[1] B. Adams, M. Gahegan, P. Gupta, and R. Hosking, "Geographic information observatories for supporting science." In Proceedings of Workshop on Geographic Information Observatories, 2014, pages 1-5.

[2] W. Bridewell, P. Langley, L. Todorovski, and S. Džeroski, "Inductive process modeling." Machine learning, vol. 71, no. 1, 2008, 1-32.

[3] R. Devillers, Y. Bédard, R. Jeansoulin, and B. Moulin, "Towards spatial data quality information analysis tools for experts assessing the fitness for use of spatial data." International Journal of Geographical Information Science, vol. 21, no. 3, 2007, 261-282.

[4] F. T. Fonseca and M. J. Egenhofer, "Ontology-driven geographic information systems." In Proceedings of the 7th ACM international symposium on Advances in geographic information systems, 1999, pages 14-19, ACM.

[5] A. U. Frank, "Metamodels for data quality description." In M. Duckham, E. Pebesma, K. Stewart, A.U. Frank editors, Data Quality in Geographic Information-From Error to Uncertainty, 1998, pages 142–158. Springer.

[6] M. Gahegan and B. Adams, "Re-envisioning data description using Peirce's pragmatics." In M. Duckham, E. Pebesma, K. Stewart, A.U. Frank editors, Geographic Information Science, Lecture Notes in Computer Science, vol. 8728, 2014, pages 142–158. Springer.

[7] M. Gahegan and W. Pike, "A situated knowledge representation of geographical information." Transactions in GIS vol. 10, no. 5, 2006, 727–749.

[8] R. Glushko, "Foundations for Organizing Systems." The Discipline of Organizing, 2013, pages 1-45, MIT Press.

[9] C. Granell, R. Lemmens, M. Gould, A. Wytzisk, R. de By, and P. van Oosterom, "Integrating semantic and syntactic descriptions to chain geographic services." Internet Computing, IEEE, vol. 10, no. 5, 2006, 42-52.

[10] K. Janowicz, B. Adams, G. McKenzie, and T. Kauppinen, "Towards Geographic Information Observatories." In Proceedings of Workshop on Geographic Information Observatories, 2014, pages 1-5.

[11] K. Janowicz, S. Scheider, and B. Adams, "A geo-semantics flyby." In Reasoning Web. Semantic Technologies for Intelligent Data Access, 2013, pages 230-250, Springer.

[12] M. I. Jordan, "Graphical models." Statistical Science vol. 19, no. 1, 2004, 140–155.

[13] W. Kuhn, "A functional ontology of observation and measurement." In GeoSpatial Semantics, 2009, pages 26-43, Springer.

[14] W. Lam and F. Bacchus, "Learning Bayesian belief networks: An approach based on the MDL principle." Computational intelligence, vol. 10, no. 3, 1994, 269-293.

[15] C. L. Palmer, M. H. Cragin, P. B. Heidorn and L. C. Smith, "Data curation for the long tail of science: The case of environmental sciences." In Third International Digital Curation Conference, Washington, DC, 2007, pages 1-5.

[16] D. Ribes, "Ethnography of scaling, or, how to a fit a national research infrastructure in the room." In Proceedings of the 17th ACM Conference on Computer Supported Cooperative Work & Social Computing, CSCW '14, 2014, pages 158–170. ACM, New York, NY, USA.

[17] R. Y. Wang and D. M. Strong, "Beyond accuracy: What data quality means to data consumers." Journal of management information systems, vol. 12, no. 4, 1996, 5-33.

[18] A. N. Whitehead, Process and Reality: An Essay in Cosmology. New York, Social Science Book Store, 1929.

[19] T. Tiropanis, W. Hall, J. Hendler, and C. de Larrinaga, "The web observatory: A middle layer for broad data." Big Data, vol. 2, no. 3, 2014, 129–133.

[20] K. McKelvey and F. Menczer, "Design and prototyping of a social media observatory." In Proceedings of the 22nd international conference on World Wide Web companion, 2013, pages 1351–1358. International World Wide Web Conferences Steering Committee.

[21] X. Gao, et al., "Supporting a social media observatory with customizable index structures: Architecture and performance." In Cloud Computing for Data-Intensive Applications, 2014, pages 401–427. Springer.

[22] R. Simpson, K. R. Page, and D. De Roure, "Zooniverse: observing the world's largest citizen science platform." In Proceedings of the companion publication of the 23rd international conference on World wide web companion, 2014, pages 1049–1054. International World Wide Web Conferences Steering Committee.

[23] K. O'Hara, N. S. Contractor, W. Hall, J. A. Hendler, and N. Shadbolt, "Web science: understanding the emergence of macro-level features on the world wide web." Foundations and Trends in Web Science, vol. 4, no. 2-3, 2013, 103–267.

[24] S. Lawrence, "Context in web search." IEEE Data Eng.Bull., vol. 23, no. 3, 2000, 25–32.

[25] L. Finkelstein, et al., "Placing search in context: The concept revisited." In Proceedings of the 10th international conference on World Wide Web, pages 406–414. ACM, 2001. [27] S. Lawrence. Context in web search. IEEE Data Eng. Bull., vol. 23, no. 3, 2000, 25–32.

[26] D. Kelly and J. Teevan, "Implicit feedback for inferring user preference: a bibliography." In ACM SIGIR Forum, volume 37, 2003, pages 18–28. ACM.

[27] T. Joachims, L. Granka, B. Pan, H. Hembrooke, and G. Gay, "Accurately interpreting clickthrough data as implicit feedback." In Proceedings of the 28th annual international ACM SIGIR conference on Research and development in information retrieval, 2005, pages 154–161. ACM.

[28] E. Agichtein, E. Brill, and S. Dumais, "Improving web search ranking by incorporating user behavior information." In Proceedings of the 29th annual international ACM SIGIR conference on Research and development in information retrieval, 2006, pages 19–26. ACM.

[29] Z. Dou, R. Song, and J.-R. Wen, "A large-scale evaluation and analysis of personalized search strategies." In Proceedings of the 16th international conference on World Wide Web, 2007, pages 581–590. ACM.

[30] G. Adomavicius and A. Tuzhilin, "Context-aware recommender systems." In Recommender systems handbook, 2011, pages 217–253. Springer.

[31] X. Shen, B. Tan, and C. Zhai, "Implicit user modeling for personalized search." In Proceedings of the 14th ACM international conference on Information and knowledge management, 2005, pages 824–831. ACM.

[32] J. Teevan, S. T. Dumais, and E. Horvitz, "Personalizing search via automated analysis of interests and activities." In Proceedings of the 28th annual international ACM SIGIR conference on Research and development in information retrieval, 2005, pages 449–456. ACM.

[33] D. Carmel, et al., "Personalized social search based on the user's social network." In Proceedings of the 18th ACM conference on Information and knowledge management, 2009, pages 1227–1236. ACM.

[34] A. Hannak, P. Sapiezynski, A. Molavi Kakhki, Krishnamurthy, D. Lazer, A. Mislove, and Wilson, "Measuring personalization of web search." In Proceedings of the 22nd international conference on World Wide Web, 2013, pages 527–538. International World Wide Web Conferences Steering Committee.

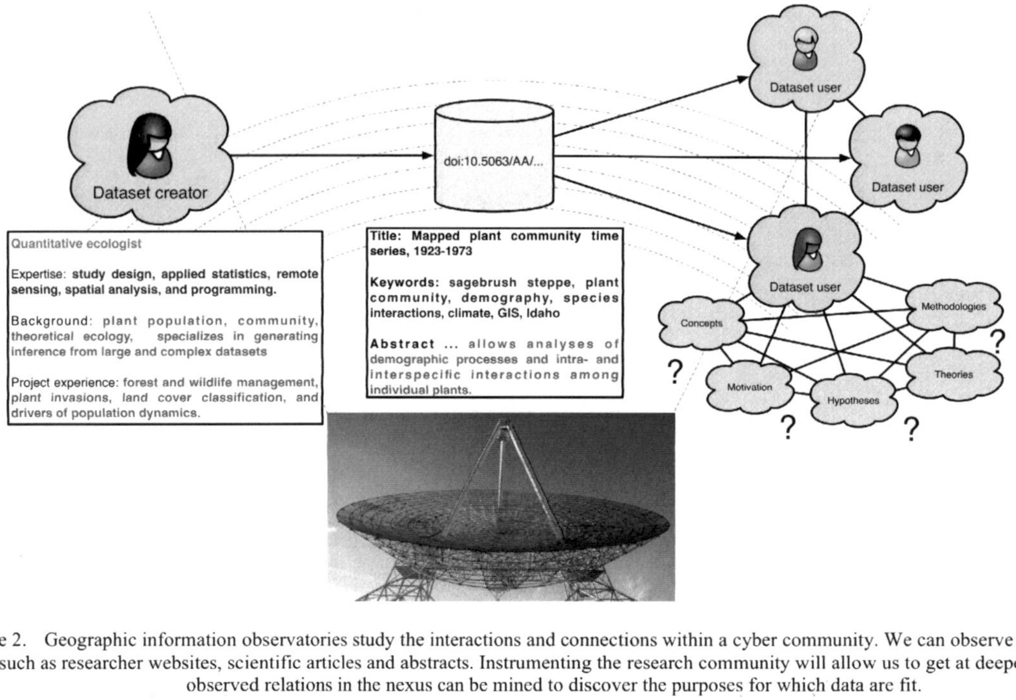

Figure 2. Geographic information observatories study the interactions and connections within a cyber community. We can observe unstructured data sources such as researcher websites, scientific articles and abstracts. Instrumenting the research community will allow us to get at deeper relationships. The observed relations in the nexus can be mined to discover the purposes for which data are fit.

Improving Spatial Data Supply Chains

Learnings from the Manufacturing Industry

Lesley Arnold
Department of Spatial Sciences
Curtin University, Western Australia
Cooperative Research Centre for Spatial Information
Email: lesley.arnold@curtin.edu.au

***Abstract* - Government spatial data holdings are the basis of many Geographic Information System applications and users have come to depend on this information for reliable evidence-based decision making. However, many spatial data supply chains, particularly those that span multiple organisations and different levels of government, are not effective and users are having to make do with out-of-date and incomplete information. The manufacturing industry has a long history in supply chain analytics and well established models. This paper discusses five traits that have generated production efficiencies in manufacturing and can be applied to effectively produce spatial data that is *fit for purpose* in the user's context. The successful supply chain traits are: a) formalized extended supply chain strategies; b) information metrics and measures; c) closed loops; d) traceability; and e) the ability to effectively communicate quality and purpose to end-users. A supply chain framework incorporating these traits is proposed for extended networks.**

Keywords-Spatial Data; Supply Chain Management; Strategy; Traceability; Fit for Purpose.

I. INTRODUCTION

Supply Chain Management (SCM) is a field of study that has evolved since the 1980's [1], principally in the manufacturing industry. SCM encompasses the planning and management of all activities involved in sourcing and procurement, conversion, and logistical management activities. Importantly, it also includes coordination and collaboration with channel partners, which can be suppliers, intermediaries, third-party service and technology providers, and customers [2].

Incremental innovations and best practice SCM solutions have evolved over time and there are opportunities for the spatial sciences to take advantage of these developments to address some of the deficiencies in extended (nationwide and cross-agency) spatial data supply chains that involve multiple suppliers, producers and consumers. Five areas where the spatial industry can improve performance are:

1) Channel partners collectively adopt a Supply Chain Strategy. *The strategy needs to be cognisant of the business models of each supply chain participant as well as the end-user in the extended supply chain. Currently, spatial data management often operates in a business vacuum and is not tied to the needs of end-users in cross-agency supply/demand systems. A supply chain strategy is critical to delivering value to the end-user, reducing costs and fostering innovation in extended supply chains.*

2) Use of Information Metrics and Measures *for capacity planning, financial management, just-in-time delivery and end-user satisfaction. These metrics and measures are implemented within a SCM system to deliver on the supply chain strategy. SCM systems seamlessly integrate functions and provide communication between organisations participating in the entire spatial data supply chain.*

3) Making the most out of Closed Loop Supply Chains *(information backhauls) as a method for strategically sourcing spatial data. Capturing reverse material flows either through process automation or crowdsourcing has potential to enable real-time spatial data supply chains.*

4) Formalise supply chain traceability *through regulations and standards to track and control spatial data, and ensure products are produced responsibly and to the required quality. From a business perspective it is about understanding who is using data and how; and from a consumer perspective it is about being able to understand the risks inherent in data for decision making.*

5) Being able to communicate 'Fit for Purpose' *to end-users so they have the knowledge about product suitability. The manufacturing industry does this well through easily understood rating systems, while, the spatial industry relies on a user's understanding of complex metadata.*

This paper discusses these five manufacturing supply characteristics (Sections 3 to 7) in terms of their effectiveness and applicability to spatial data supply chains. Prior to this, Section 2 presents the functional concepts with the objective providing a common understanding of supply chain terminology. This is important as one of the limiting factors in establishing supply chain theory in the spatial domain is the lack of a controlled vocabulary for ontology development. Finally, Section 8 introduces the high-level Supply Chain Framework, which adopts the lessons learned from the manufacturing industry and is being used to generate an ontology to understand the interrelationships between supply chain domains.

II. SUPPLY CHAIN FUNDAMENTALS

The term supply chain stems from the manufacturing industry. It is gaining common usage across the wider spatial sector where it is used to describe the flow of raw spatial data through to the end-user as a product. However, the supply chain concepts, terminology and theory that are ingrained in manufacturing [3], are not entrenched in the acquisition and management of spatial information and its delivery as a product, particularly in extended supply chains, such as cross-agency networks and National Spatial Data Infrastructures.

A simplified network is illustrated in Figure 1 to explain the concepts, terms and relationships between organisations (supply chain nodes) that need to work in synergy to store, process and create component parts of a product that are progressively aggregated and combined to deliver a product or service to an end-user further along the supply chain. The nodes represent locations where value-added operations occur and include supplier, producer and distributor nodes, as well as mixed nodes. Arrows represent the linear flow of spatial data and customer information. These are referred to as supply chain links. Flows between the nodes are two-way and consist of material flows (data) and information flows (customer requirements and product feedback). There are typically several supplier and customer tiers (1…*n*). These tiers are illustrated in Figure 1.

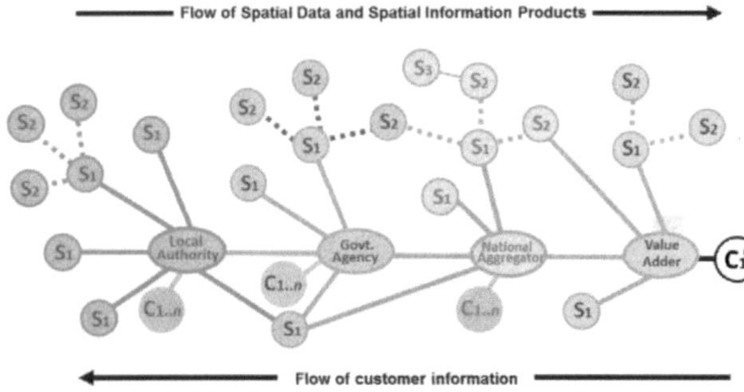

Figure 1. An example of a linear spatial data supply chain that is characterised by nodes and links; data flows and information flows.

Supply chains integrate supply (push production) and demand (pull production) management within and across organisations. *Push* production is based on forecast demand and *pull* production is based on actual or consumed demand.

In the spatial industry, the *push* approach is a business response to anticipated customer demands, long range forecasts or forecasting based on previous sales (downloads and/or views) and the maturation of the market with respect to geospatial understanding.

In contrast, the pull approach is typically a query driven approach. End-users require knowledge about location to answer a question or to visualize patterns and complex relationships so that they can make an informed decision. This is the subject of rapid spatial analytics.

This paper focusses on push supply chains, drawing on comparisons with the manufacturing industry to reduce production costs and attain operational excellence.

Push supply chains are generally non-linear [4] and there are few models that capture the web of multiple networks and relationships required to understand spatial data lineage from its initial capture through to its transformation and delivery as a product or knowledge service further downstream.

The inherent complexity and convoluted nature of digital supply chains is highlighted in a recent study into Australian geocoded addressing [5]. Geocoded addresses (verified or otherwise) are pushed (or dragged) along various pathways from one supply chain participant to another (Figure 2). While an authoritative 'primary' pathway exists for address data, many government departments, hospitals, education institutions and businesses collect address information directly from home occupiers using online forms or over the counter. These address data sets may enter the primary supply chain at any point where residual value is deemed to be recyclable. However, data integration is essentially manual and few data sets incorporate verification processes. Therefore, time delays and the potential for human error may compromise the value of this data to the consumer.

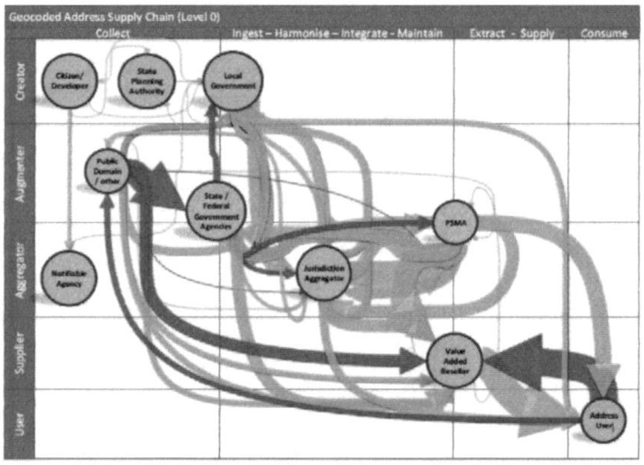

Figure 2. Geocoded Address Supply Chains are non-linear and inherently complex [5].

To achieve improvements in productivity, quality and timeliness, extended supply chains need to be more streamlined. While, automation of processes is a key enabler, so too is understanding the interrelationships between participants engaged in supply chain activities and their business motivation.

III. SUPPLY CHAIN STRATEGY

To achieve a sustainable extended spatial data supply chain, the lesson from the manufacturing industry is to formulate an overarching supply chain strategy [6] to deliver on the goals and vision of all the participants in the business of developing nationwide data products. This initially requires an agreed business strategy for the extended supply chain [7].

The business strategy establishes the overall direction that organisations participating in the supply chain collectively aspire to. It includes decisions about what products and services are to be offered, identifies market segments, the timing of product releases and whether products are to be sold or made freely available.

In contrast to the business strategy, the supply chain strategy is the mechanism by which organisations formalize how they work with supply chain partners (suppliers, distributors, customers, and the customers' customers) to deliver on the business strategy [7].

As a whole, the manufacturing industry is good at preparing and implementing comprehensive supply chain strategies. There is a plethora of research in this field; most aimed at driving down operational costs and maximizing efficiencies.

In the spatial industry, anecdotal evidence suggests there are few formally documented extended supply chain strategies. The supply chain strategies that do exist are typically those of individual organisations and their customers and are therefore not necessarily relevant to the needs of the end consumer of a nationwide data product developed downstream. There is often no business incentive to work in the national interest. In the manufacturing industry, this silo approach has shown to result in sub-optimization of the supply chain as a whole and, as a consequence supply chain performance suffers and end-user needs are not met [8].

Spatial infrastructures (globally) tend not to be aligned with a business strategy and therefore formulating a supply chain strategy becomes a difficult task. Without a business strategy, the network of suppliers, producers and distributors have no clear direction on the type of product to generate and where to focus effort; nor how their intellectual property will be protected and what mechanism will be used to measure and generate a return on investment.

National Spatial Data Infrastructure (NSDI) strategies often serve as the business strategy for governments and are a useful starting point for a nationwide spatial data supply chain strategy. The NSDI strategy captures the core purpose for implementing a NSDI and the aspirations of the nation in using spatial technologies for improving and sustaining social, economic and environmental development. Common NSDI goals are for spatial information that is:

- An accurate nationwide representation of the landscape.
- Available in a variety of forms and accessible through multiple channels.
- Used widely in response to emerging business opportunities.
- Easily integrated with economic, social and environmental geographies for evidence-based decision making.
- Produced efficiently and according to sustainable principles.
- An enabler for economic growth and social wellbeing.

However, while NSDI strategies exist they are often not supported by financial models and capacity plans, nor a clear understanding of the market and end-user needs. These aspects are required to engage, incentivize and obligate supply chain participants. As a consequence, nationwide supply chain strategy often goes unaddressed through lack of commitment.

This happens for two main reasons. Firstly, its execution requires a high degree of organisation and collaboration between many suppliers, producers and distributors, and this is difficult without a clear business strategy. Secondly; the core problem is to balance supplies against demand across several nodes and a sound financial investment model is required. This is important. Supply chain participants will have various existing business models and different financial investment perspectives – their imagery suppliers will expect to be paid for services and value-adding activities will need to be resourced. As such, the national spatial data supply chain needs to comprise an overarching business strategy and model that satisfies the business objectives of all participants in the extended supply chain and provides the goals, measures and value proposal of the national data product.

Currently, national supply chain frameworks suffer because participants often have no clear understanding of what business model they are contributing to - commercial, commercial 'free' or public good. With an agreed business model the national supply chain strategy has a blueprint from which to operationalize and sustain national product objectives.

IV. Information Metrics and Measures

SCM systems are embedded in the manufacturing industry. They integrate logistics management and manufacturing operations to coordinate processes and activities. SCM systems are used to link major business functions and workflows within and across organisations into a cohesive and high-performing business model. SCM systems use information metrics to drive business success, provide information about the performance of the overall supply chain and to identify problem areas.

The spatial industry can draw from this experience. Performance measures, targets and quality standards can be used to keep track of spatial data investment decisions and monitor progress towards achieving business objectives for national products.

A review of literature indicates SCM systems are not utilized in the spatial sector, and yet, there are some compelling reasons to do so. The primary benefit is to fulfil end-user demands through the most efficient use of resources. SCM systems provide oversight for capacity planning, financial management and just-in-time delivery through forecasting and production monitoring. In this way, the potential impact caused by a change in operation at one supply chain node can be evaluated ahead of time to understand possible repercussions along the entire supply chain and implications for service delivery [3].

Spatial data supply chain metrics that focus on performance across the entire supply chain can be used to better understand:

- *Carrying costs:* to measure how much it costs to store data over a given period of time.
- *Production Costs:* to measure efficiency and provide a benchmark for process improvement.
- *Warehouse turnover:* to measure how often and which products are sold/downloaded in a given year.
- *Order Tracking:* to monitor the status of requests for data and updates and associated turnaround times.
- *Inventory to sales ratio:* to measures the ratio of in-stock items, such as imagery, versus the amount of data being used/orders being filled.
- *Product performance:* to measure the rate of consumer satisfaction i.e. product returns, and quality/usability issues.
- *Units per transaction:* to measure the average number of units purchased/downloaded to establish a baseline with which to compare future targets.

Measures and metrics for spatial data, both financial and non-financial, are generally available from individual organisations along the supply chain. This information is usually of an operational and discrete nature and is not an indication of the performance of the entire supply chain. In addition, there is usually no interrelationship between the strategic measures of success of each supply chain participant.

The majority of manufacturing performance frameworks, such as the one proposed by [9], are not suited to spatial data supply chains. While SCM systems utilize diagnostics software to show exactly where bottlenecks occur in manufacturing and where data quality improvement is required, the focus is on building physical products with large inventories and complex transportation logistics. A new approach is required for the spatial domain that captures the value activities at each node, such as integration, generalization and level of accessibility.

Information Management System (IMS) metrics are more aligned with spatial data management. IMS have adopted eSupply chains in which supply chain participants are interconnected via internet technologies at technical, application and business management levels. Similarly to spatial infrastructures, the objective of IMS is to improve the effectiveness of decision-makers by getting the right information, to the right people, in the right format, at the right time [9].

A study by [10] evaluated six eSupply chain performance measures: (a) web-enabled service metric; (b) data reliability metric; (c) time and cost metric; (d) e-response metric; (e) invoice presentation and payment metric; and (f) e-document management metric. The researchers surveyed 120 companies. Results indicate that while companies believe these metrics are important, the challenge was to measure them.

There is an opportunity to re-examine these eSupply chain metrics in light of spatial information supply chain needs. In the spatial industry, supply chain metrics are not well documented with the exception of spatial data quality metrics [11] [12] [13]. However, quality is only one aspect of measure for a spatial data supply chain and there is still much work to be done in this area.

V. Closed Loop Supply Chains

The manufacturing industry has adopted opportunities for backhauling in transport logistics to reduce supply chain costs through collaboration and partnership. Often referred to as closed-loop supply chains [14], backhauls additionally transport items in the reverse direction from customers (usually retailers) to the depot (or warehouse). An example is a supplier of gas canisters. Full canisters are delivered to the customer and empty canisters collected at the same time – saving transportation costs and time.

The concept of backhauls (or reverse material flows) are not new to the spatial industry. Crowdsourcing and trusted partnerships have potential as viable strategic data sourcing solutions for maintainers of large geographic datasets. They have the ability to reduce costs (updates are free), improve data currency (updates are timely), and improve the overall accuracy of information (updates stem from local knowledge).

Crowdsourcing has not been seriously adopted by government mapping agencies where there are concerns about integrating data from potentially unreliable sources into authoritative data sets. Yet, vendors of navigation systems have embraced crowdsourcing to update their mapping base. Google Maps goes one step further. It displays crowdsourced traffic conditions along major routes by calculating vehicle speeds from the GPS-determined locations transmitted from 'opted in' mobile phone users. Both methods essentially create a closed-loop supply chain. There is significant opportunity for innovation in this area using volunteered GPS vehicle tracing to record map updates and errors.

Research is the key to increasing the uptake of reverse information flows and falls into four areas:

- resolving the *trust* problem [15];
- data harvesting to collect and verify information rapidly;
- integration of crowdsourced and authoritative data; and
- community engagement strategies to stimulate reverse information flows.

The efficiency and effectiveness of supply chains can be improved by embracing the backhaul concept. Benefits are cost avoidance for data maintenance in the longer term, better engagement with end-users and the community, and increased potential for product innovations.

VI. Supply Chain Traceability

Spatial information products are being used to save lives, prepare for natural disasters, mitigate environmental damage, form legal judgments on land boundaries and make significant economic decisions, such as where to locate infrastructure, source minerals and direct social services.

The importance of this information implies that data products are produced using scientifically proven reproducible methods. Yet this is not necessarily the case. Currently, there is no legal requirement or standard that imposes traceability practices on spatial data products.

As the spatial industry considers outsourcing parts of the spatial data supply chain, consideration needs to be given to tracking products and suppliers, using methods that are reproducible, and incorporate elements of traceability into metadata standards.

The ease with which data can be copied and transformed has made it increasingly difficult to determine the origins of a piece of data and therefore, its legitimacy for a particular usage. Supplier and product auditing needs to go beyond direct relationships with first-tier suppliers.

Understanding where data comes from and how it is created and by whom is important to:

- End-users in determining if data are fit for their purpose.
- Consumers who want to know if data has been produced in an environmentally and socially responsible manner.
- Decision makers needing to know the risks inherent in using particular data and thus their level of accountability.
- Data producers in identifying the need for product recalls and understanding the end-user/consumers of their products.

The drivers for supply chain traceability are similar to those encountered in the manufacturing industry. Challenges include [16]:

- Regulatory pressures and consumer demand for responsibly sourced and produced goods and services.
- Tracking and controlling materials and the processes applied on those materials to create finished products.
- Proactively managing product recalls (or data errors) with near real-time corrective actions.
- Improving customer safety and consumer satisfaction when using products.
- Managing product quality and reducing costs associated with nonconformance.

However, the spatial industry has no automated and fool proof solutions for:

- Backward Traceability: tracing back to the data source.
- Forward Traceability: tracking the end-users of data products.
- Component Traceability: tracking the component parts that makeup the end data product.
- Process Traceability: tracing what processes have been applied to data in the finished product.

A. Backward Tracability

The manufacturing and clothing industry has adopted backward traceability as a means of demonstrating a company's corporate social responsibility. Incidents, such as the 2013 Savar building collapse in Bangladesh, where more than 1,100 workers died because of unsafe conditions, have led to widespread discussions about corporate social responsibility across global supply chains. Law makers are increasingly legislating that manufacturers disclose where raw materials are sourced [17] [18], particularly if sourced from war-torn countries where revenue is funding violent military groups [19].

In the food industry, companies are increasingly sourcing directly from farmers or trusted aggregators rather than purchasing crops that have passed through several layers of collectors. The drivers for this change include concerns about food safety, child labor and environmental sustainability. The aviation industry standards require traceability to ensure the authenticity of parts, aircraft maintenance history and approved supplier identification [20].

From a spatial data industry perspective, CRCSI research is examining methodologies to trace data provenance along the supply chain and be able to present this knowledge in a way that allows end-users to make informed decisions on whether the information is suitable for their purpose. Currently, there are few models that address spatial data provenance from both a detailed metadata and lineage perspective. The CRCSI research is seeking to develop an ontology that goes beyond traditional metadata models that only capture the *who/what/when/why* of information. The provenance model will incorporate process knowledge at the various stages of a data product's lifecycle and include quality measures [21].

B. Forward Traceability

Forward traceability is mandatory in some industries, such as car manufacturing, food and beverage and pharmaceuticals. Forward tracking distribution is necessary in case a product has to be recalled. In 2009 Toyota recalled eight car models and put a halt to production, China recalled 170 tons of melamine-tainted milk powder in 2010 and Unilever United States recalled peanut butter due to potential salmonella contamination [16].

Anecdotally, product recall risk is low in the spatial sector. Nonetheless, being able to trace data usage and consumers will become more important. Today, web portals are extensively used to distribute spatial data but few sites require users to register their details online. This makes it difficult to keep track of who is using data, what their needs are in terms of future product design, and how to let them know that product updates are available.

C. Component and Process Traceability

Radio Frequency Identification (RFID) has revolutionized component and process traceability. RFID technology is embedded in many industries including baggage handling, livestock management, toll collection, theft prevention systems and automated production systems [22]. RFID tags are used to automatically identify and track products, materials and parts along the supply chain. The radio-frequency identification (RFID) market is expected to rise from $8.89 billion to $27.31 billion by 2024 (23) based on 2014 figures.

The equivalent of the RFID in the spatial domain is the persistent Global Unique Identifier (GUID) or Global ID. The Global ID is a unique identification code permanently assigned to a piece of data (database record) so that information about the data element can be easily retrieved. The importance of the Global ID is that it can be used to unambiguously track a data element through its lifecycle.

However, the value of the Global ID has not been fully exploited in spatial data supply chains. Part of the problem is that organisations that collect and manage spatial data generally only store feature identifiers (ID) that are unique to their systems. This means that when data are integrated from more than one provider there is a risk that the 'system' feature ID will be the same. While IDs can be reassigned, the ability to track data and its lineage along the supply chain, is significantly reduced. Industry needs to consider the application of the GUID in terms of supply chain efficiency (GUIDs support automatic update propagation) and understanding the fit for purpose nature of a data product (GUIDs support provenance modelling).

VII. COMMUNICATING FIT FOR PURPOSE

Users of spatial information can be faced with a choice of multiple datasets, each containing the information required, but the question is 'which one is fit for their purpose?' Traditionally, the spatial industry has used descriptive metadata to describe datasets – contact information, coverage, accuracy and recommended purpose are all included in the metadata. However, similar data sets have similar metadata and therefore the choice of which one to use becomes difficult. For example, if an end-user searches for a data set containing road information in a particular area, the following may be retrieved:

- polygonised roads in a cadastral dataset;
- highways only, in a road authority dataset; and
- a topologically correct road network in a topographic dataset.

Understanding which product is best for their needs requires experience or subject matter knowledge. The manufacturing industry appreciates that consumers do not necessarily have the information to understand whether a product will suit their needs or not. Manufacturers have embedded methods in production processes to let the consumer know if a product is fit for their purpose. The systems used are based on production standards and their compliance with legislation. For example, toy manufacturers include age suitability on packaging, food producers include nutrition panels and the hotel industry has a 'star' quality rating system. These methods act as a purchasing guide for consumers. They build an expectation that a product will be satisfactory for a given purpose. For example, television codes are a guide to whether or not a program is suitable for a given type of audience based on predetermined criteria (Figure 3). This approach builds consumer confidence as the codes are regulated through a recognised code of practice. There are also systems that put the quality ratings in the hands of the consumer. The internet has become well-entrenched as a vehicle for consumers to rate their experience of a product.

Product: Television Program	**How do I know if a television program is suitable?**	
	A rating and recommendation to audiences based on age and level of maturity; that considers level of violence, coarseness of language, nudity, horror, adult themes, medical practises etc.	G Suitable for all
Standards:	**Australian Commercial Television Code of Practice**	

Figure 3. Rating System for Television Programs [24].

The spatial industry does not have a ratings approach for data quality and usability, and has traditionally relied on consumers' understanding of spatial metadata as a means of interpreting whether a data product is fit for a particular purpose or not. However, metadata is often not reliable because it is out-of-date and often incomplete. Interpretation is generally only through descriptive metadata that is at best a subjective interpretation from the perspective of the data custodian. As a result the metadata approach is not user-friendly as it does not consider the needs of the consumer or their viewpoint when determining if a product is suitable. A different approach is required. For example, the food labelling industry in Australia is considering including 'walking time' kilojoule 'burn off' to help consumers make sense of nutrition panels that are difficult to interpret for weight loss programs [25].

Current approaches to providing 'fit for purpose' advice for consumers of spatial information products and services are not adequate. Organisations that move down the fit for purpose track will typically address a single business objective. In addition, the spatial industry has typically relied on self-regulation and many organisations and businesses have adopted their own standards rather than a national data quality standards approach.

There are inherent difficulties in establishing criteria that can be applied across a single data set due to the varying degrees of quality. Data elements are often sourced from multiple suppliers and have been subject to different processes.

In moving towards a new approach the spatial industry needs to firstly, identify and classify the purposes for which spatial information products are used; and secondly, develop a set of criteria with which spatial data products can be rated so that they can be assigned a fit for purpose code. Criteria would be based on:

- Data standards and quantity measures, such as currency, completeness, integrity, accuracy.
- Origin, including method of capture and equipment used.
- Lineage, such as the transformation and processing methods applied.

VIII. Supply Chain Framework

A Supply Chain Framework has been developed as a guide to formulating supply chain strategy (Figure 4). The framework adopts learnings from the manufacturing industry: where supply chain strategy is the key mechanism by which producers formalize how they work with their supply chain partners (suppliers, distributors, customers, and the customers' customers) to deliver on business strategy.

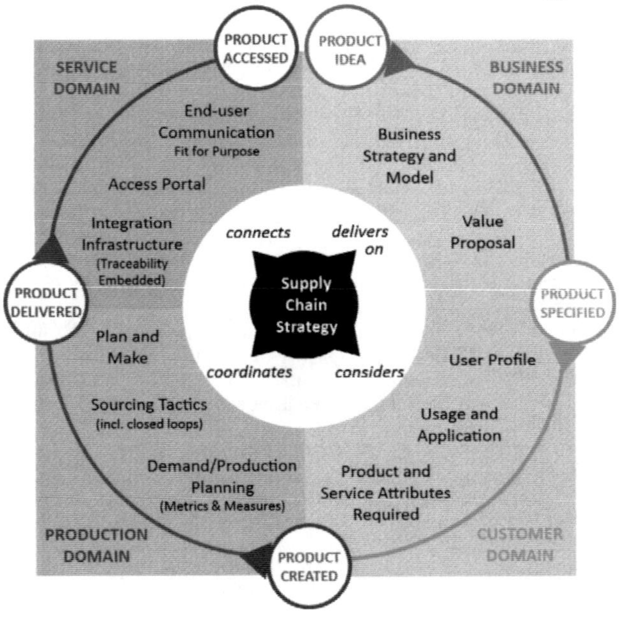

Figure 4. Spatial Data Supply Chain Framework

The framework applies to individual nodes (organisations) along the supply chain where value is created for first tier consumers. It also applies to the overarching strategy required for extended supply chains. The aim of the latter is for all supply chains participants to cooperate in a way that provides value for the end-user.

The Supply Chain Framework considers the life cycle of a data product from its inception as a product idea through to its delivery to customers. The framework includes four domains; business, customer, production and service domains.

A detailed version of the framework (supply chain ecosystem) is currently being used to develop an ontology to link supply chain components across extended supply chains to create value for the end-user.

A. *Business Domain:*

The supply chain strategy delivers on the business outcomes required including return on investments and business incentives. These are generally specified in the business strategy along with the collective vision, mission and goals of the supply chain partners. The supply chain strategy also considers the value proposition to the end-user. This stems from the collective efforts and activities of the supply chain partners.

B. *Customer Domain:*

The supply chain strategy considers the end-user requirements, the factors influencing product usage behavior and the design criteria that will create the most value for the end-user, such as timeliness, content, coverage, semantics and accuracy.

C. *Production Domain*

The supply chain strategy coordinates the external forces that impact on demand planning, data sourcing complexity and the types of transformation processes required to make a data product. A compliance framework is required to support interoperability including data and technology standards, quality measures and metrics, and custodian roles and responsibilities. Collaboration with supply chain partners is a key component to sustaining production in the longer term; as are closed loop systems that capture additional product value through process integration. Future partnerships between nodes are likely to be characterised by digital collaborative environments and automated information flows.

D. *Service Domain:*

The supply chain strategy focuses on connecting people to products and services. It considers the integration of component products from multiple sources to create standard offerings as well as tailored solutions. A policy framework is required to manage open access to data products balanced with individual privacy, copyright and intellectual property considerations. These aspects are more complex in extended supply chain networks. Communicating product suitability will require a rating system that is meaningful in the end-user's context.

IX. Conclusion

Spatial Data Supply Chains have evolved over time to become complex networks that are difficult to visualise and manage. The relationships between suppliers, producers and consumers in extended (or national/cross-agency) supply chains are difficult to formalise, and understanding the origin of a piece of data is often challenging. The ease with which data can be copied and transformed by individual supply chain participants is creating inherent problems.

A new approach is required to improve the way spatial data products are produced and distributed. The objective is to automate tasks to deliver productivity improvement, cost savings, timeliness and improved data quality. The approach is essentially to strengthen the *push* supply chain model. The second viewpoint is to improve the experience of the end-user by more effectively communicating the purpose for which the data product is intended to simplify consumer decision-making when faced with multiple data sets to choose from.

Drawing on manufacturing supply chain experiences, it is clear that the underpinning issue is to create effective supply chain strategies. The supply chain strategy formalizes how supply chain partners (suppliers, distributors and end-users) work together to deliver on an agreed vision and business goals and provide the incentive to participate and driving opportunities for efficiency and innovation

Supply chain metrics can then be used to drive business success, provide information about the performance of the overall supply chain and to identify ongoing problem areas. As a pre-requisite, having an understanding of existing supply chain costs will better direct where process improvement is most critical.

A more strategic approach to data sourcing is an industry imperative in driving down costs and increasing end-user engagement. Closed loop supply chains are one such mechanism that has potential to deliver efficiencies and increase community participation. Supply traceability combined with methods to communicate the 'fit for purpose' nature of spatial products has the potential to improve the experience of consumers when tapping into spatial data holdings.

In many cases, it will be obvious what spatial information can be used for. However, consumers are becoming far savvier about what spatial information products and services are available, and are applying this information to increase business acumen.

The next step in this research is to formulate a supply chain ontology to examine the interrelationships between business strategy, customer requirements, spatial data workflows, metrics and measures, and data access from both a supplier and consumer perspective. The aim is to develop best practice extended spatial data supply chain strategies.

ACKNOWLEDGMENT

This work has been supported by the Cooperative Research Centre for Spatial Information (CRCSI), whose activities are funded by the Australian Government's Cooperative Research Centre Programs.

REFERENCES

[1] L. Dawei, "Fundamentals of Supply Chain Management, [Online]. e-Book available at http://www.pdfdrive.net/fundamentals-of-supply-chain-management-e299250.html, retrieved March 2016.

[2] J. Wolfe, "The Nature of Supply Chain Management Research: Insights from Content Analysis of International Supply Chain Management Literature from 1990 to 2006.," Gabler Edition Wissenschaft, Ed. 1, 2008, pp. 11-12, ISBN 978-3-8349-0998-5.

[3] J. Coyle, C. Langley, R. Lovak, and B. Gibson, "Supply Chain Management: A Logistics Perspective," Cengage Learning, Business and Economics, Edition 9, pp720.

[4] H. Min and Z. Gengui, "Supply Chain Modeling: Past Present and Future," in Computers and Industrial Engineering, Pergamo, vol 43, pp 231-249.

[5] CRCSI (2015) Optimising the Supply Chains for Geocoded Addressing in Australia, Current State Supply Chains, Final Review 6.0, March 2015, [Online] Available at http://www.crcsi.com.au/assets/Resources/Geocoded-Address-Supply-Chain-Review-Final-V6.0-Final.pdf, accessed June 2015.

[6] A. Karunaratne, L.M. Chi H.S. Min, and S. Sorrooshian, "Supply Chain Strategy" in International Journal of Innovative Ideas, vol. 12(4), pp 7-7 November 2012.

[7] H.D. Perez, "Supply Chain Roadmap: Aligning Supply Chains with Business Strategy', ISBN:149420049X, pp. 186, 2013

[8] S. Homberg, "A Systems Perspective on Supply Chain Measurements", International Journal of Physical Distribution and Logistics Management, vol. 30, Number 10, 2000 pp. 847-878

[9] A. Gunasekaran, C. Patel, and E. Tirtiroglu, "Performance Measures and Metrics in a Supply Chain Environment", International Journal of Operations and Production Management, vol. 21, Issue 1/2, 2008 pp 71-87.

[10] M. Sambasivan, Z.A.Mohamed, and T. Nandan "Performance Measures and Metrics for e-Supply Chains", Journal of Enterprise Information Management, vol. 22 Iss: 3, 2009, pp.346 – 360

[11] J. Xia "Metrics to Measure Open Geospatial Data Quality, in Issues in Science and Technology, Winter 2012. [Online] Available at http://www.istl.org/12-winter/article1.html, retrieved March 2016.

[12] H. Veregin,"Data Quality Parameters, Geographical Information Systems", Chapt 12, 2009, [Online] Available at http://www.geos.ed.ac.uk/~gisteac/gis_book_abridged/files/ch12.pdf, retrieved February 2015.

[13] R. Devillers, Y. Bedard, R. Jeansoulin and B. Moulin "Towards Spatial Data Quality Information Analysis Tools for Experts Assessing the Fitness for Use of Spatial Data", International Journal of Geographical Information Science, vol. 21, Issue 3, pp. 261-282.

[14] F. Schultmann, M. Zumkeller. and O. Rentz 'Modeling Reverse Logistic Tasks Within Closed-loop Supply Chains: An example from the automotive industry", in European Journal of Operational Research, vol. 171, Issue 3, 16 June 2006, pp. 1033-1050.

[15] P. Goodhue, H. McNair, F. Reitsma "Trusting Crowdsourced Geospatial Semantics" in The Internatiponal Archives of the Photgrammetry, Remote Sensing and Spatial Information Sciences, vol. XL-3/W3, 2015, pp. 25-28.

[16] G. Hedge 'Global Traceability: Key Drivers, Challeges and Solutions for Today's Manufacturers", SAP Insider 2011, [Online] Available at http://sapinsider.wispubs.com/Assets/Articles/2011/July/Global-Traceability-Key-Drivers-Challenges-And-Solutions-For-Todays-Manufacturers, accessed March 2015.

[17] Australian Government, "Food Standards Code", 2015 [Online] A available http://www.foodstandards.gov.au/code/Pages/default.aspx, accessed April 2015.

[18] SAI Platform "Guidelines for Sustainable Agriculture Supply Chains of SAI Platform", 2013 [Online] Available at http://www.saiplatformaust.org/media/W1siZiIsIjIwMTQvMDkvMDkvNHpqN2RodjU0eF9TdXN0YWluYWJpbGl0eV9HdWlkZWxpbmVzXzEzMTEyMDEzLnBkZiJdXQ/Sustainability%20Guidelines%2013112013.pdf?sha=c64f9015d7f40dda, accessed March 2015

[19] M. Stronmer "Conflict Minerals: Yet Another Supply Chain Challenge", Ideas and Insights, 2012 {Online] Available at https://www.atkearney.com/paper/-/asset_publisher/dVxv4Hz2h8bS/content/conflict-minerals-yet-another-supply-chain-challenge/10192, accessed April 2015.

[20] Civil Aviation Safety Authority "Documents for the Supply of Aeronautical Products", Revised July 2010, [Online] Available at https://www.casa.gov.au/sites/g/files/net351/f/_assets/main/download/caaps/airworth/42w_1.pdf, accessed April 2015.

[21] CRCSI "Spatial Infrastructures Research Strategy" 2013 [Online] Available at

http://www.crcsi.com.au/library/resource/spatial-infrastructures-research-strategy, accessed June 2015.

[22] K. Pramatari "RFID-enabled Traceability in the Food Supply Chain, Industrial Management and AMP Data Systems", March 2007, [Online] Available at http://www.researchgate.net/profile/Katerina_Pramatari/publication/220672527_RFID-enabled_traceability_in_the_food_supply_chain/links/0f31752e3867dcf59c000000.pdf, accessed April 2015

[23] R. Das and P. Harrop (2014) "RFID Forecasts, Players and Opportunities in 2014-2024". IDTechEx Report 2014, [Online] Available at http://www.idtechex.com/research/reports/rfid-forecasts-players-and-opportunities-2014-2024-000368.asp, accessed January 2015.

[24] Australian Government "Australian Classification" [Online] Available at http://www.classification.gov.au/Pages/Home.aspx, accessed December 2015.

[25] Daily Mail Australia "Popular snacks may soon be labelled to show how many minutes WALKING will be needed to burn off the calories if you eat them", 2015 [Online] Available at http://www.dailymail.co.uk/news/article-3193903/Australian-food-soon-labelled-minutes-WALKING-needed-burn-calories.html, accessed August 2015.

Spatial Data Supply Chain Provenance Modelling for Next Generation Spatial Infrastructures Using Semantic Web Technologies

Muhammad Azeem Sadiq
Department of Spatial Sciences, Curtin University
Cooperative Research Center for Spatial Information
Perth, Australia
Email: Muhammad.sadiq@postgrad.curtin.edu.au

David McMeekin
Department of Spatial Sciences, Curtin University
Cooperative Research Center for Spatial Information
Perth, Australia
Email: d.mcmeekin@curtin.edu.au

Lesley Arnold
Department of Spatial Sciences, Curtin University
Cooperative Research Center for Spatial Information
Perth, Australia
Email: l.arnold@curtin.edu.au

***Abstract*—This research addresses spatial data supply chain provenance issues using semantic Web technologies to resolve knowledge gaps when disseminating spatial data products. Two models from the World Wide Web Consortium (W3C) and the Open Provenance Group for general data on the Web do not satisfy geospatial end-user needs. The Open Geospatial Consortium (OGC) has investigated the W3C PROV model for spatial datasets. Issues identified are the lack of provenance captured at the feature and attribute level, and for time series, data set series, representation and presentation interfaces, and elements at different levels. In order to answer user queries comprehensively, a geospatial provenance model in conjunction with semantic technologies has been identified as a potential solution to increase a user's trust in datasets and processes. This is important as raster dataset provenance, time series conflation processes and incremental updates have not been addressed. This has created a critical gap between provenance currency and the believability of geospatial datasets.**

Keywords-spatial data supply chain; spatuial data provenance; semantic Web; ontology; trust; processess and services.

I. INTRODUCTION

This research focusses on the needs of next generation spatial infrastructures. It explores different aspects of spatial infrastructures with a view to improving our understanding and management of data provenance along the spatial data supply chain, including end-user trust and believability. This research aims to produce a geospatial data provenance model called GEOPROV. It will investigate and implement semantic Web techniques to aid the user in assessing the results of comprehensive queries by linking provenance features with other information available from spatial systems. The objective is to improve the accessibility and usability of spatial data for Australia and New Zealand, in the first instance, but the techniques created will be generic and applicable to global use.

The main objectives of this research are: (1) detailed exploration of the requirements for geospatial provenance models; (2) development of a comprehensive spatial data supply chain provenance model for spatial information that is applicable to all feature types of spatial datasets including vector and raster datasets; (3) exploration and development of techniques to present provenance information to users for their assessment in an understandable form, via geospatial interfaces; and (4) exploration, development and enhancement of the proposed model through real case studies from relevant industry collaborators.

This paper is organized as follows: Section II describes the purpose of work performed; Section III identifies and discusses different provenance models and current work on provenance. In Section IV, the importance of work is followed by a detailed research methodology in Section V. In the last section, current findings, open issues and future directions are discussed.

II. BACKGROUND

The Cooperative Research Centre for Spatial Information (CRCSI) Program 3, Spatial Infrastructures, seeks to improve the organization, access and use of spatial data in Australia and New Zealand [1]. The research program has embraced advanced Semantic Web Technologies and Artificial Intelligence as a means of improving spatial data supply chains [1].

III. CURRENT RESEARCH

A. Current provenance models

The Open Geospatial Consortium (OGC) and the World Wide Web Consortium (W3C) define the provenance of spatial data as "information on the place and time of origin or derivation or a resource or a record or proof of authenticity or of past ownership. The W3C PROV model is a generic provenance information standard" [2].

W3C PROV is a conceptual model for provenance that offers an elegant and flexible solution for linking provenance information to geospatial elements with the necessary semantics. It can be realized in RDF, XML, and text formats, giving alternative options for implementing the same model suggested by [2]. However, no dedicated geospatial provenance model currently exists. The OGC test bed 10 Cross Community Interoperability (CCI) thread has conducted provenance activities and provided guidelines to capture provenance information through examining PROV for geospatial data.

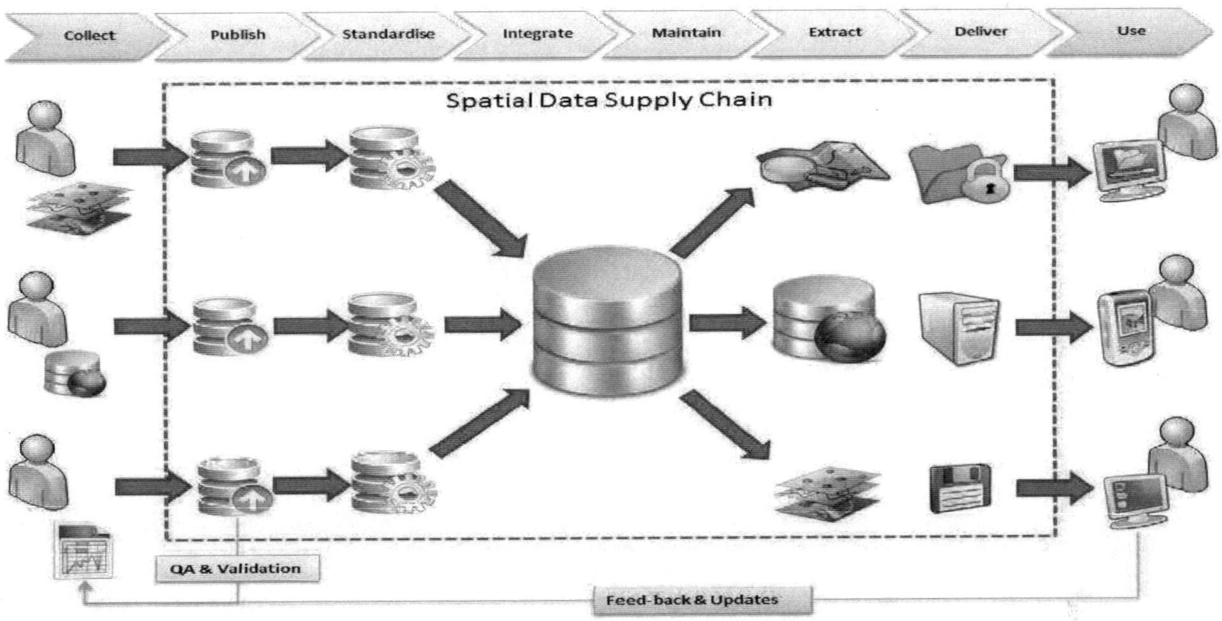

Figure 1. Spatial Data Supply Chain (van der Vlugt., 2012) [4]

B. Contemporary Research

In a Geospatial Web Service environment, data are often disseminated and processed widely and frequently, and often in an unpredictable way. This means that it is important to have a mechanism for identifying original data sources. Geospatial data provenance records the derivation history of a geospatial data product in [3]. It is important for evaluating the quality of data products, tracing workflows, updating or reproducing scientific results, and in evaluating geospatial data products' reliability and quality. As a consequence, geospatial data provenance has become a fundamental issue in establishing Spatial Data Infrastructures (SDIs).

The exchange and sharing of geospatial data provenance in a distributed information environment requires an interoperable model for provenance in [3]. The rationale for designing the provenance model in this way is to combine the W3C PROV with the ISO 19115 metadata standards. This will enrich the model with domain specific details and allow domain specific representations to be translated into an interoperable form for exchange on the Web. He et al. in [3] argue that fine-grained provenance modelling for geospatial data could be achieved by borrowing from existing modelling approaches for geospatial data such as feature, coverage and observation.

One representation of a spatial infrastructure is a supply chain that involves processes from data collection to production. An alignment study of SDSCs proposed the model shown in Figure 1 in which a number of stages are identified. It shows the variability in data from suppliers and products used by consumers. Of importance are the feedback loops that are needed for quality assurance and performance monitoring [4].

An attempt has been made to develop standards for provenance tracking in spatial analytical workflows. A prototype using spatial weights has been developed by [5] for metadata and provenance for spatial analysis. They are currently in collaboration with other researchers to refine these standards and extend them as a broader set of spatial analytical services. In scientific workflows, the most valuable service is the automatic capture of sufficient provenance data to establish trust and potentially allow other researchers to reproduce a result. Scientific workflows have emerged as de facto models for researchers to process, transform and analyse scientific data. Workflow management systems provide researchers with many valuable and time saving features, from cataloguing, workflow activities and Web services, to visual authoring and monitoring [6].

A current CRCSI project is concerned with geocoded address optimisation. Each valid address in Australia is required to have one or more geocoded locations for emergency services, and efficient delivery of mail and other services. At a Geocoding Address Workshop (held in Canberra, July 2014), stakeholders highlighted the need for geospatial provenance for geocoded addressing spatial data supply chains.

As the products are published, managers and scientists will have easy access to consistent baseline information of Australia to holistically monitor and predict the impact of natural and man-made changes on the Australian environment. We presumed that temporal Provenance is an active research area that has generated complex findings.

This is because as the original data source is updated, the integrated dataset will also be updated. Similarly, an integrated dataset may be updated if a new version of the integration algorithm becomes available.

The integration process is usually re-executed or the updates may be done routinely or as required. To manage these scenarios, Harth et al. in [7] have developed different approaches to temporal provenance for geospatial data and derived integrated spatial products. Changes in spatial datasets with time are crucial as well as continuous capturing of provenance for each process.

C. *Provenance in spatial infrastructure*

Spatial data provenance is often difficult to trace in a spatial infrastructure. Regardless of the application domain, data are collected and manipulated by a wide range of users, with distinct interests and applications, using their own vocabularies, work methodologies, models, and sampling needs. We observed that in particular there is a huge effort to improve the means and methodologies to capture process and disseminate geospatial data. In real life situations, provenance information of geospatial data is used to decide pre-processing procedures, storage policies and even data cleaning strategies, with direct impact on data analysis and synthesis policies in [8].

The next generation of Spatial Infrastructures will need the capability to integrate and federate geospatial data that are highly heterogeneous. Adams et al. in [9] discuss new data that can come with variable, loosely defined, and sometimes unknown provenance, semantics and content. They further explain that the geospatial datasets that we might wish to combine could be highly heterogeneous. They will be represented in many forms, will have been generated by a variety of producers using different processes and may have originally been intended for purposes that are different from their present use. The explicit consideration of provenance into Spatial Infrastructures is needed because of massive datasets and complex functionality involved. Wang et al. in [10] state that Geographical Information Systems (GISs) are widely used for manipulating geographically referenced data and supporting spatial analysis and modelling.

Gill in [11] researched intelligent semantic workflows for complex computations and data processing at a large scale, providing assistance in setting up parameters and data, validating workflows created by users, and automating the generation of workflows from high-level user guidance. Harth et al. in [7] report their experiences with integrating geospatial datasets using Linked Data technologies. They describe NeoGeo, an integration vocabulary, and an integration scenario involving two geospatial datasets.

Despite significant advances in computational infrastructure, many environmental scientists are hampered by the resource intensive task required to set up their analysis process because data comes in daily from their sensors [12]. Data preparation is time-consuming: scientists (1) gather data from multiple sources and sensors, (2) clean the data, (3) normalize it so that data from different sources is represented using the same units and formats, and (4) integrate it and configure it according to the requirements of their models and simulation software.

D. *The semantic Web approach*

Provenance is seen as an important aspect of the Web that becomes crucial in Semantic Web research. Research described in [13] addresses the many questions raised about the Semantic Web in the context of automated applications. "Modelling Provenance of DBpedia Resources Using Wikipedia Contributions" by Orlandi et al. in [14] presents an approach for adding provenance information about the statements in DBpedia by connecting these statements to the Wikipedia edits they are derived from. This provenance information is subsequently exposed as Linked Data using several existing provenance ontologies.

The use of provenance for information is recommended by Artz et al. in [15] to support trust decisions, as is the automated detection of opinions as distinct from objective information. They provide an overview of existing trust research in computer science and the Semantic Web and argue that trust has another important role in the Semantic Web.

IV. SIGNIFICANCE

A provenance model is needed for geospatial data as no model currently exists. This research will investigate the generation of a provenance model, called GEOPROV building on the work of the W3C and the Open Provenance Group. Based on GEOPROV a comprehensive provenance application will be developed which will extract, capture and store provenance information in an intelligent way that it can be queried semantically.

V. RESEARCH METHODS

Experts from industry will be engaged as industry supervisors to assist with aligning the research with the needs and requirements of stakeholders. Workshops have been conducted with the land survey commission from the Surveying and Spatial Sciences Institute of Western Australia and in conjunction with their comments the ontology details have been created. During the ontology design, trust, quality, lineage, history and authoritative attributes of datasets have been considered as the building blocks. On the basis of these elements of provenance, different decision metrics will be built to rank and further analyse datasets for decision making processes. Progress review workshops will be conducted quarterly with stakeholders. Regular visits will be arranged and close working relationships will be maintained. Below are the major activities which are on-going:

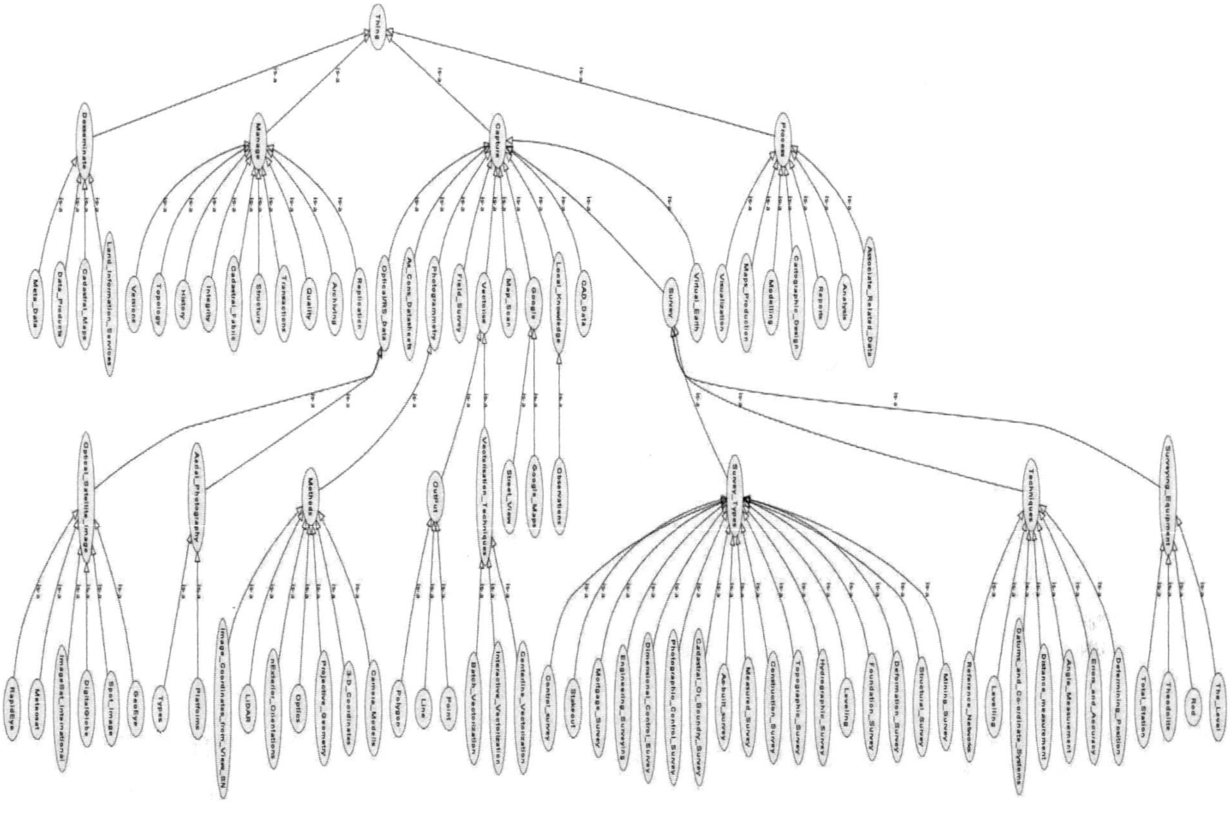

Figure 2. Land Administration Provenance Ontology Model

A. Use case development

With the close consultation of stakeholders, use cases will be developed and modelled in a comprehensive way to provide a common standard for the Australian and New Zealand geospatial industry.

Use cases are being explored with the Public Sector Mapping Agency (PSMA), Department of Land, Water and Planning (DELWP Victoria), Landgate (WA), as well as Land Information New Zealand (LINZ), and Geoscience Australia. As a result of this consultation process, the final design of GEOPROV will be developed and as a common standard.

Based on final design, GEOPROV tool will be developed to extract, store and visualize provenance of spatial datasets and will be tested by GIS teams of land administration departments across Australia and New Zealand.

B. Use case 1

A land administration subdomain provenance ontology structure has been developed. During the land administration process, data may be collected using several surveying techniques and methods. Different types of equipment are used and at various levels of sophistication. For example the popular Total Stations verses simple handheld GPSs are used to capture locations. The use of optical remote sensing techniques is also used to obtain location information that is produced by different organisations using different accuracy and modalities. For example, aerial photography is often a combination of many platforms and techniques. The nature and requirement of data capture is domain specific. Google Street View and Maps are handy sources of information for visual ground verification. All these methods, equipment and techniques are included in the ontology defined as type, resolution, calibrations, orientations, optics, geometry, combinations and principles. Capturing all these characteristics is important for determining feature accuracy and suitability for further use in the land administration life cycle (Figure 2).

C. Use case 2

Integrating road network data across State jurisdiction level may have result in anomalies due to the different standards currently used to collect the datasets. The conceptual design process is presented in Figure 3. When a linear feature road, river or any utility infrastructure are collected by different organisation and using a diverse range of standards, tools and methods, they may not be aligned at State borders. Automating such processes can provide several benefits as compared to the manual process. Matching source features with corresponding adjacent features quality may be questionable if errors or uncertainties will not be audited in post processing as in Figure 4.

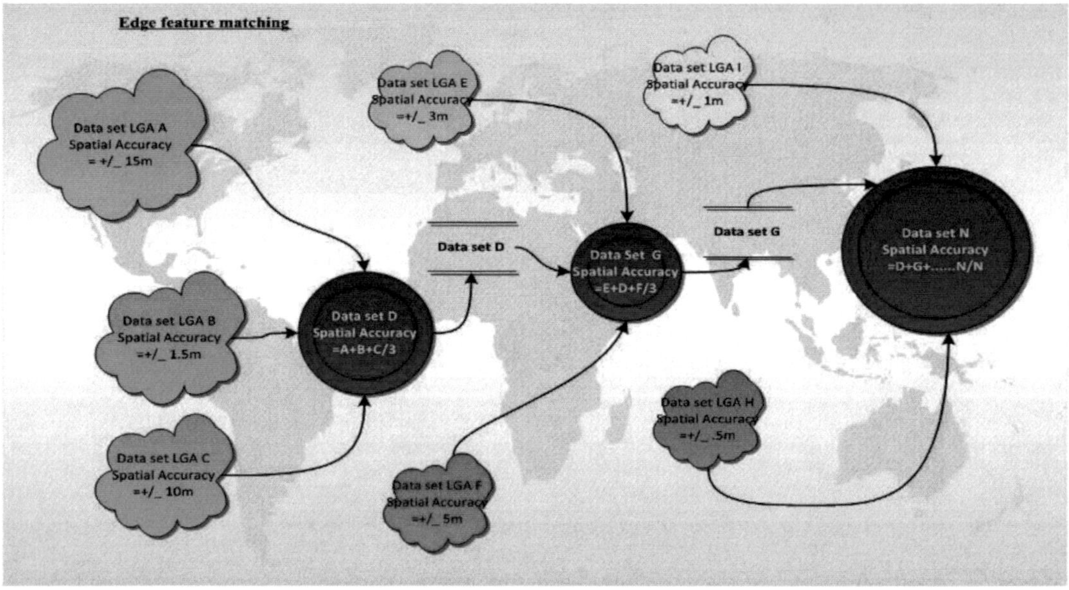

Figure 3. Edge feature matching

D. Encoding and mapping GEOPROV

A modular approach will be investigated initially and may be used depending upon the nature of the geospatial processes involved. Elements of the provenance model will be encoded, relationships will be defined, and geospatial data will be mapped. GEOPROV will make use of ontologies and rules. These will be explored and developed as part of this research. Open source tools such as Protégé, Pellet and others will be evaluated for usability.

W3C PROV ontology classes, properties, and constraints will be used to represent and allow the interchange of provenance information. Using this ontology, provenance records can be encoded in RDF triples. The OGC defined geospatial terms will be mapped to GEOPROV in RDF. Relationships between features, their geometric and non-geometric attributes will be defined through the latest version of an ontology Web language, namely OWL-2 and RDF.

E. Engineering Design Experiments

All use cases will be tested with the developed solution combined with a linked data approach. Provenance will be linked with other information available in the system to make query results more comprehensive. Research will also develop a weighted matrix approach to enable a user to determine the fitness for purpose of datasets for selected use cases. Besides this, the performance of queries will be studied as well as issues with storage, redundancy and application architectures. Geospatial provenance model requirements will be explored and defined and validated through stakeholder consultation and use cases investigated. There may be different query requirements based on the business need of each organization and governance level. These requirements will be input for testing provenance model effectiveness.

GEOPROV will be applicable to all feature types of spatial datasets including vector and raster datasets. The GEOPROV physical geospatial data provenance model will be developed in UML and Protégé along with a conceptual and functional business model applicable to all feature types of spatial datasets and different levels of granularity for geometric and non-geometric attributes including vector and raster.

VI. Conclusion

Spatial data supply chains (SDSC) for next generation spatial infrastructures require extensive investigation to address several contemporary issues and challenges that are hampering innovation and the use of spatial information across industry sectors. SDSCs consist of multiple value chains. Each value chain has heterogeneous geo-processes, methods, models and workflows that combine to generate, modify and consume spatial data as shown in Figure 5.

The integration and processing of multiple datasets gives rise to questions about trust, quality, fitness for purpose, currency and the authoritativeness of data. This is because multiple datasets originate from heterogeneous sources, and different geo processes have been executed to reach the final product. Users have different data requirements and therefore knowing how data is collected and at what level of accuracy, provides knowledge about what it can be used for leading to increased user confidence. With the advent of semantic Web technologies, new methods for exploring and understanding the provenance of spatial data have become possible. However, there are few models that address data provenance and none that adequately cater for spatial information management and the dissemination of data to users.

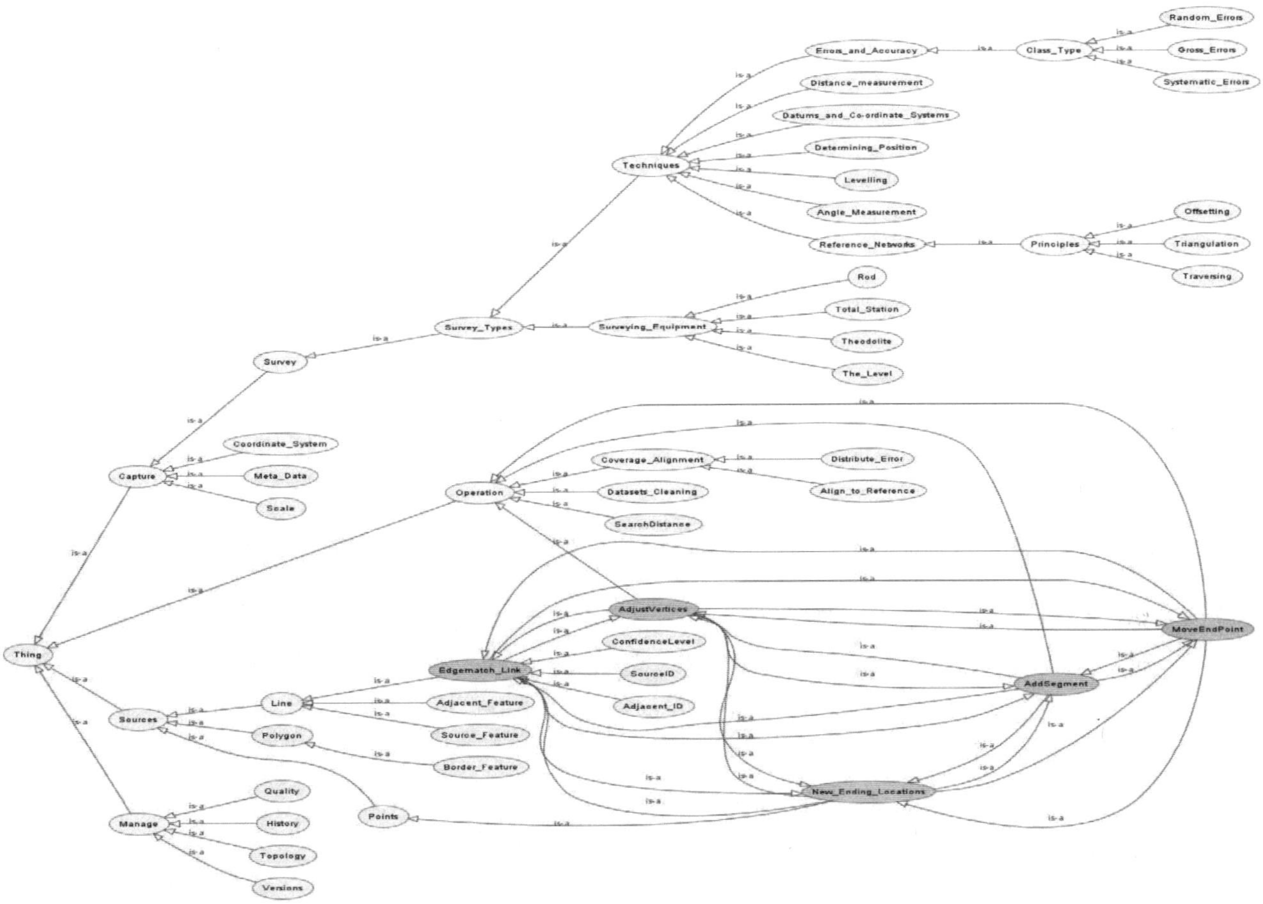

Edge matching provenance ontology model

A comprehensive provenance model for the spatial domain in Australia and New Zealand is an industry imperative. Understanding provenance is crucial to capturing information about spatial features, such as who/what/when/how/why it has been generated. This information is needed to support well informed and reliable evidence-based decision making. In addition, geospatial provenance models related to spatial data storage, scalability, robustness and query performance are yet to be examined.

Currently, GEOPROV is under development. Use cases have been produced. As result of which a Land Administration subdomain model in Figure 6 has been developed and ontology produced. The sub domain provenance model is still to be tested.

Besides this, a model for temporal and spatial provenance at feature and attribute level is under development (Figure 7). Ontologies and relationships between classes and subclasses have been defined. To achieve feature and instance level spatial provenance an edge matching line feature Web processing service is being modeled when two or more line features from heterogeneous source aligned and merged together to produce as new feature or manipulation of existing features.

This may result in the addition of new vertices, and shift vectors that may change the position of existing edges. This is a typical use case for survey data that is merged to form multiple sources to form a single cadastral dataset. The question is, what are the best techniques to enable a user to query, understand and analyse provenance information to determine trust in the data and whether it is fit for purpose? For example, a weighted matrix of provenance values may be appropriate, similar to hotel star rating. Alternatively, a user may want data at a specific accuracy and use other provenance information, such as a Web service having graphical charts for different levels of accuracy and thus trust. One representation can be the retrieval of provenance information by selecting a specific feature on the screen by querying the triples stored. The model developed will answer provenance information at feature and attribute level. For example if two features are merged, information will be captured and can be retrieved to answer queries about how the information was generated in the first instance. It will support requests about data, type of source, entities, processes, characteristics and agents.

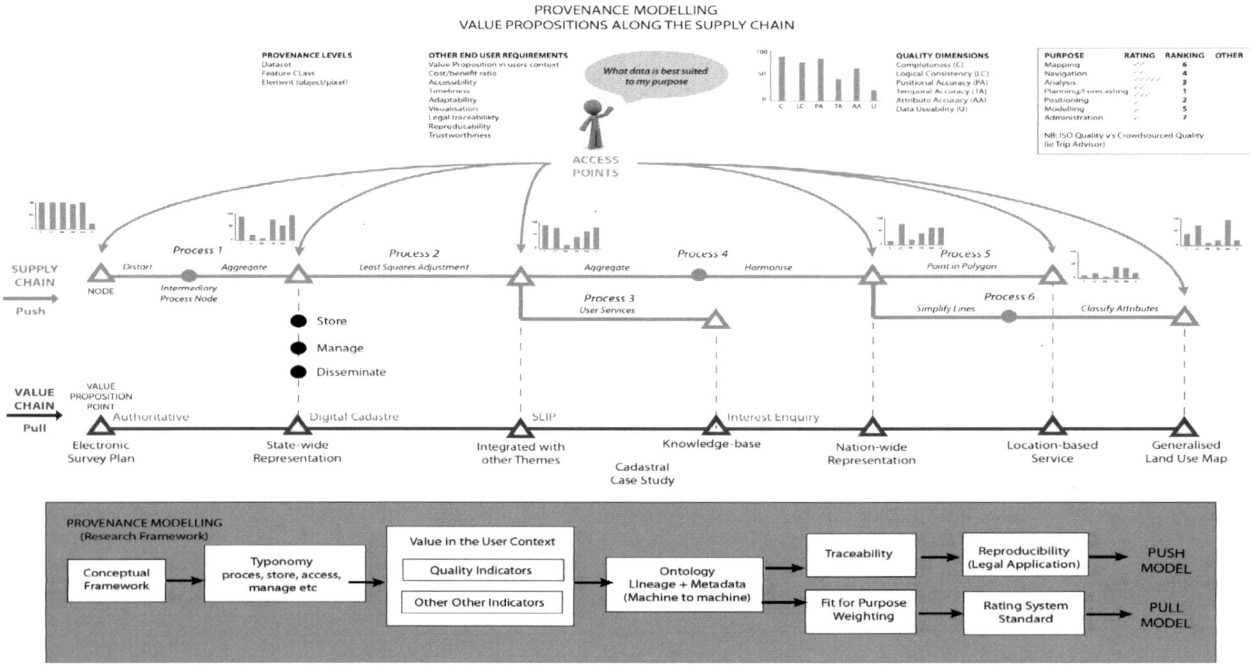

Figure 5. Supply chain provenance modeling

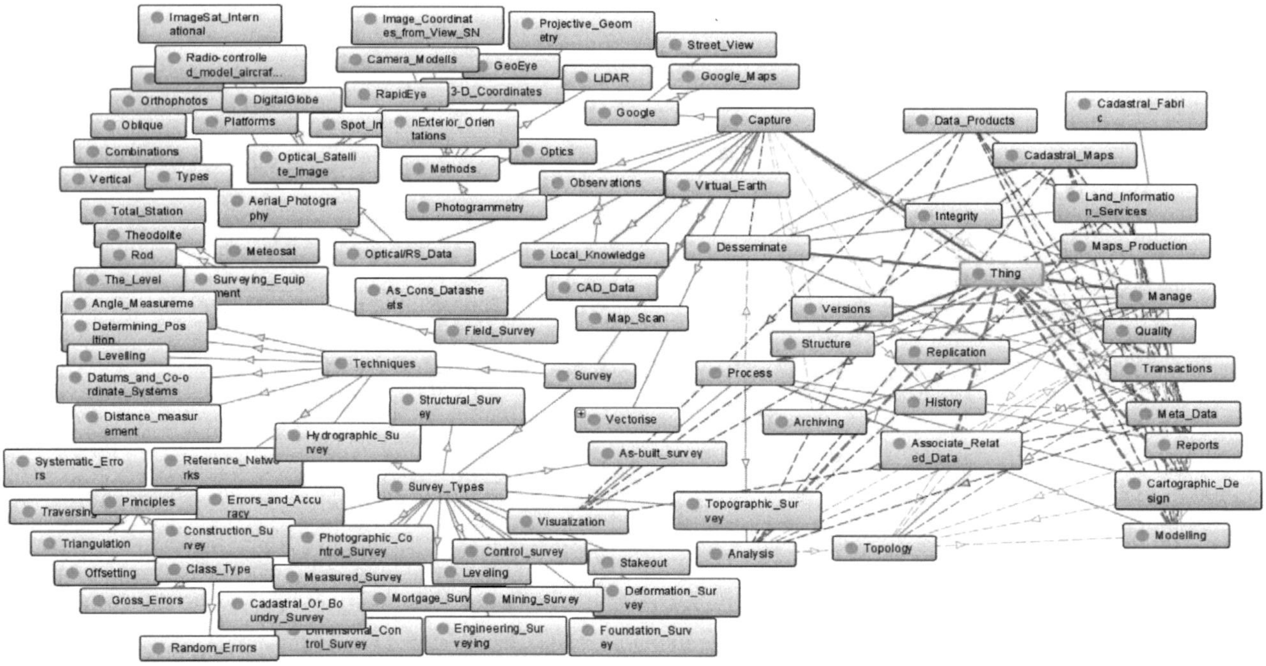

Use case 1 ontologies and relationships

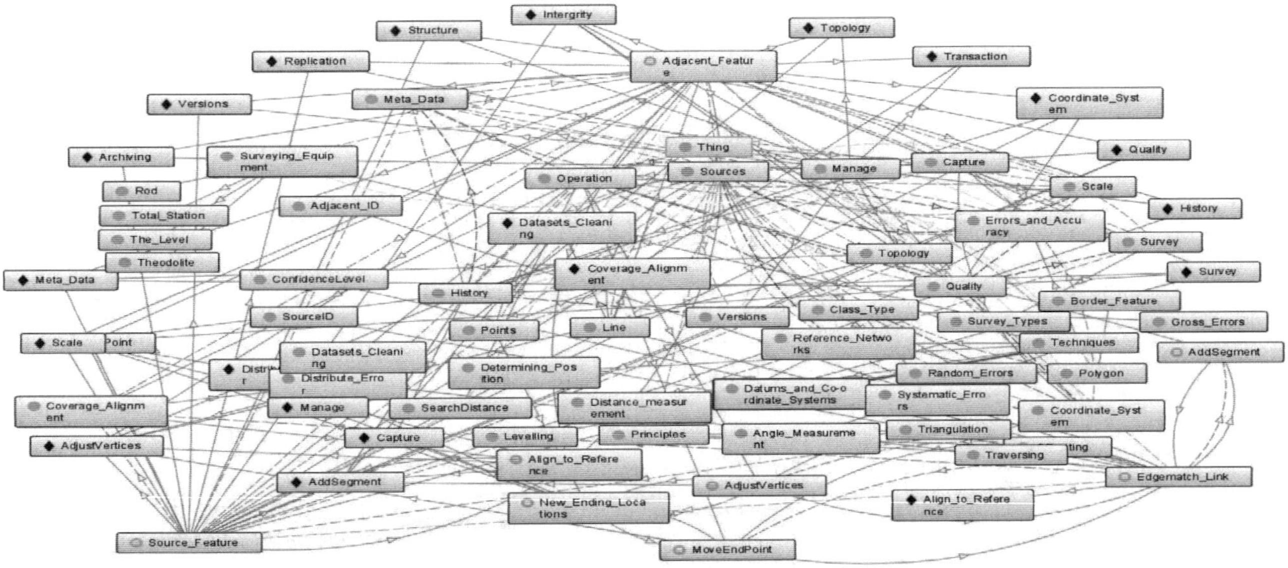

Figure7. Use case 2 ontologies and relationships

The implementation resulted from this study is a working prototype based on Python. This tool will extract metadata in XML format, it will extract properties of data sets and save them to a text files on physical location. Currently, it is extracting and collecting spatial properties of features and feature classes, their grouping schemes, data paths, counts, location, scale, data frame properties information. The next step is to capture spatial data workflow level provenance.

REFERENCES

[1] G. West, "Research Strategy Spatial Infrastructure, (Program 3)". Updated. Retrieved: March 2016, from www.crcsi.com.au/Resources/ Research/ P3-final-Research-Strategy.aspx

[2] J. Maso, C. G., Y. Gil, and B. Prob, "OGC® Testbed 10 Provenance Engineering Report OGC Public Engineering Report" (pp. 1-87): Open Geospatial Consortium.

[3] L. He, P. Yue, L. Di, M. Zhang, and L. Hu, "Adding Geospatial Data Provenance into SDI-A Service-Oriented Approach". IEEE Journal of Selected Topics in Applied Earth Observations and Remote Sensing. Feburary 2015, vol. 2, pp. 926-936. doi: 10.1109/JST ARS.2014. 2340737

[4] D. McMeekin, and M. Van der Vlugt, "CRC-SI – Alignment Study of Spatial Data Supply Chains. Version 2.0". Retrieved: March 2016, from http: //crcsi.com.au/getattachment/dabdca57-d7e7-4080-97cc-36db9a89173a/. aspx

[5] L. Anselin, S. J. Rey, and W. Li,. "Metadata and provenance for spatial analysis: the case of spatial weights". International Journal of Geographical Information Science. 15 May 2014, pp.1-20. doi: 10.1080/13658816.2014.917313

[6] R. Barga, et al. "Provenance for Scientific Workflows Towards Reproducible Research". IEEE Data Eng. Bull., 2010, vol. 33(3), pp. 50-58.

[7] A. Harth, and Y. Gil, "Geospatial Data Integration with Linked Data and Provenance Tracking". Paper presented at the Linking Geospatial Data, London. 2014, Iss. 54, pp. 1-5.

[8] J. Malaverri, E. Medeiros, C. B., and R. C. Lamparelli, "A provenance approach to assess the quality of geospatial data". Paper presented at the Proceedings of the 27th Annual ACM Symposium on Applied Computing. 2012, pp. 2043-2049.

[9] B. Adams, and M. Gahegan, "Emerging data challenges for next-generation spatial data infrastructure". S. Winter and C. Rizos (Eds.), Research@Locate'14, Canberra, Australia, Vol. 1142, April 7-9, 2014, pp. 118-129. Retrieved: Feburary 2016 from http://ceur-ws.org/Vol-1142/paper13.pdf

[10] S. Wang, A. Padmanabhan, J. D. Myers, W. Tang, and Y. Liu,. "Towards provenance-aware geographic information systems". Paper presented at the Proceedings of the 16th ACM SIGSPATIAL international conference on Advances in geographic information systems, Irvine, California. 2008 Nov 5, pp. 70.

[11] Y. Gill, "Intelligent Workflow Systems and Provenance-Aware Software". Paper presented at the Ames, D.P., Quinn, N.W.T., Rizzoli, A.E. (Eds.), Proceedings of the 7th International Congress on Environmental Modelling and Software, San Diego, California, USA. 2014, p. 91.

[12] Y. Gil, et al. "Mind Your Metadata: Exploiting Semantics for Configuration, Adaptation, and Provenance in Scientific Workflows". In L. Aroyo, C. Welty, H. Alani, J. Taylor, A. Bernstein, L. Kagal, N. Noy and E. Blomqvist (Eds.), The Semantic Web – ISWC 2011, Vol. 7032, pp. 65-80.

[13] Y. Gil, and P. Groth, "Using provenance in the Semantic Web". Web Semantics: Science, Services and Agents on the World Wide Web, 31 July 2011, Vol. 9(2), pp. 147-148.

[14] F. Orlandi, and A. Passant, "Modelling provenance of DBpedia resources using Wikipedia contributions". Web Semantics: Science, Services and Agents on the World Wide Web, 21 April 2011, Vol. 9(2), pp. 149-164.

[15] D. Artz, and Y. Gil, "A survey of trust in computer science and the Semantic Web". Web Semant., 15 March 2007, Vol. 5(2), pp. 58-71. doi: 10.1016/j.Web sem.2007.03.002

Geographical General Regression Neural Network (GGRNN) Tool For Geographically Weighted Regression Analysis

Muhammad Irfan, Aleksandra Koj, Hywel R. Thomas, Majid Sedighi
Geoenvironmental Research Centre,
School of Engineering, Cardiff University
Cardiff, UK
emails: {MuhammadI2, KojA, ThomasHR}@cf.ac.uk, majid.sedighi@manchester.ac.uk

Abstract—**This paper presents a new geographically weighted regression analysis tool, based upon a modified version of a General Regression Neural Network (GRNN). The new Geographic General Regression Neural Network (GGRNN) tool allows for local variations in the regression analysis. The algorithm of the GRNN has been extended to allow for both globally independent variables and local variables, restricted to a given spatial kernel. This mimics the results of Geographically Weighted Regression (GWR) analysis in a given geographical space. The GGRNN tool allows the user to load geographic data from the Shapefile into the underlying neural networks data structure. The spatial kernel can be either a fixed radius or adaptive, by using a given number of neighboring regions. The Holdout Method has been used to compare the fitness of a given model. An application of the tool has been presented using the benchmark working-age deaths in the Tokyo metropolitan area, Japan. Standardized residual maps produced by the GGRNN tool have been compared with those produced by the GWR4 tool for validation. The tool has been developed in the .Net C# programming language using the DotSpatial open source library. The tool is valuable because it allows the user to investigate the influence of spatially non-stationary processes in the regression analysis. The tool can also be used for prediction or interpolation purposes for a range of environmental, socioeconomic and public health applications.**

Keywords- GGRNN; GRNN; GWR; ANN; Spatial Kernel.

I. INTRODUCTION

The GGRNN tool is part of a Spatial Decision Support System (SEREN-SDSS) developed by the Geoenvironmental Research Centre of Cardiff University. SEREN-SDSS has been designed and developed for geoenergy and geoenvironmental applications. It facilitates the decision making process by combining several Multicriteria Decision Analysis (MCDA) and Artificial Neural Network (ANN) techniques [1]. The GGRNN tool utilises and extends the capabilities of GRNN in order to facilitate local spatial variations in regression analyses.

GRNN was first presented by Spetch [2]. GRNN are powerful function approximations, capable of modelling linear and non-linear relationships in data despite being very simple in their structure and operation [3].

GRNNs have been considered in this research because unlike some of the other type of ANNs, GRNNs do not operate as a "black box". Rather, they predict the values at an unknown location on the basis of its proximity to known location in terms of the selected independent variables. Additionally, because of its structure, it is easier to incorporate spatial parameters as one of the independent variables to support local variation in the regression analysis.

The paper is organised as follows. Section II covers the structure, empirical formulation and algorithmic details of the training of the GRNN. Section III describes the nature, operation and different variations of GWR analysis. Section IV presents the GGRNN introduced in this research. Section V highlights the development and operation of the GGRNN tool used here to carry out the GGRNN analysis. Section VI covers the validation of the proposed GGRNN tool. Results obtained using the proposed GGRNN tool, are provided in Section VII together with a comparison of its results against the GWR4 tool. Section VIII summarizes conclusions and future work.

II. GENERAL REGRESSION NEURAL NETWORKS

GRNNs have the capability to predict, interpolate and undertake regression analysis. It is a useful tool when the relationship between dependant and independent variables is unknown and complex. It supports both linear and non-linear relationships.

GRNNs have been used in a number of applications. For example, a GRNN has been used to predict rainwater runoff in two small sub-catchments of Tiber River Basin in Italy using rainfall and soil moisture information at different soil depths [4]. The GRNN prediction was found to be satisfactory in relation to the actual runoff, with coefficient of determination, R_2, equal to 0.87 [4].

Similarly, three different types of neural networks have been used to predict and classify the per-capita Ecological Footprint (EF) of 140 nations [5]. These neural networks are Multi-layer Perceptron Neural Networks (MLPs), Probabilistic Neural Networks (PNNs) and GRNNs. The results reveal that neural networks outperform traditional statistical methods used for this application [5].

GRNNs can also be utilised in finding the most useful set of variables that can be used in an analysis. For example, GRNNs have been used in [6] for the determination of the most appropriate variables to forecast chlorine in preventing the spread of waterborne diseases.

A. *Structure of GRNN*

GRNNs are very simple in their structure and have the following four layers of neurons: a) Input Layer, b) Pattern Layer, c) Summation Layer, d) Output Layer.

Figure 1 shows the general structure of a GRNN with these four layers, as originally suggested by [2]. A GRNN can approximate a function and estimate the value of a dependent variable from a set of independent variables.

The Input Layer contains as many neurons as there are variables in the input dataset. The input data points are presented to the Input Layer which simply feeds them into the Pattern Layer. Each input data point is then stored in the Pattern Layer. The number of neurons in the Pattern Layer is equal to the total number of data points. The value of the dependent variable (Y) at the prediction point is calculated based on the difference between the values of independent variables at the prediction point and their respective values at other points at which the independent variables are known. The Summation Layer computes the numerator and denominator terms for Equation 1, by using the difference factor of the independent variables (at known and unknown location) and the dependent variable (at known location). The last layer is called the Output Layer where the value of function $\acute{Y} = f(x)$ is computed using (a).

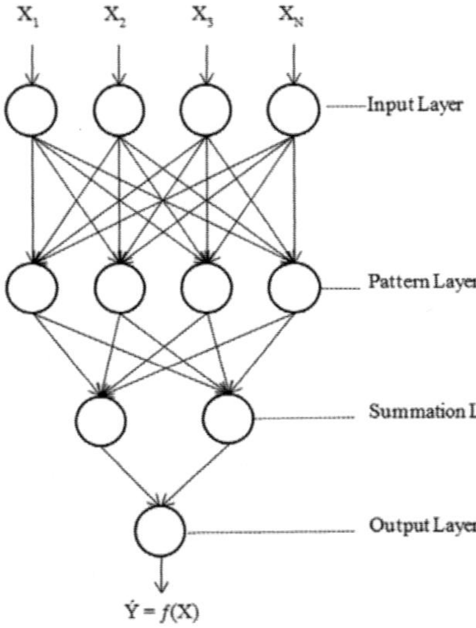

Figure 1. Structure of General Regression Neural Network [2]

The mathematical formulation to implement GRNN is straightforward and similar to probability distribution function. The output function of the GRNN is given as [2]:

$$\acute{Y} = f(x) = \frac{\sum_{i=1}^{n} Y^i \ exp^{(-D_i^2/2\sigma^2)}}{\sum_{i=1}^{n} \ exp^{(-D_i^2/2\sigma^2)}} \quad (1)$$

where $\acute{Y}$ is the estimated value of the dependent variable at the unknown location, Y^i is the value of dependent variable at known locations and D_i is a scalar term that accounts for the differences between the prediction point and the training sample for all independent variables (dimensions) and is calculated as[2]:

$$D_i^2 = (X - X^i)^T(X - X^i) \quad (2)$$

The distance between the prediction point and a training sample defines the influence of that training sample in the calculation of $f(x)$ (the dependent variable $\acute{Y}$). If this distance is small, the term $exp^{(-D_i^2/2\sigma^2)}$ increases and is exactly one for a difference of zero. A larger value of this term means the known value of dependent variable at this training sample will have more influence in the calculation of the dependant variable at the prediction point. If the distance is large, the value of the term $exp^{(-D_i^2/2\sigma^2)}$ decreases, tending to zero for very large distances. Such sample points will provide no contribution to the estimation of dependent variable at the prediction location. The predicted output is bounded between the maximum and minimum known values of the dependent variable [2].

B. *Smoothing parameter sigma (σ)*

The σ parameter can have single or multiple values for different variables (dimensions) in an input dataset. If a single value is used, it is very important to standardise the independent variables so that they have a mean of zero and a standard deviation of one. Without standardisation of the independent variables, a single σ value will cover different distances in each dimension and the value of D_i^2 will not represent the actual difference between the training sample and the prediction point [2]. A smaller σ value will result in a localised regression analysis, i.e., only the sample points that are very close to the prediction point in terms of their distances on different axis (domains) will contribute to the calculation of dependent variable. A larger σ value results in a more globalised regression where almost the entire set of data samples contributes to the calculation of the dependent variable. In this latter case, results are very close to the mean value of the dependent variable for the entire set of sample points.

C. *Holdout Method for training of GRNN*

GRNNs require supervised training and the selection of the most suitable value for the smoothing parameter, σ, is very important to obtain reliable results [2]. The Holdout Method is a useful and common method for the selection of σ [2]. Figure 2 explains the Holdout method algorithm in a flow chart.

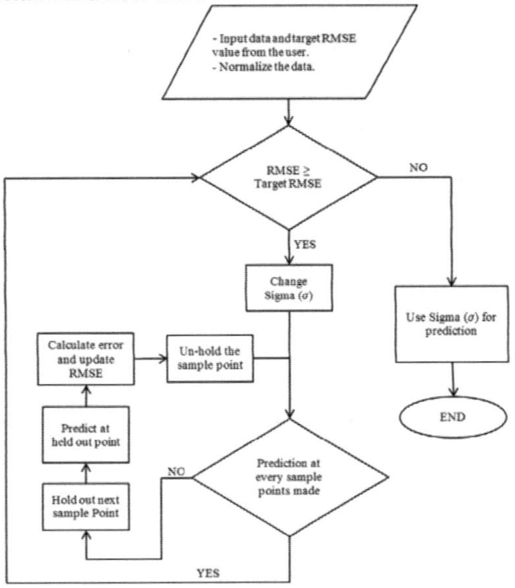

Figure 2. Flow chart of the Holdout Method for the selection of sigma

In the Holdout Method, only one training sample is selected from the training set at a time and the value of Ŷ is predicted at this sample point using the rest of the samples [2]. The predicted value is compared with the actual value and the difference is used in the calculation of mean squared error [2].

III. GEOGRAPHICALLY WEIGHTED REGRESSION (GWR)

Geographically weighted regression (GWR) models can be used to understand and analyse spatially varying relationships between dependent and independent variables [7]. A conventional GWR regression model is represented by the following equation [7]:

$$Y_i = \sum_k \beta_k(u_i, v_i) x_{k,i} + \varepsilon_i \quad (3)$$

where Y_i, $X_{k,i}$ and ε_i are the dependent variable, *kth* independent variable, and the error term at location *i(u, v)* respectively. β_k is locally varying coefficient at the *i*th location. Another variation of GWR model is where some of the independent are treated as global while others are restricted to vary locally. In such models a user given spatial kernel defines the area in which the local variables are analysed. Such models are called semi-parametric or mixed GWR are normally represented by [7]:

$$Y_i = \sum_k \beta_k (u_i, v_i)\, x_{k,i} + \varepsilon_i \; + \; \sum_l \gamma_l\, z_{l,i} + \varepsilon_i \quad (4)$$

where $z_{l,i}$ is the lth independent variable that is treated globally and has a fixed coefficient γ_l.

GWR or mixed GWR functions can be applied using Gaussian, Poisson, and logistic regression models. The models give better regression results and enhanced understanding of the relationship between different parameters, whether global or local [7].

IV. GEOGRAPHICAL GENERAL REGRESSION NEURAL NETWORK (GGRNN)

The GGRNN presented in this study extends the basic GRNN model described earlier in Section 2. This extension of the original GRNN algorithm allows for local variation in the relationship between different parameters. The influence of local and global variables are computed separately and then summed together. The difference is in the calculation of the term D (Distance) if spatial distance is used as independent variable as explained earlier. Also for the locally independent variables, the influence is calculated only within the given neighbourhood in contrast to the global variables for which the locations are involved.

In order to define the neighbourhood for local variations, two different techniques are used:

A. *Fixed spatial kernel*

In this technique, a user defined spatial kernel, e.g., 15km, is used to select the neighbouring geographical regions (features). These features are used for the computation of the influence of the local dependent variables only. The influence of global variables is calculated in the normal manner from the entire study area.

B. *Spatially adaptive kernel*

If the spatially adaptive kernel technique is used, the user selects the number of neighbouring areas to define the kernel, within which the influence of the local parameters is calculated. Since the geometries of the administrative boundaries (e.g. districts) are asymmetric, a fixed number of neighbouring areas will result in a varying spatial kernel, hence the naming of this technique.

C. *Spatial distance as independent variable*

The use of an appropriate neighbourhood size is important for the model to fit the data properly. Different iterations and comparison of the standardised error can help in the identification of the appropriate neighbourhood size. However, if it is not clear what type and size of kernel is to be used, the GGRNN tool also provides a mechanism to use spatial distance between different areas as one of the independent variables for the prediction of the dependent variable. As discussed earlier in Section 2, the neighbouring areas of the prediction location will have a greater influence in the calculation of the dependent variable. The distance between two geographical features (areas) is calculated using (5) based on the centroids of either feature [7]:

$$D_{spatial} = \sqrt{(X_i - X_j)^2 + (Y_i - Y_j)^2} \quad (5)$$

V. GGRNN TOOL

The GGRNN tool has been developed in the .Net C# programming language using the DotSpatial open source library. Figure 3 shows the user interface of the GRNN based prediction tool.

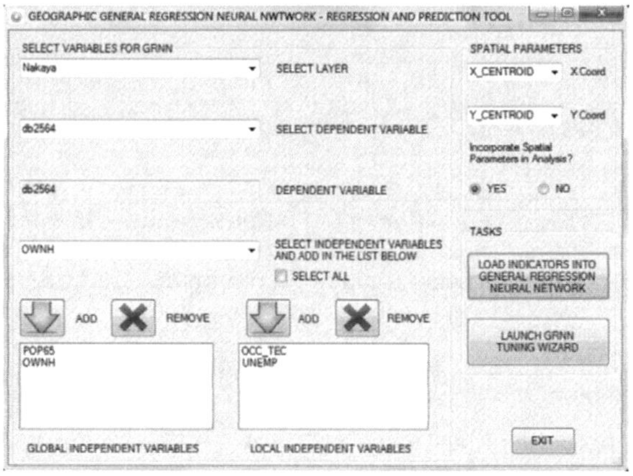

Figure 3. GUI of GGRNN based prediction and regression analysis tool

The user first selects the GIS layer (Shapefile) containing the indicators. The user identifies the dependent, global and local independent variables, and loads the data into the GRNN tool. The user can select whether or not to use spatial distance in the analysis.

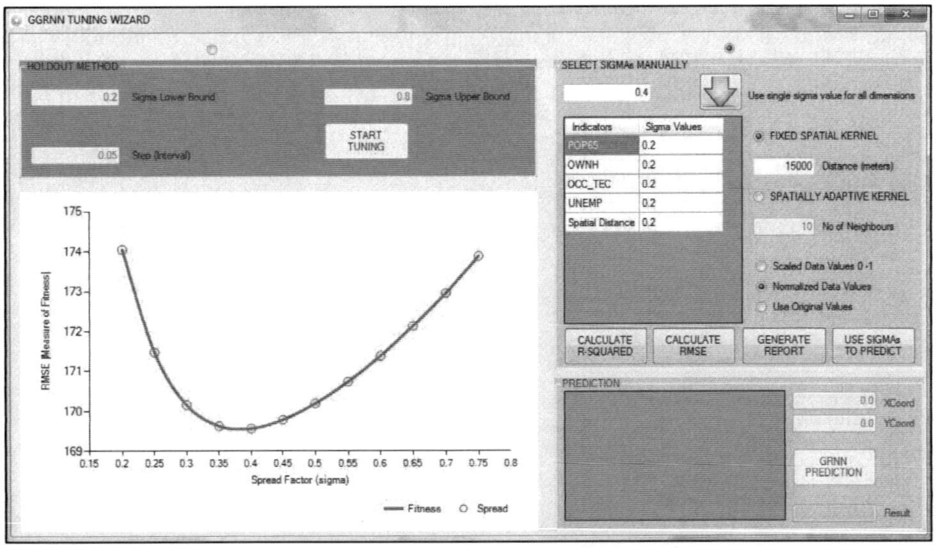

Figure 4. GRNN sigma tuning and prediction tool

Once the data is incorporated in the neural network, a tuning wizard is launched helping the user to select best sigma (σ) parameters for the analysis. The tuning wizard utilises the Holdout Method for the calculation of the Root Mean Square Error (RMSE).

The user can give upper and lower bounds for the sigma parameters and a step (interval) to calculate the RMSE using the Holdout method. The system plots the RMSE values against the corresponding sigma spread factors as shown in Figure 4.

Either the actual, scaled or normalised data values can be used for the calculation of RMSE for a given set of σ values. The user can assign the same σ parameters for all the independent variables if the data is normalised or scaled. However, if the original data values of the independent variables are used for the estimation of the dependent variable, then it is important to assign the sigma values with care. This is important as some of the variables may have a different spread and range of data values as compare to the others and using a similar sigma value can adversely affect the results.

Adopting spatial parameters in the regression analysis in GRNN is similar to the Geographically Weighted Regression (GWR) suggested by [7]. If spatial parameters are included in the analysis then the tool provides two different methods to identify a specific number of neighbouring geographical features to be used for the prediction analysis. These two methods are a) Fixed Spatial Kernel and b) Adaptive Spatial Kernel.

If an Adaptive Spatial Kernel is selected, only a given number (N) of neighbouring geographical features are selected for the analysis. The system first calculates the distance of each geographical feature from the prediction point. Then only N closest neighbours are selected and used in the process. However, if a fixed spatial kernel is used then all neighbouring geographical features found within the spatial kernel are selected.

In either case the smoothing parameter, sigma (σ), used for each independent variable computes the influence of each neighbouring area on the calculation of the independent variable at the prediction point. If a sigma parameter is assigned to the spatial dimension, then features closer to the prediction point will have a greater influence on this calculation. Large values of sigma parameters cause the prediction to tend to the mean value of the dependent variable in the entire study area of the given neighbourhood.

Once a set of sigma parameters has been selected with an acceptable RMSE value, the user can select to use them for the actual prediction at an unknown location. If spatial parameters were not used in the analysis, only the independent variables need to be provided by the user at the unknown location, where prediction is to be made for the dependent variable. If however, spatial parameters were used, then the user must also provide the X and Y coordinated of the centroid of the geographical feature, for which the dependent variable is to be predicted.

VI. VALIDATION

An application of the GGRNN tool is presented to compare its results with those produced by the GWR tool. A semi-parametric GWR model application has been presented to analyse the relationships between the working-age mortality and socio-economic conditions in Tokyo metropolitan area, Japan [8]. The same dataset is used in this research for two reasons:

a. The dataset is known to have local spatial variations found in parts of the study area, as explained in [8].
b. The standardised error resulting in the application of GWR and the GGRNN tool can be mapped, analysed and compared for benchmarking purpose.

The Tokyo mortality data covers the 262 municipality zones of the Tokyo Metropolitan area, Japan. The older age population and rate of house-ownership are used by

[8] as the global independent variables, whereas the other two variables are controlled locally in the regression analysis. The description of dependent and independent variables are given in Table 1 below.

TABLE 1. TOKYO MORTALITY DATASET

Variable	Description	Relationship
Working age mortality rate	Standard mortality rates for the 25–64 age group	Dependent Variable
Older population	Proportion of elderly people (aged over 64) within each zone	Independent (Global)
Own houses	Rate of house-ownership in each zone	Independent (Global)
Professional and technical workers	Proportion of professional and technical workers in each zone.	Independent (Local)
Unemployment	Rate of unemployment in each zone	Independent (Local)

VII. COMPARISON OF RESULTS

GWR version 4 has been used to analyse the Geographically Weighted Regression of working age mortality rates with socio economic conditions. A Gaussian Model has been used for the kernel analysis in both the GWR and GGRNN tools. The introduction of an offset and a local intercept variable in the GWR analysis is recommended [8]. Therefore, the two variables have been included in the GWR tool; however, the GGRNN tool doesn't have a provision for this because of the structure of its underlying neural network. In both cases the independent variables are standardised. Both fixed and adaptive kernels have been used to run the model in GWR. The recommended fixed kernel for this dataset is 15km and, for an adaptive kernel type, 50 neighbours are recommended [8]. In order to compare the results with those produced by the GGRNN tool, the most suitable sigma parameter is identified using the Holdout Method and RMSE. A sigma value of 0.4 was obtained for both adaptive and fixed spatial kernel techniques. Standardised residual maps are produced in ArcMap; the resultant maps obtained using the GWR tool, are shown in Figure 5 below.

Figure 5 shows the standardized residual maps produced by the GGRNN tool and GWR4 tool by using an adaptive kernel. The results show that the GGRNN tool has produced very similar results to the GWR4 tool using the adaptive kernel. A slight difference can be observed between the two results in the south-eastern part of the region which needs to be further investigated. A possible reason is the difference between the locally varying coefficient used in the GWR tool and the sigma parameter used in the GGRNN tool.

In the second process, both the tools have been set to use a fixed spatial kernel of 15 km. The Holdout Method used in the GGRNN tool suggests that a network model with sigma parameter of 0.4 exhibits the best fit to the dataset. The results are shown in Figure 6. It can be seen that the two tools have again produced very similar results in the case of fixed spatial kernel.

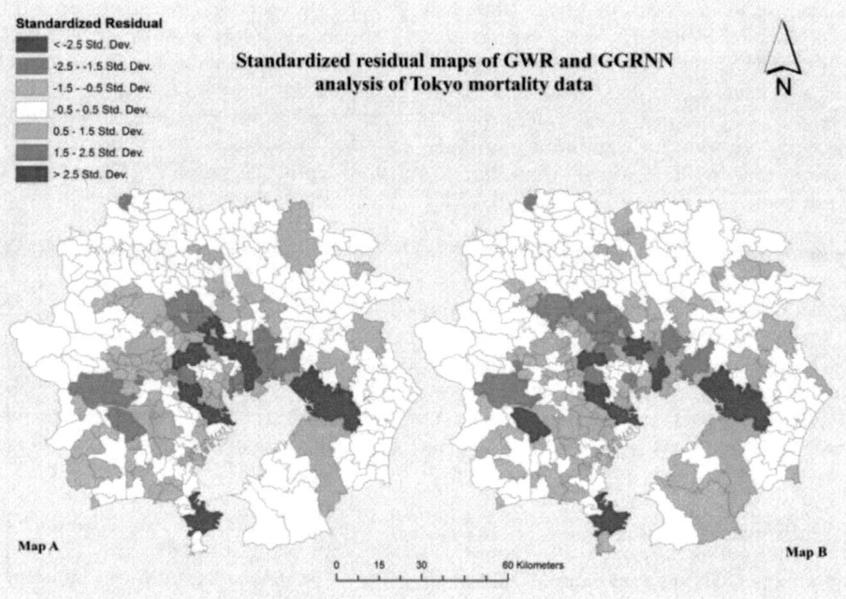

Figure 5. Standardised residual maps using adaptive Gaussian with 50 neighbours. Map A: GGRNN tool. Sigma parameter: 0.4 (for all independent variables). Map B: GWR4 tool

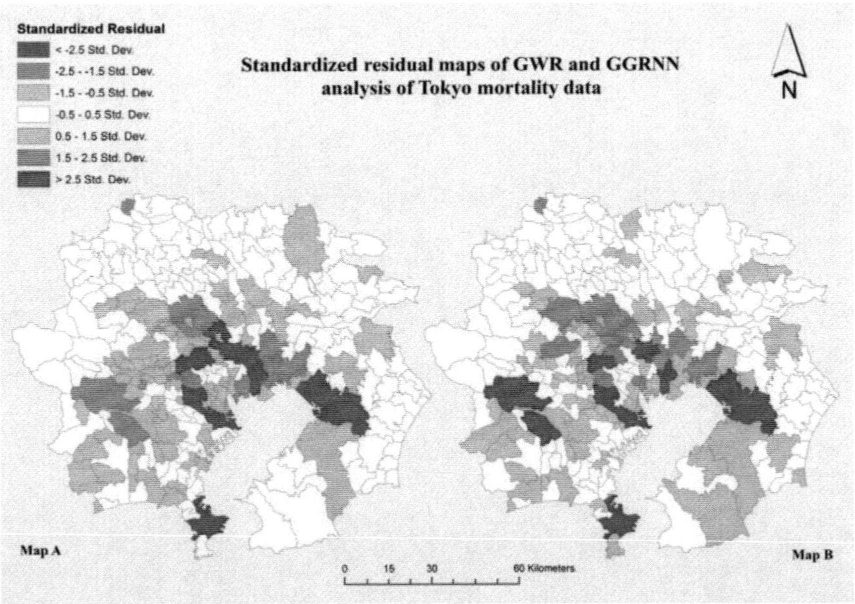

Figure 6. Standardised residual maps using fixed kernel of 15kms. Map A: GGRNN tool. Sigma parameter: 0.4 (for all independent variables). Map B: GWR4 tool

VIII. CONCLUSIONS AND FUTURE WORK

This paper presents a new regression analysis tool, based upon a modified version of the General Regression Neural Network (GRNN). The Geographical General Regression Neural Network (GGRNN) tool can be used to perform to Geographically Weighted Regression (GWR) analysis. It can be useful in understanding the underlying spatially varying relationships between dependent and independent variables and for prediction analysis. The GGRNN tool can be used in a number of environmental, socio-economic and public health applications.

The tool provides options to select the independent variables as globally fixed or locally varying. The spatial kernel can either be assigned as a fixed radius or adaptive, i.e., by assigning a given number of neighbouring regions.

The Holdout Method has been used to compare the fitness of a given model. The GGRNN tool allows the user to compare the fitness of different models by using the Holdout Method. The Holdout Method helps in selecting the most appropriate network parameters, essential for the working of a neural network. A validation of the tool has been carried out using the benchmark Tokyo mortality dataset and using the GWR4 tool. The validation results demonstrate that the GGRNN tool can be used with confidence to carry out geographically weighted regression analysis.

In future work, the performance of the tool will be tested against the GWR tool. Also, it will be tested to assess its prediction of dependent variable at unknown locations for impact assessment.

ACKNOWLEDGMENT

The work described in this paper has been carried out as part of the GRC's Seren project (GRC SEREN Project 2015) [9], which is part funded by the Welsh European Funding Office (WEFO).

REFERENCES

[1] M. Irfan, "An Integrated, Multicriteria, Spatial Decision Support System, Incorporating Environmental, Social and Public Health Perspectives, for Use in Geoenergy and Geoenvironmental Applications", Ph.D. Thesis, Cardiff University, The Wales, UK (2014).

[2] D. F. Specht, "A General Regression Neural Network", Neural Networks, IEEE Transactions on, vol. 2, 1991, pp. 568-76.

[3] N. Currit, "Inductive Regression: Overcoming OLS Limitations with the General Regression Neural Network", Computers, Environment and Urban Systems, vol. 26 2002, pp. 335-53.

[4] G. Tayfur, G. Zucco, L. Brocca, and T. Moramarco, "Coupling Soil Moisture and Precipitation Observations for Predicting Hourly Runoff at Small Catchment Scale", Journal of Hydrology, vol. 510, 2014, pp. 363-71.

[5] M. M. Mostafa, and R. Nataraajan, "A Neuro-Computational Intelligence Analysis of the Ecological Footprint of Nations", Computational Statistics & Data Analysis, vol. 53, 2009, pp. 3516-31.

[6] G. J. Bowden, J. B. Nixon, G. C. Dandy, H. R. Maier, and M. Holmes, "Forecasting Chlorine Residuals in a Water Distribution System Using a General Regression Neural Network", Mathematical and Computer Modelling, vol. 44, 2006, pp. 469-84.

[7] S. Fotheringham, C. Brunsdon, and M. Charlton, Geographically Weighted Regression: The Analysis of Spatially Varying Relationships (Wiley, 2003).

[8] T. Nakaya, A. S. Fotheringham, C. Brunsdon, and M. Charlton, "Geographically Weighted Poisson Regression for Disease Association Mapping", Statistics in Medicine, vol. 24, 2005, pp. 2695-717.

[9] GRC SEREN Project. [online] Available from: http://grc.engineering.cf.ac.uk/research/seren/ 2016.04.14

AUTHOR INDEX

Abdelmoty, A. 14
Adams, B. 131
Almeida, C. 38
Arnold, L. 137, 146
Baldo, F. 63
Baptista, C. 78
Beaumont, B. 43
Becher, S. 32
Bertazzon, S. 26
Bezdek, A. 128
Braz, F. 63
Clemens, K. 118
Close, O. 7, 43
Cools, M. 22
Daems, D. 126
Donnay, J. 109
Doytsher, Y. 93
Dries, J. 126
Fa, J. 46
Ferreira, T. 1
Frippiat, C. 7
Gahegan, M. 131
Goor, E. 126
Gulland, E. 85
Hallot, E. 43
Hong, C. 100
Hubauer, T. 32
Irfan, M. 154
Jager, M. 106
Jardim, A. 38
Jessel, J. 58
Kang, T. 100
Kasprzyk, J. 109
Kim, K. 56
Klokocnik, J. 128
Koj, A. 154
Kostelecky, J. 128
Kroll, N. 106
Latvala, P. 121
Lee, I. 56
Lee, J. 56
Lehto, L. 121
Leite, D. 78
Lisboa-Filho, J. 1
McMeekin, D. 46, 85, 146
Melo, A. 63
Mohamed, S. 14
Moncrieff, S. 46, 85
Mustafa, A. 22
Namikawa, L. 38
Noskov, A. 93
Oliveira, J. 38
Park, D. 56
Park, S. 100
Poelmans, L. 43
Rahajaniaina, A. 58
Ramanana, H. 58
Reed, T. 85
Reini, J. 121
Rosim, S. 38
Ruckemann, C. 69
Saadi, I. 22
Sadiq, M. 146
Sarjakoski, T. 121
Scheibel, G. 63
Sedighi, M. 154
Silva, T. 78
Solomakhina, N. 32
Stankov, K. 76
Stempliuc, S. 1
Stephenne, N. 7, 43
Suß, S. 106
Teller, J. 22
Thomas, H. 154
Tolton, B. 76
West, G. 46, 85